CHICAGO PUBLIC LIBRARY
BUSINESS / SCIENCE / TECHNOLOGY
400 S. STATE ST.   60605

```
QK          Belousova, L. S.
86
.A1         Rare plants of the
B45            world.
1992

$85.00
```

RUSSIAN TRANSLATIONS SERIES

1. K.Ya. Kondrat'ev et al. (editors): *USSR/USA Bering Sea Experiment*
2. D.V. Nalivkin: *Hurricanes, Storms and Tornadoes*
3. V.M. Novikov (editor): *Handbook of Fishery Technology*, Vol. 1
4. F.G. Martyshev: *Pond Fisheries*
5. R.N. Burukovskii: *Key to Shrimps and Lobsters*
6. V.M. Novikov (editor): *Handbook of Fishery Technology*, Vol. 4
7. V.P. Bykov (editor): *Marine Fishes*
8. N.N. Tsvelev: *Grasses of the Soviet Union*
9. L.V. Metlitskii et al.: *Controlled Atmosphere Storage of Fruits*
10. M.A. Glazovskaya: *Soils of the World* (2 volumes)
11. V.G. Kort & V.S. Samoilenko: *Atlantic Hydrophysical Polygon-70*
12. M.A. Mardzhanishvili: *Seismic Design of Frame-panel Buildings and Their Structural Members*
13. E.'A. Sokolenko (editor): *Water and Salt Regimes of Soils: Modeling and Management*
14. A.P. Bocharov: *A Description of Devices Used in the Study of Wind Erosion of Soils*
15. E.S. Artsybashev: *Forest Fires and Their Control*
16. R.Kh. Makasheva: *The Pea*
17. N.G. Kondrashova: *Shipboard Refrigeration and Fish Processing Equipment*
18. S.M. Uspenskii: *Life in High Latitudes*
19. A.V. Rozova: *Biostratigraphic Zoning and Trilobites of the Upper Cambrian and Lower Ordovician of the Northwestern Siberian Platform*
20. N.I. Barkov: *Ice Shelves of Antarctica*
21. V.P. Averkiev: *Shipboard Fish Scouting and Electronavigational Equipment*
22. D.F. Petrov (Editor-in-Chief): *Apomixis and Its Role in Evolution and Breeding*
23. G.A. Mchedlidze: *General Features of the Paleobiological Evolution of Cetacea*
24. M.G. Ravich et al.: *Geological Structure of Mac. Robertson Land (East Antarctica)*
25. L.A. Timokhov (editor): *Dynamics of Ice Cover*
26. K.Ya. Kondrat'ev: *Changes in Global Climate*
27. P.S. Nartov: *Disk Soil-Working Implements*
28. V.L. Kontrimavichus (Editor-in-Chief): *Beringia in the Cenozoic Era*
29. S.V. Nerpin & A.F. Chudnovskii: *Heat and Mass Transfer in the Plant–Soil–Air System*
30. T.V. Alekseeva et al.: *Highway Machines*
31. N.I. Klenin et al.: *Agricultural Machines*
32. V.K. Rudnev: *Digging of Soils by Earthmovers with Powered Parts*
33. A.N. Zelenin et al.: *Machines for Moving the Earth*
34. *Systematics, Breeding and Seed Production of Potatoes*
35. D.S. Orlov: *Humus Acids of Soils*
36. M.M. Severnev (editor): *Wear of Agricultural Machine Parts*
37. Kh.A. Khachatryan: *Operation of Soil-working Implements in Hilly Regions*
38. L.V. Gyachev: *Theory of Surfaces of Plow Bottoms*
39. S.V. Kardashevskii et al.: *Testing of Agricultural Technological Processes*
40. M.A. Sadovskii (editor): *Physics of the Earthquake Focus*
41. I.M. Dolgin: *Climate of Antarctica*
42. V.V. Egorov et al.: *Classification and Diagnostics of Soils of the USSR*
43. V.A. Moshkin: *Castor*
44. E'.I. Sarukhanyan: *Structure and Variability of the Antartic Circumpolar Current*
45. V.A. Shapa (Chief editor): *Biological Plant Protection*
46. A.I. Zakharova: *Estimation of Seismicity Parameters Using a Computer*
47. M.A. Mardzhanishvili & L.M. Mardzhanishvili: *Theoretical and Experimental Analysis of Members of Earthquake-proof Frame-panel Buildings*
48. S.G. Shul'man: *Seismic Pressure of Water on Hydraulic Structures*

(continued)

RUSSIAN TRANSLATIONS SERIES

49. Yu.A. Ibad-zade: *Movement of Sediments in Open Channels*
50. I.S. Popushoi (Chief editor): *Biological and Chemical Methods of Plant Protection*
51. K.V. Novozhilov (Chief editor): *Microbiological Methods for Biological Control of Pests of Agricultural Crops*
52. K.I. Rossinskii (editor): *Dynamics and Thermal Regimes of Rivers*
53. K.V. Gnedin: *Operating Conditions and Hydraulics of Horizontal Settling Tanks*
54. G.A. Zakladnoi & V.F. Ratanova: *Stored-grain Pests and Their Control*
55. Ts.E. Mirtskhulava: *Reliability of Hydro-reclamation Installations*
56. Ia.S. Ageikin: *Off-the-road Mobility of Automobiles*
57. A.A. Kmito & Yu.A. Sklyarov: *Pyrheliometry*
58. N.S. Motsonelidze: *Stability and Seismic Resistance of Buttress Dams*
59. Ia.S. Ageikin: *Off-the-road Wheeled and Combined Traction Devices*
60. Iu.N. Fadeev & K.V. Novozhilov: *Integrated Plant Protection*
61. N.A. Izyumova: *Parasitic Fauna of Reservoir Fishes of the USSR and Its Evolution*
62. O.A. Skarlato (Editor-in-Chief): *Investigation of Monogeneans in the USSR*
63. A.I. Ivanov: *Alfalfa*
64. Z.S. Bronshtein: *Fresh-water Ostracoda*
65. M.G. Chukhrii: *An Atlas of the Ultrastructure of Viruses of Lepidopteran Pests of Plants*
66. E.S. Bosoi et al.: *Theory, Construction and Calculations of Agricultural Machines*, Vol. 1
67. G.A. Avsyuk (Editor-in-Chief): *Data of Glaciological Studies*
68. G.A. Mchedlidze: *Fossil Cetacea of the Caucasus*
69. A.M. Akramkhodzhaev: *Geology and Exploration of Oil- and Gas-bearing Ancient Deltas*
70. N.M. Berezina & D.A. Kaushanskii: *Presowing Irradiation of Plant Seeds*
71. G.U. Lindberg & Z.V. Krasyukova: *Fishes of the Sea of Japan and the Adjacent Areas of the Sea of Okhotsk and the Yellow Sea*
72. N.I. Plotnikov & I.I. Roginets: *Hydrogeology of Ore Deposits*
73. A.V. Balushkin: *Morphological Bases of the Systematics and Phylogeny of the Nototheniid Fishes*
74. E.Z. Pozin et al.: *Coal Cutting by Winning Machines*
75. S.S. Shul'man: *Myxosporidia of the USSR*
76. G.N. Gogonenkov: *Seismic Prospecting for Sedimentary Formations*
77. I.M. Batugina & I.M. Petukhov: *Geodynamic Zoning of Mineral Deposits for Planning and Exploitation of Mines*
78. I.I. Abramovich & I.G. Klushin: *Geodynamics and Metallogeny of Folded Belts*
79. M.V. Mina: *Microevolution of Fishes*
80. K.V. Konyaev: *Spectral Analysis of Physical Oceanographic Data*
81. A.I. Tseitlin & A.A. Kusainov: *Role of Internal Friction in Dynamic Analysis of Structures*
82. E.A. Kozlov: *Migration in Seismic Prospecting*
83. E.S. Bosoi et al.: *Theory, Construction and Calculations of Agricultural Machines*, Vol. 2
84. B.B. Kudryashov and A.M. Yakovlev: *Drilling in the Permafrost*
85. T.T. Klubova: *Clayey Reservoirs of Oil and Gas*
86. G.I. Amurskii et al.: *Remote-sensing Methods in Studying Tectonic Fractures in Oil- and Gas-bearing Formations*
87. A.V. Razvalyaev.: *Continental Rift Formation and Its Prehistory*
88. V.A. Ivovich and L.N. Pokrovskii: *Dynamic Analysis of Suspended Roof Systems*
89. N.P. Kozlov (Technical Editor): *Earth's Nature from Space*
90. M.M. Grachevskii and A.S. Kravchuk: *Hydrocarbon Potential of Oceanic Reefs of the World*
91. K.V. Mikhailov et al.: *Polymer Concretes and Their Structural Uses*
92. D.S. Orlov: *Soil Chemistry*
93. L.S. Belousova & L.V. Denisova: *Rare Plants of the World*

RARE PLANTS
OF
THE WORLD

Rare Plants of the World

L.S. Belousova
L.V. Denisova

RUSSIAN TRANSLATIONS SERIES
93

A.A. BALKEMA / ROTTERDAM / BROOKFIELD / 1992

Translation of:
Redkie rasteniya mira, Lesnaya
Promyshlennost', Moscow

© 1992 Copyright reserved

Translator :   Dr. B.R. Sharma
Technical Editor : Dr. G.N. Dixit
General Editor :   Dr. V.S. Kothekar

ISBN 90 6191 482 5

Distributed in the USA and Canada by: A.A. Balkema Publishers,
Old Post Road, Brookfield, VT 05036, USA

Preface

The International Strategy for the Conservation of Nature and Natural Resources was adopted in 1980. One of the major problems identified under it is the conservation of the vast diversity of biological species irrespective of their economic importance, as their disappearance is irreversible. The problem of conservation of the species diversity has two aspects: identification of the endangered species which require immediate protection measures and development and implementation of a system of such measures. The preparation of international and national red data books as well as books of rare and endangered plants of different countries and regions corresponds to the former objective. Simultaneously, measures are being developed and used for conservation of plants; special areas with different levels of protection are being established (national parks, prohibited forests, reserves, etc.), legislative measures are being taken at the national and other levels, and international agreements are being signed. A large number of the species is included in the appendix to the Convention of International Trade in Endangered Species of Wild Flora and Fauna in Commerce (CITES). This Convention, adopted in 1973 in Washington, bans trade in some species (Appendix 1) without special permission and restricts and regulates trade in other species (Appendix 2).

This book is compiled to illustrate the present status of the problem of conservation of the diversity in plants in different parts of the world. While collecting material for the book, the authors used the list of rare and endangered plants of different regions and states in Europe, Australia, New Zealand, the USSR, the United States, and other countries and numerous publications on identification and study of individual taxa that need protection, as well as the International Red Data Book of Plants prepared by a special commission on endangered plant species under the International Union for the Conservation of Nature and Natural Resources (IUCN). According to this commission, about 20,000 species of higher plants, or about 10 per cent of the world's flora, need protection at present. Only 250 taxa are included in the Red Data Book of the IUCN.

The present book provides a brief description of about 2000 rare and endangered plants. The plants are described under different families in alphabetic order. A brief description of the family is presented only at its first mention. The rarity categories of the IUCN have been used for characterization of species: extinct or probably extinct, endangered, vulnerable (decreasing in population strength), and rare.

The material on Asia, North America, Australia, and Oceania was prepared by L.S. Belousova, and the material on Europe, Africa, Central and South America, and the islands in the Atlantic and the Indian oceans by L.V. Denisova.

Contents

Preface	v
1. Europe	1
2. Asia	71
3. Africa	125
4. North America	167
5. Central and South America	208
6. Australia	229
7. Oceania	281
8. Islands of the Atlantic and Indian Oceans	316
Index of Latin Names of Plants	334

1
Europe

5 This continent is included in the Holarctic floristic kingdom; a larger part of the continent is in the circumboreal region, and a smaller part in the Mediterranean region. Europe has a fairly complex topography (a series of mountain systems and valleys at different altitudes) and a variety of climate from mild humid Atlantic to fairly hot Mediterranean and continental. All this has a great influence on the composition of its flora and vegetation. Besides, a large part of the continent has experienced repeated glaciation, which has also affected its flora. Several vegetational zones are located in Europe: tundra, taiga, broad-leaved forests, forest-steppes, steppes, and subtropical. The mountain regions are particularly rich in flora.

Europe is economically the most exploited continent, and its vegetation has been altered by humans to a great extent. For instance, the zone of agricultural lands appeared in place of the steppe zone; the relics of steppes have remained only on the protected territory (national parks, prohibited forests, rivers) or in small areas at inconvenient places (e.g., steep slopes, along the ravines). The broad-leaved forests have been replaced by small-leaved forests over large areas. As a result of reclamation work, many common plants of the past in water bodies and marshy habitats have disappeared or are close to extinction.

In all the European countries, depletion of flora is being observed, especially during the recent decades. In France, for example, 4 of the 73 endemic flowering plant species are extinct (or probably extinct), 7 are endangered, 10 are vulnerable, and 23 are rare. Some non-endemic species have also disappeared, for example, several Mediterranean orchids. Many aquatic plants have become rare as a result of reclamation work. In Yugoslavia, out of 135 endemic species, one is endangered with extinction, 5 are vulnerable, and 84 are rare. In Poland, about 10 per cent of the vascular plant flora disappeared during the last 100 years.

At present, laws or special regulations have been enacted in all countries to protect the rare plants. In some countries, the lists of protected species are fairly large. National Red Data Books are being prepared in

England, the USSR, Germany, and several other countries. The Commission on Rare Species of the IUCN has prepared and periodically reexamined the list of endemic, rare, or endangered plants of Europe.

Since it is impossible to describe all the endangered and rare plants of Europe, we shall restrict ourselves to a brief description of the members of individual families and genera. In several cases, we have not listed the reasons for reduction of one or another species; however, the most important reason is change in habitat due to various types of economic exploitation of the land. The characteristics of the endangered and rare plants of Europe growing on the territory of the USSR (with a few exceptions) are also not included. Information on these plants can be obtained from the books on rare and endangered plants of the USSR and its individual regions (e.g., Ukraine, Crimea) published earlier.

FAMILY AMARYLLIDACEAE

The family Amaryllidaceae has about 1000 species, which are widely distributed in the arid regions of the tropics and subtropics. The maximum species diversity is found in South Africa (the Cape region), South and Central America, and the Mediterranean countries. They are mainly perennial, herbaceous, bulbous plants with regular flowers, arranged in umbelliferous inflorescence, enclosed in a membranous cover before blooming. Perianth is simple, petaloid; there are 6 stamens; pistil consists of 3 fused carpels. The fruit is a capsule or berry. Many species of amaryllis, clivia, crinum, and snowdrops are very ornamental, several plants contain alkaloids, and some are poisonous. The population of the plants of this family is declining because of their collection as well as cultivation of land.

The genus *Leucojum* (snowflake) includes bulbous plants with linear, lobed leaves, distributed in the Mediterranean and broad-leaved forest regions of Europe and North Africa. Among them many species are ornamental and widely cultivated. Translated from ancient Greek, the generic name means 'white violet'. Three European species of snowflake are faced with extinction.

Long-leaved snowflake [*L. longifolium* (Gay ex Roemer) Gren. et Godron] is a plant with an inflorescence of 7–10 mm diameter. Leaves are 12–25 × 1–2.5 mm, narrow-linear, and appear after blooming. The leafless flowering shoot is 15–27 cm long, the umbel consisting of 1–4 white flowers. The plant flowers during spring. It is found on stony dry slopes in Corsica. Plant population is decreasing due to collection for bouquets.

Nice snowflake (*L. nicaeënse* Ard.), an endemic plant that needs protection, is a perennial with 15–20 mm bulb and 10–30 × 1.5–2.5 mm long, narrow-linear leaves, developing after blooming. Stem is 5–18 cm long, weak, usually with one, rarely with 2–3 white flowers; seeds are black. The plant flowers during spring. Its distribution is limited to the exposed rocks in Provence from Nice (southern France) to the border with Italy, possibly, partly extending into Italy. Many populations are already extinct or extremely reduced.

Valencian snowflake (*L. valentinum* Pau) externally resembles the Nice snowflake, but differs in that it flowers in late summer or autumn. It is found in the eastern region of Spain (northern Valencia), and there are reports of its occurrence in northwestern Greece.

The daffodil genus (*Narcissus*) includes the most ancient cultivated ornamental plants. The large number of cultivated forms of daffodil are a result of hybridization with many wild species. The genus has about 60 species, mainly distributed in southern Europe and the Mediterranean region. They are all annual bulbous plants with 2–4 linear leaves and long leafless stems. The flowers are fairly large, often scented, with campanulate or cup-shaped corolline corona at the throat; the limb and corona are often of different color. Three endemics of Spain, one of Portugal, and two of France belong to the rare daffodils of the world. The number of species is declining because of collection and destruction of habitats.

Cuatrecasian daffodil (*N. cuatrecasisii* Casas et al.) is a plant with broad, straight leaves, width 2.2–4.5 mm, with 2 dorsal keels and smooth margins. Flowers are 22–30 mm in diameter, solitary, rarely in pairs. Tube length is 12–15 mm; perianth lobes 9–12 × 7–11 mm; and corona 4–6 × 8–10 mm with entire or crenate margins. The plant grows on limestone rocks in the mountains of central and southern Spain.

Sessile daffodil [*N. hedraeanthus* (Webbt et Colm. Heldr.)] is a plant with dark brown, 15–10 mm [sic] bulbs; leaves are solitary, 6–12 cm × 1 mm, dark green, almost straight. Stem is ascending or procumbent, shorter than the leaves. Flowers are pale yellow, light green; tube length is 10–15 mm; perianth leaves are 8–13 × 3–4.5 mm, oblong, obtuse; corona is 10–15 mm, obconical. The plant grows in mountain meadows and stony soils in the southeastern and southern part of central Spain.

The distribution of **long-spathe daffodil** (*N. longispathus* Pugsley) is in the Sierra de Casorla mountains (Spain). This is a plant with bulbs of 30 mm diameter, leaves 40–60 cm or longer and 10–15 mm broad, and stem 30–175 cm long (plant height depends on water level). Flowers are solitary or paired; tube length is 10–15 mm; perianth lobes are 25–32 mm, yellow in the middle, loosely spreading corona is 25–30 mm long, darker than perianth, crenate. The plant grows along the banks of mountain streams.

The **scaberulous daffodil** (*N. scaberulus* Henriq.), an endemic of central Portugal (valley of the Mondegu River), is a rare species. The leaves are ascending, 1–3.5 mm broad, with 2 keels and minutely rough margins. Flowers are 12–17 mm in diameter, solitary, or in groups of 2–3; floral tube is 12–17 mm long; perianth lobes are 4.5–7 × 3.5 mm; corona is 2–5 × 5–7 mm with entire, crenate margins. The plant grows along rocky places.

The **French daffodil** (*N. tazetta* L.) is possibly the most beautiful species. Plant height is 30–50 cm, diameter of bulbs is 2–5 cm. Leaves are 4–6, linear, flat, with a keel, gray-green, almost equal to the stem in length. Inflorescence is umbellate, consisting of 3–18 flowers. Floral tubes are greenish, cylindrical, up to 2 cm long; perianth lobes are horizontally divergent, ovate, white; corona is cup-shaped, golden-yellow (Plate 1). Blooming begins in December and continues till May. The plant grows in humid meadows near the Mediterranean Sea in France. At one time the French daffodil was very common and found in large numbers. Now its population has reduced and continues to decline, as does that of its subspecies. It has been cultivated since 1557.

The rare species **calciphilous daffodil** (*N. calceolus* Mendonça) is found in the crevices of limestone rocks in western and central Portugal. Its leaves are 2–5 mm broad, straight, with 2 keels. Flowers are 1–4, usually 17–25 mm in diameter; perianth is 4–8 × 6–9 mm (Plate 1).

FAMILY APIACEAE (UMBELLIFERAE)

The Apiaceae family mainly consists of herbaceous plants of the temperate belt, very rarely trees and shrubs (in tropics and subtropics). Usually, they are perennial or biennial plants, often with a vertical and thick rhizome (sometimes obliquely ascending). Leaves are alternate, with sheaths. Flowers are pentamerous, bisexual, rarely unisexual, arranged in complex umbels. Fruit is usually dry (cremocarp), divided into 2 medicarps. The family has about 300 genera and more than 3000 species, among which many species have essential oils, have edible fruit, or are poisonous.

The **longwort angelica** (*Angelica heterocarpa* Lloyd), indigenous to Portugal, is threatened with extinction. It is an almost leafless perennial plant with a height up to 200 cm; leaves are bipinnate; leaflets are 10 × 3 cm, ovate-lanceolate. Bracts are either absent or a few in number. The plant suffers due to land reclamation work.

Hare's ear (*Bupleurum kakiskalae* Greuter) has been included in the Red Data Book of the IUCN. It is a monocarpic perennial. In vegetative condition the plant lives up to 12 years and produces a solitary, unbranched, strong woody stem, 12 cm long and about 1 cm thick, emerging from a thick rosette of 15–30 oblanceolate leaves. The reproductive branches, up to 1 m high, emerge from the rosette of the previous year. Inflorescence is sparse, profusely branched, with umbels consisting of 4–6 rays; bracts and bracteoles are 5–9 in number, unequal, leafy, obtuse, ligulate or lobate. Flowers are yellow.

At present, the only habitat of hare's ear known is the Island of Crete, in the Nom Hanya mountains, where it grows in the crevice of a vertical rock at the height of 1450 m. The population is at critical level: flowering was not recorded from 1966 to 1971, and only 4–5 plants flowered in 1971. Because of a small number of flowering plants, there is danger of inbreeding. The plant can be preserved by introducing it into cultivation. Scientifically, it is interesting as an isolated and only species in the genus with such a life cycle.

Three other species of longwort are endangered and 2 species have been included in the category of rare species.

The eryn16 genus (*Eryngium*) includes 230 species, distributed in tropic and temperate latitudes of both hemispheres, predominantly in arid regions; some species are ornamental. Two European species are endangered with extinction.

White-flowered eryngo (*E. spinalba* Vill.) is a perennial with a very strong, 20–35 cm tall stem. The basal leaves are leathery, 5–8 × 5–10 cm long, roundish, palmately divided, with 3–5 imparipinnate segments, dentate, with awns. Petioles are twice as long as lamina. Inflorescence is usually bluish, with 10 oval-cylindrical heads, 4–6 × 2–3 cm in size. Involucral bracts are 6–9 cm long, linear-lanceolate, often 3-fid with 15–30 spinelike teeth. The plant grows on dry stony places in the southwestern Alps (France, Italy).

The **alpine eryngo** (*E. alpinum* L.), close to extinction, is a perennial with a height of 30–70 cm. Its basal leaves are soft, with lamina 8–15 × 5–13 cm, ovate, triangular-cordate at the base, irregularly serrated, with very long petioles. Inflorescence is light blue, with 1–3 drooping ovate-cylindrical heads of 2–4 × 1.5–2 cm size. Bracts number more than 25 and are 3–6 cm long, many of them pinnate, with numerous crestate-roundish lobes and long and tender awns. Bracteoles are entire or tricuneate (Plate 1). The plant is found in meadows and grassy swales, usually on lime soils in the Alps, Jura, and mountains of western and central Yugoslavia; it was once found in the western Carpathians.

Sea holly (*E. maritimum* L.) has been placed under protection in many European countries, as its population is declining. It grows on sandy and stony sea coasts.

The **giant fennel** (*Ferulago capillaris* Link ex Sprengel Cout.), endemic to southern Portugal, has possibly already disappeared. It is a light green plant with almost cylindrical grooved stem. Leaflets are filiform; bracts are setose, straight; petals are yellow. Fruits are flattened and winged.

Cow parsnip or **hogweed** (*Heracleum minimum* Lam.) is placed in the group of vulnerable plants. It is a rhizome-producing perennial species with smooth, sometimes pubescent stems, up to 30 cm tall. Leaves are glabrous, bipinnate or tripartite. Umbels are 6–9 cm in diameter, petals are white or pinkish. Fruits are broadly elliptical. The plant grows only in the French Alps.

Laserwort (*Laserpitium longiradiatum* Boiss.) is endangered. It is a perennial with fibrous rhizome and cylindrical stem, 60–150 cm tall. Leaves are bipinnate, with lobes 18–30 mm long. Fruit is narrow-winged. The plant is indigenous to southern Spain, growing in the Sierra Nevada.

The genus *Lereschia* is represented in Europe by a single species: the Thomas lereschia (*L. thomasii* Boiss.). This rhizome-producing perennial has mainly basal leaves, ternate, with rhombic segments, entire or serrated margins in the lower half and dented or lobed in the upper half. Umbels are irregular, arranged like leafless petioles. The plant is found along the humid places in Italy (Calabria).

Albanian ligusticum or **bladder-reed** (*Ligusticum albanicum* S. Jávorka) is an almost glabrous plant with nonfibrous rhizome. Stem is hollow, 10–50 cm tall, with 1–2 branches; leaves are triangular, bipinnate or tripinnate, lobes are narrow linear; bracts number up to 7, involucral bracts are often longer than the umbels. The plant is indigenous to northern Albania.

Ligusticum lucidum Miller ssp. *huteri* (an endemic of the Balearic Islands) and *L. corsicum* J. Gay (an endemic of Corsica) are also rare in Europe.

Balearic naufrage (*Naufraga balearica* Constance et Cannon), the only species in the genus, is endemic to Baleares (1–2 locations known on Majorca). It is a small, turfy annual with a height of 2.5–4 cm, forming pads of leafy rosette at the tips of short stolon. The leaves have long petioles, 3–5 small ovate or oblongate-ovate 1.5–5 mm long leaflets; petioles have an auriculate membranous base. The leaves (on stems) have 3 leaflets, with papery-white stipules, arranged in groups of 2–4. The umbels appear in the leaf axils and consist of 1–8 small white flowers. Among other umbellifers, it is distinguished by its fruit and well-developed stipules.

Naufrage grows on steep, humid, marine, and permanently shaded cliffs facing north and forms dense groups in the upper part of slopes and in eroded areas. It was first found in 1962, and again in 1969. Possibly, it is still preserved on the coastal inaccessible cliffs. The Balearic naufrage is one of the few species of island plants with relatives in Aus-

PLATE 1. 1—Calciphilous daffodil; 2—French daffodil; 3—cultivated parsley; 4—alpine eryngo.

PLATE 2. 1—Small-flowered centaury; 2—ornamental telekia; 3—ragus centaury; 4—subacauline berardia.

tralia, New Zealand, and Chile. It is included in the Red Data Book of IUCN.

Possibly, **dropwort** (*Oenanthe conioides* Lange) has already disappeared. It was earlier found in the valley of the Lower Elbe (FRG) and in Belgium. It is a plant with a rough, straight stem. The leaves are bipinnate or tripinnate, leaflets are ovate or semicircular with cuneate base and small lobes. Cauline leaves are bipinnate, with ovate lobes. Umbels are opposite the leaves; rays number 6–15.

The genus *Petagnia* is represented by a single species in Europe, which is endangered. This is the **sanicle-leaved petagnia** (*P. saniculifolia* Guss.), growing along the margins of streams in northern Sicily, a perennial plant with strong rhizome and a stem up to 50 cm tall. Basal leaves have long petioles, pentapartite; lobes are serrated; cauline leaves are almost sessile, tri- or pentapartite. Inflorescence is a compound umbel with central female or bisexual flowers surrounded by 2–4 male flowers. Bracts and bracteoles are small.

The parsley genus (*Petroselinum*) includes biennial or annual herbaceous plants with unipinnate or tripinnate leaves and yellowish-green or almost white petals. The genus has 5 species, which have been found in Europe and along the entire Mediterranean region. One species is widely cultivated as a food plant. The **cultivated parsley** [*P. segetum* (L.) Koch.], a tender bluish plant with a height up to 10 cm, is endangered. Stem is hard, round, with scattered branches. Leaves are linear-oblong, pinnate; leaflets are 3–10 mm long, ovate, serrated or lobate, with thick margins. Umbels have 2–5 highly unequal rays. Bracts number 2–5, petals are white (Plate 1). This plant is indigenous to western Europe and is found from the Netherlands and England to Portugal and the eastern part of central Italy.

Close to extinction is **burnet saxifrage** (*Pimpinella bicknellii* Briq.), a perennial plant with height up to 50 cm and strong stem. Leaves are biternate and pinnatisect; leaflets are ovate, lobate-serrate; rays number 5–7. Flowers are white. The plant is an endemic of Baleares (Majorca Island).

The meadow saxifrage genus (*Seseli*) is represented by perennial or biennial plants with 2–4 cuneately divided leaves. Bracts are numerous; calyx lobes are small; petals are white, light yellow, or pinkish. Four species are rare, 2 of them endemic to Bulgaria. **Degen's saxifrage** (*S. degenii* Urumoff) is a glabrous perennial with a height up to 100 cm. Leaves are ternate pinnate with $30-40 \times 0.5$ mm filiform lobes; rays number 12–20; bracts are pubescent and fused at the base; petals are white. The plant grows along dry limestone deposits in the northern regions of Bulgaria.

FAMILY ASPLENIACEAE

The family Aspleniaceae includes small deciduous or evergreen ferns with rhizomes covered with hard scales. Sori are oblongate or linear.

The spleenwort genus (*Asplenium*) includes about 680 species, distributed throughout the world. It is a plant with short, erect rhizomes, covered with dark, oblong-triangular, linear-oblong or filiform scales. Leaves are agglomerate; petioles are usually dark; sori are elliptical or linear, covered.

The **Jahandies spleenwort** [*A. jahandiezii* (Litard.) Rouy.] is included in the Red Data Book of IUCN. It is a small fern with short scaly rhizome, bearing at its tip a rosette of almost straight, pale green, linear-elliptical, pinnate, 3–10 cm long leaves on 4–25 mm long, green petioles. Leaflets are 5–10 mm long, oblong, unequally dented. Spores are in numerous covered sporangia. Sori are arranged in 2 rows on the lower surface of the leaves. The plant is found in the Verdon ravine over a distance of 12 km and in adjoining ravines of the Provence Alps (southern France). It grows on humid shady limestone rocks at the height of 400–700 m above m.s.l. together with the widespread species *A. trichomanes*. Of the known populations, 2 have already disappeared, and the populations of others decreased.

FAMILY ASTERACEAE

The large family of Asteraceae consists of more than 20,000 species, widely distributed throughout the world. They are particularly numerous in the hilly regions, steppes, and deserts. The family is distinguished by a large variability of living forms. It includes herbs, semishrubs and rarely shrubs (in temperate latitudes), creepers and trees (in tropics and subtropics). The members of this family have small flowers, which are arranged in heads of diameter varying from a few millimeters to 30–50 cm. Flowers in the head are bisexual, tubular, ligulate, or of other shape; fruit is an achene.

Many plants of this family are of great economic importance as sources of food, essential oils, latex, and medicine.

Downy sowthistle [*Andryala levitomentosa* (Nyár.) Sell.], threatened with extinction, is an endemic of Rumania. It is found in the eastern Carpathians, 80 km toward the southern border with the USSR. It is known only in the Piatra mountains from an area of about 150 m^2, where it forms 3 spots, each not more than 30 m^2. This is the only species of the Mediterranean genus which has penetrated so far into the north, a relict of the Tertiary period. Downy sowthistle is a perennial herbaceous plant with lignified, strongly branched, 6–20 cm high stem, covered with glandular, stellate, and long simple hairs. Numerous leaves, 10–100 mm long and 10–25 mm broad, form a rosette at the base, varying from almost roundish to ovate or broadly elliptical in shape, with winged petioles covered with stellate and simple hairs; cauline leaves are small and linear. A single stem has 1–2 heads of 15–20 mm diameter, each covered with hairy lanceolate involucral bracts. Flowers are yellow, ligulate; ligules are divided into 5–7 hairy segments at the apex. Achene is 1.5 mm long, obconical; pappus is longer than the corolla tube; peduncle has fimbriate scales, covering the flowers. Seeds have not been found; possibly, the plant reproduces only vegetatively.

Downy sowthistle grows on deep mountain cliffs facing south or southeast, at the height of 1600–1700 m above m.s.l. on metamorphic crystalline rocks, shales, and gneiss with high carbonate content. Downy sowthistles have been included in the Red Data Book of the IUCN.

Norwegian cat's foot (*Antennaria nordhageniana* Rune et Rǫnning.) is a herbaceous perennial not taller than 6 cm. Lower surface of leaves, flowering shoot, and lower half of stipules are reddish. Heads number 1–2; flowers are white or yellow. The plant grows in patches of thawed snow in the mountains of northern Norway, where it is known from several places at 70°N. Possibly, it is also found in Finland. The plant is threatened with extinction.

Eight European species of the camomile (dogfennel) genus (*Anthemis*) need special protection measures. These are annual or perennial herbs, sometimes semishrubs, typical for the Mediterranean regions.

A vulnerable species, indigenous to southeastern France, is **Gerard's camomile** (*A. gerardiana* Jordan), a perennial plant with 20–40 cm tall stems, usually straight, unbranched, or with 1–3 branches. Stems are surrounded by vegetative shoots, which are 1/10 to 2/9 of the stems. Leaves are 1.5–4 cm long, bipinnate, fleshy, green, covered with compressed hairs. Head is 25 mm in diameter; thalamus is conical; involucre is hemispherical. The plant grows at stony places in Provence and Rhone Alps and is disappearing due to increasing use of lands for agriculture.

The wormwood genus (*Artemisia*) consists of more than 400 species, mainly distributed in the temperate regions of the northern hemisphere. These are perennial, biennial, and annual herbs or semishrubs with erect or ascending, sometimes prostrate stems, and alternate, usually divided leaves. Numerous minute heads are borne in inflorescences of various types.

The **Granada wormwood** (*A. granatensis* Boiss.), indigenous to Spain, is threatened with extinction. It is a dense, pad-like, silvery-silky perennial up to 18 cm tall. Leaves are pentapartite, petiolate, with trilobed segments; cuneate head is solitary or in a terminal raceme, broadly hemispherical, 4–7 mm in diameter, with 80 very small flowers. Bracts are ovate-lanceolate, acute, reddish in the center. Corolla is dark, its lobes covered with dense short hairs. The plant mainly reproduces vegetatively. Its distribution is limited over 6 km^2 in the Sierra Nevada at the height of more than 3200 m above m.s.l. The plant grows on acidic shales along with many other endemics.

The Granada wormwood has become very popular during the last decades; it is collected for preparing wormwood tea and tinctures. This has resulted in a sharp reduction of its population. It has poor survival under cultivation. It is included in the Red Data Book of IUCN. A few protected reserves have been created in Spain to conserve this plant, and the seeds are being preserved in the seed bank in Madrid.

Possibly, the **insipid wormwood** (*A. insipida* Vill.) is already extinct. It is a 10–70 cm tall, hairy perennial with long rhizome and numerous vegetative branches. The heads are borne in a short panicle. The plant was found on the rocks in the southwestern Alps (France); it has not been found during recent years.

Molinier's wormwood (*A. molinieri* Quezel, Barbero et Loisel) is a rare endemic of France, a fragrant semishrub with a height not more than 60 cm, and ovate heads borne in a dense spicate inflorescence. Each head bears 10–15 flowers. The plant grows in southern France, near the town of Flassan* (Var).

Pyrenean aster (*Aster pyrenaeus* Desf. ex DC.) is a perennial 40–90 cm tall, with strong, erect stems and oblanceolate, sessile leaves. Heads are solitary or in groups of a few in a loose comb. The plant is ornamental. Its distribution is restricted to the western and central Pyrenees (France). It grows on rocks in grassy communities. It is disappearing due to destruction of habitats and collection.

Bairnwort daisy (*Bellis bernardii* Boiss. et Reuter), a perennial, rare endemic of France, grows in humid grassy places among rocks in the Corsican mountains and usually forms large groups. Leaves are not more than 10 × 6 mm, glabrous, entire or with 1–3 distantly placed teeth on each side. Diameter of flower head is up to 12 mm; ligules are 4–6 mm long.

Subacauline berardia (*Berardia subacaulis* Vill.) is a perennial herb, with short, densely pubescent stem. Leaves are oblongate, entire or slightly serrated and incised at the base, petiolate. Head is hemispherical. Bracts are unequal, in 3–4 rows, densely pubescent. Flowers are bisexual and cream to pale yellow or pinkish (Plate 2). The plant grows on rocks

---

*Some place names could not be verified, as they were transliterated from the original language into Russian.

and along stony waste, above 1500 m above m.s.l. in the southwestern Alps (France and Italy).

The **suffruticose marine calendula** [*Calendula suffruticosa* Vahl ssp. *maritima* (Guss.) Meikle] is found in the western part of Sicily. This is a plant with prostrate stems, densely glandular pubescence, and fairly fleshy leaves, which are entire or sparsely serrated. Heads are 3–8 cm in diameter, flowers are ligulate and tubular, yellow. The plant grows on pebbles and sands along the sea coast. It is endangered.

The cauline thistle genus (*Carlina*) comprises biennial and perennial plants, sometimes with highly reduced stems, rarely shrubs with basal or alternate, serrate or pinnately lobed, pinnate or bipinnate leaves with prickly teeth or lobes. The distribution of this genus is the Mediterranean region, Europe, and the temperate regions of Asia.

The **Dian cauline thistle** [*C. diae* (Rech. f.) Meusel et Kästner] is included in the Red Data Book of IUCN. It is a dense, white, tomentose dwarf shrub with several woody branches; numerous short, fruitless (vegetative) branches are covered with 5–8 cm long, lanceolate leaves. Flowering shoots are rarely foliate, 40–60 cm high, terminating in 1–4 heads, each up to 15–35 mm in diameter, with multirowed involucre; outer involucral lobes are leafy, 10–15 mm long, entire; inner lobes are similar to petals, bright yellow, glossy, hard, serrated, 10–16 mm long. Flower is tubular, with yellow corolla and a hairy pappus.

The distribution of this species is limited to a few islands in the Mediterranean Sea. Two varieties are known. One is found on Nicos Dia Island (eastern part of the Mediterranean Sea), with an area of 16 km^2. Cauline thistles were found here in large numbers in 1962 after the restoration of *Phrygana* as a result of 10 years' ban on cattle grazing. However, after the introduction of Cretan ibex on the island, which multiplied fast, the main population of cauline thistle was reduced to 3 plants, and a few more plants survived on cliffs on the eastern part of the island. In general, the entire population is threatened with extinction. The other variety has been reported from the northeast of the Isle of Crete, from the Dragonada and Gianisada islands and at the Sidheros peninsula.

The Dian cauline thistle grows on steep limestone rocks facing the sea, together with Turnefor's woodruff, grape hyacinth, and other endemics. The plant is ornamental. It is of scientific interest as a relict of the Tertiary period, belonging to the primitive subgenus *Lyrolepis*.

Cirsioid cauline thistle (*C. cirsioides* Klok.) has a small area in the USSR (right bank of the Dnepr, Carpathians) and Poland (the Lublin elevation). It is a biennial, herbaceous plant, 20–40 cm tall, with simple densely pubescent stem. Leaves are petiolate, pinnatisect or compound pubescent. A large solitary head, 7–10 cm in diameter, is found at the tip. Corollas are yellowish brown, membranous. The plants are often collected for dry bouquets. They grow in well-lit forests, along dry ravines and steppe slopes. **Cotton thistle-leaved cauline thistle** (*C. onopordifolia* Bess. ex Szafer *et al.*) also has a limited area, in the USSR (Volyn-Podol'sk plateau) and Poland. Distributed sporadically and with a few plants, it grows along dry steppe slopes facing south in the outgrowths of shrubs. It is a perennial acauline plant with a rosette of pinnate-lobed leaves densely pubescent on the upper surface and fluffy on the lower surface. Head is large, 2 cm in diameter, solitary, located among the rosette of leaves.

The reserves of cauline thistles are shrinking due to collection of plants for bouquets as well as destruction of their habitats during reclamation of land for agriculture. Both species of cauline thistle are included in the Red Data Book of the USSR; in Poland, they are protected by law.

The centaury genus (*Centaurea*), predominantly a Mediterranean genus, has more than 500 species.

In Europe, about 30 centaury species require special protection. Spain is particularly rich in rare centauries, where it is proposed to conserve 9 endemic species. The **small-flowered centaury** (*C. micracantha* Dufour) is an annual with erect branched stem, 20–50 cm tall, with discontinuously winged branches. Leaves are ciliate-pubescent, serrated; lower leaves are pillate, upper leaves are lanceolate. Heads are arranged in corymbs. Flowers are purple (Plate 2). The plant is distributed in southern Spain.

In Yugoslavia, the endemic **ragus centaury** (*C. ragusina* L.) is protected. It is a white tomentose plant, 30–60 cm tall. Leaves are basal, their petiolate segments ovate to oblongate, obtuse. Involucre is 20–25 mm in diameter, spherical, bracts are whitish-tomentose, with triangular brownish processes and a terminal spine curved backward. Flowers are pale yellow (Plate 2). The plant grows along coastal rocks and shores in the western part of the country.

Corymbose centaury (*C. corymbosa* Pourret) is a biennial with erect, branched, 10–30 cm long stems. Leaves are green, glandular-punctate, lower leaves are bipinnate. Involucre is 18–20 × 8–10 mm, ovate; its bracts are blackish in the middle, with turned tips and membranous margin. Flowers are purple; pappus is 3–3.5 mm long, equal to achene. The plant grows on the elevations in southern France (Languedoc-Roussillon), near Narbonne. Found very rarely, its population is decreasing due to collection. The plant is ornamental.

The thistle genus (*Cirsium*) includes more than 200 species. Some of them are weeds or growing in large numbers in degraded meadows. Ornamental and melliferous species are also found. Two Spanish endemics are considered rare.

Bourgean thistle (*C. bourgaeanum* Willk.) is a biennial, 80–110 cm tall, with fibrous roots and stems branched in the upper part. Leaves are almost flat, elliptical-lanceolate, pinnate; leaflets are bi- or trilobate with 2–3 mm long, tender cusps. Heads are in groups of 2–8 at the tip of the stem and branches. Involucre is 10–12 × 7–9 mm; bracts are erect, glabrous; upper bracts are obtuse, middle bracts have a membranous-ciliate process. The plant grows in marshes near Avilés.

Valentian thistle (*C. valentinum* P. Porta et G. Rigo) is a perennial, 20–70 cm tall, with branched leafless stems. Leaves are oblongate-lanceolate, pinnate, with long hairs on both sides, arranged along the veins. Bracts are triangular, lobate, with a 4–8 mm long spine. Head is solitary, flowers are purple.

The **hawk's beard crepis** [*Crepis suffreniana* (DC.) Lloyd], an annual, 3–35 cm tall with numerous basal branches, is threatened with extinction. The basal leaves are lobate, serrated; cauline leaves are lanceolate, acute, and sessile. Involucre has 10–12 linear outer and 10–16 linear-lanceolate inner bracts. Distribution is confined to the southern and western regions of France and Italy.

The artichoke genus (*Cynara*) includes 10–11 species, distributed in the Mediterranean region and the Canary Islands.

Algarbian artichoke (*C. algarbiensis* Coss.) is a whitish-pubescent plant, 10–50 cm tall; stems are erect; leaves are elliptical and lanceolate, short-petiolate with web-like coating, bluish-green on the upper side, white pubescent on the lower side, and spiny along the margins. Bracts are greenish or purple. Crown is purple to dove-blue. A very rare species, endemic to Portugal, it is found on cultivated lands in the south of the country.

A very rare endemic of Austria, **leopard's bane** (*Doronicum cataractarum* Widder), has been included in the Red Data Book of IUCN. This is a rhizomatous perennial, 130 cm tall. Basal leaves are glabrous, cordate, with long petioles; cauline leaves are smaller and narrower, sharply serrated. Head is 4–7 cm in diameter. Ligulate flowers are glandular-pubescent. Achenes are of 2 types: those developing from ligulate florets are usually glabrous, and those developing in tubular flowers are glandular-pubescent.

Leopard's bane is found in Koralpe on a small area in the western part of the central Alps (southwest of Graz). It grows along the banks of streams, on moist rocks, in shady places at the height of 1250–2000 m above m.s.l. Possibly, this plant is a relict of the Tertiary period, as it is found together with other relicts. It has poor competitiveness, and it suffers from cattle grazing and digging of plants for collection and transfer to ornamental gardens. At present, about two-thirds of its area is under legal protection; collection of not only plants but also individual plant parts is banned.

The genus *Hymenostemma* is represented by a single species in Europe: the **false chamomile hymenostemma** [*H. pseudanthemis* (Kunze) Willk]. This is an annual with slightly pubescent stem, 10–20 cm tall, simple or branched. Leaves are 5–20 × 2–3 mm, crestate-pinnate. Leaf segments are oblong, obtuse, alternate. Disc is solitary, 10–20 mm in diameter. Bracts are in 2–3 rows, unequal; inner bracts have a broad membranous margin. Thalamus is conical. Ligulate white flowers are yellowish at the base; tubular flowers are yellow. Achenes are oblongate-ovate with 5–6 whitish ridges. The plant grows along dry shady places in southwestern Spain (Cadiz province).

The *Jurinea* genus includes perennial herbs or semishrubs, often with reduced stem. Leaves are simple or pinnatisect. Heads are multiflorate; flowers are all bisexual, developing into fruits. Corolla is straight or slightly obliquely quinquepartite, narrowed at the base. About 300 species of this genus are distributed in Eurasia from Atlantic to eastern Siberia, in northwest, middle and also central Asia.

Tsar Ferdinand jurinea (*J. tzar-ferdinandii* Davidoff) is a perennial with a height of 15–30 cm and thick woody rhizome. Basal leaves are 4–15 cm long, with entire, rolled margins, strigose on the upper surface, lower surface densely pubescent. Heads number 4–8, are obconical. Bracts are narrow, pointed, membranous along the margin, glabrous or densely pubescent, purple along the margin, white or pale green. Pappus is 2–4 times as long as the achene. The plant is found along the limestone slopes. It is endemic to Bulgaria, threatened with extinction.

Font-Quer jurinea (*J. fontqueri* Cuatrec.) is a perennial with scaly rhizome. Stems are up to 4 cm tall, leafy. Leaves are ovate, densely tomentose on the lower surface and puberulous or almost glabrous on the upper surface. Heads number 3, with diameter of 3–4 cm. Bracts are lanceolate, pointed, puberulous, green. Flowers are bisexual. The plant grows along limestone slopes at the height of 2000 m above m.s.l. in the mountains of southern Spain.

Plumb lettuce (*Lactuca livida* Boiss. et Reuter) is a prickly biennial. Lower leaves are oblongate-spatulate, entire or lobate, narrowed into a long petiole; upper leaves have curved margins and sagittate base. Inflorescence is a panicle, heads have 25 yellow flowers; achenes are elliptical, narrow-winged, black, with 5–6 ridges. The plant grows along shaded rocks in the Toledo mountains (Spain) and is found very rarely.

Long-dented lettuce (*L. longidentata* Moris) is a bluish-green biennial, up to 100 cm tall. Lower leaves are obovate to lanceolate, or pinnate-crenate, narrowing into petiole; upper leaves are ovate, deeply incised, dented, acute, with amplexicaul base. Inflorescence is in the form of a pyramidal panicle with ascending branches. Heads have 5–6 yellow flowers. Achenes are 7–9 mm long, oblongate, black, with 7–11 ridges. The plant grows along limestone rocks of Sardinia.

Small-head lamyropsis [*Lamyropsis microcephala* (Moris) Dittrich et Greuter] is a perennial with a height of 20–50 cm, with whitish and densely pubescent stems. Leaves are broad oblongate-lanceolate, pinnate, ciliate along the veins; leaflets have 2 or 3 deeply incised, narrow triangular lobes, terminating in hard spines. Heads are solitary or a few are found in terminal raceme, surrounded by upper leaves. Flowers are whitish. The plant grows along stony rocks in Sardinia and is endangered.

Deer-horn launaea [*Launaea cervicornis* (Boiss.) Font-Quer et Rothm.] is a spiny dwarf shrub, 5–20 cm tall, with dense branches. Leaves are mainly basal, serrated to pinnately divided, their lobes triangular, pointed. Heads are 8–10 × 3–4 mm, borne on lateral branches. Bracts are 1–1.5 mm long, outer ones with apical appendages. Achenes are cylindrical, often curved. The plant grows on rocks, mainly near the sea. It is endemic to Baleares and is a rare plant.

The edelweiss genus (*Leontopodium*) includes perennial herbs, sometimes with highly reduced stems, fruitless rosette or bundles of leaves and simple straight leafy stems. Leaves are linear to lobate, on the lower side usually more pubescent and light-colored. About 30 species of the genus are distributed in Eurasia, mainly in the high mountains and mountain-steppe belts. The **alpine edelweiss** (*L. alpinum* Cass.) is a plant with short and thick ascending, simple or shortly branched rhizome, covered with dark brown remnants of leafy sheaths. Flowering shoots are solitary or sometimes in tufts, leaves are grouped into fruitless rosettes. Numerous bracts are narrowly ovate or triangular-ovate, acute or obtuse, snow-white tomentose, forming a red multiradiate 'star' with 2–2.5 cm diameter; rarely the inflorescences have a few 'stars' (Plate 3). The plant grows in the mountains of Europe (Carpathians, Alps), on rocks, cornices, and exposed areas in the alpine and subalpine belts. A rare ornamental plant, its population is decreasing due to destruction by tourists. It is being protected in all European countries.

The daisy genus (*Leucanthemum*) includes about 20 species, distributed predominantly in the mountain regions of southern and central Europe. Only one species has a wide Eurasian area of distribution. Many plants are ornamental.

Corsica daisy [*L. corsicum* (Less.) DC.] is 20–60 cm tall, with simple, sometimes pubescent stem. Basal and lower cauline leaves are lobate or pinnatisect, middle cauline leaves are pinnatilobate to bipinnatisect. Heads are 2–3 cm in diameter. Bracts have brownish-black margin. The plant grows on the rocks and stony slopes in Corsica. It is endangered.

The genus *Leuzea* is represented by rhizomatous herbaceous plants with simple fine ridges and stems slightly swollen at the tip, terminating in large, almost spherical heads. Flowers are bisexual, with tubular corolla which is divided into linear lobes, consequently, broadened pinnatisect in the upper half. Achenes have long feathery, soft, snow-white, multirowed pappus. The genus has 3 species, distributed in the western Mediterranean region. All of them are ornamental and valuable medicinal plants.

Long-leaf leuzea (*L. longifolia* Hoffmanns et Link.) is 20–80 cm tall, with stems leafless at the top. Leaves are whitish-tomentose on the lower surface, lower leaves are lanceolate, entire, with bent lyrate lobes at the base; cauline leaves are lyrate-pinnate. Head is 3 cm in diameter, spherical; middle bracts are glabrous, with a brownish appendage; corolla is purple. Achenes are rectangular. The plant is found in moist aggregation of shrubs. It is an endemic of Portugal.

Rhaponticoid leuzea (*L. rhaponticoides* Graells) is 100–150 cm tall. Leaves are pinnate, scattered, densely pubescent on the lower surface, upper leaves are lobate. Involucre is 6–8 cm in diameter, spherical; middle bracts are scattered-tuberculate, with reddish-brown appendages. Flowers are violet-purple. Achenes are 4.5 mm long, ovate, dark brown, with whitish long pappus. It is a rare species, found in the hilly forests of northeastern Portugal and in central Spain.

The genus *Palaeocyanus* is represented by a single species, **crassifoliate palaeocyanus** [*P. crassifolius* (Bertol.) Dostál]. It is a naked evergreen shrub, usually up to 1 m in height. Leaves are succulent, scapulate, entire, 5–10 × 1.5–2 cm, mostly in rosettes. Heads are 2–3 cm in diameter, located on long crestate stems, each with a sheath of entire bracts; flowers are tubular, purple; achenes are 6–8 mm long, naked, whitish-purple. The plant grows in cracks on vertical coastal cliffs, on oligocenotic and monocenotic coralline along the southern coast of Malta and on Gozo Island. Ten locations are known, and the largest among them does not have more than 500 plants. The areas suitable for its growth in Malta have reduced, but reduction in its population has not yet been reported, although there are very few young plants. This is because larvae of flies parasitize in the ovaries. This is a rare plant, included in the Red Data Book of IUCN. Since 1971, it is the national flower of Malta.

Lavender cotton (*Santolina oblongifolia* Boiss.) is an aromatic, silky dwarf shrub. Vegetative shoots are long, semiprostrate; reproductive shoots are 15–30 cm long, ascending, rarely leafy on the lower side and leafless in the upper. Leaves are oblongate-scapulate, flat, sometimes pinnatisect, with obtuse lobes; upper leaves on the reproductive shoots are entire and pointed, all petiolate. Spathe is 15–18 mm broad, hemispherical; flowers are bright yellow. The plant grows in the western and central hilly regions of Spain and is endangered.

Among the rare species of Spain are also the endemic **sticky cypress** (*S. viscosa* Lag.), found in the southeast of the country, and **elegant cypress** (*S. elegans* Boiss. ex DC.), distributed in the Sierra Nevada.

The fleawort genus (*Senecio*) includes perennial or annual-biennial herbs, in the tropics often shrubs or trees with pinnate or rarely bi- and tripinnate or entire alternate leaves, usually with serrated margins. The heads are solitary or grouped in different types of inflorescence. The marginal flowers are female and ligulate, arranged in a single row, bright-colored, often absent. Middle flowers are bisexual, tubular. The genus has about 2000 species distributed all over the earth surface, especially in South Africa, Mediterranean, and temperate zones of Asia and America.

Cambridge fleawort (*S. cambrensis* Rosser) is a floccose perennial (very rarely annual), up to 50 cm with erect branched stems. Lower leaves are petiolate; middle and upper leaves are auriculate-amplexicaul, irregularly pinnatifid with lanceolate dented lobes. Inflorescence is branched; heads are numerous, 12 mm in diameter, broadly cylindrical. Ligulate flowers are yellow, ovate, drying immediately after flowering. The plant is one of the few endemics of England, is found along roads and wastelands in northern Wales, and is endangered.

Lopez fleawort (*S. lopezii* Boiss.) is a perennial with a height of 30–100 cm and short rhizomes. Stem is erect, simple, sparsely pubescent. Leaves are sharply decreasing in size, becoming very sparse toward the upper part of the stem. Basal and lower cauline leaves are 20–30 × 2.5–7 cm, with scattered curling pubescence, entire or serrated margin, petiolate; upper cauline leaves are linear-lanceolate, sessile. Heads number 1–8, are up to 40–55 mm in diameter, grouped in sparse comb. Spathe is 10–14 mm in diameter, with curling pubescence. Flowers are ligulate, yellow. The plant grows in forests, shady places in southwest Spain and Portugal. It is endangered.

Saw-wort (*Serratula bulgarica* Acht. et Stoy.) is distributed in northeastern Bulgaria and eastern Rumania. Plant height is 40–100 cm, with weak, erect stems. Basal leaves are lanceolate-ovate, serrated; cauline leaves are lanceolate, long-cuneate, deeply or finely serrated at the base, upper leaves deeply divided. Head is of diameter up to 30 mm, campanulate, solitary. Outer bracts are broadly ovate with membranous appendages; inner bracts are acute, with hairy membranous margin. Flowers are pink-purple.

The genus *Telekia* comprises 2 species distributed in Europe and Asia Minor. The **ornamental telekia** [*T. speciosissima* (L.) Less.] is a pubescent perennial up to 50 cm tall, with simple stem and a head. Lower leaves are sessile, cordate. Bracts are lanceolate (Plate 2). The plant is found on the rocks and mountains of northern Italy, between lakes Lugano and Gardo.

FAMILY BORAGINACEAE

The members of the family Boraginaceae are annual and perennial herbs, rarely shrubs or trees. Usually, they are covered with hairy bristles. Leaves are alternate, entire, exstipulate. Flowers are bisexual, regular, usually arranged in helicoid inflorescences, which themselves form a complex inflorescence. Calyx is fused, quinquepartite or pentalobate. Corolla is pentalobate. Fruit splits into 4 (rarely 2) nutlets. Sometimes the fruit is a stone or capsule. Many species of this family contain alkaloids and are used as dye or food plants.

The alkanet genus (*Anchusa*) is represented by the herbs with lanceolate sessile leaves and paniculate inflorescences of helicoids.

Curly alkanet (*A. crispa* Viv.) is endangered. It is an annual or biennial with bristles; stem is 10–35 cm, prostrate or ascending. Leaves are 50–170 × 5–10 mm. Helicoids are long, loose; length of calyx is 5 mm, and with fruit up to 8 mm, divided into lanceolate obtuse lobes over one-third to one-half of its length. Corolla is blue, tube 4–5 mm long, limb 5 mm in diameter. Nuts are 2 × 3 mm, ovate. The plant grows on coastal sands on the islands of Sardinia and Corsica.

The genus *Buglossoides* includes annual or perennial herbs and dwarf shrubs. Their inflorescence is an apical umbel. Corolla is dove-blue or purple, funnel-shaped with 5 long stripes of hairs on the inner side. The fruit is a nut, sometimes tuberculate.

Glandular buglossoid [*B. glandulosa* (Velen.) R. Fernandes] is a coarse-pubescent and glandular annual with numerous erect shoots. The basal leaves are stipulate, narrowed into a petiole; cauline leaves are oblongate-scapulate to linear, serrated. Umbels are located on short peduncles. Calyx lobes are linear-setose. Corolla is 6–8 mm in diameter, white. Nut is ovate. This endangered plant is found in Bulgaria and southeastern Rumania.

Calabar buglossoides [*B. calabra* (Ten.) I.M. Johnston] is a perennial with weak, creeping stems 30–80 cm tall. Flowering shoots are unbranched, arising from a rosette of leathery elliptical leaves of the previous year. Umbel has a few flowers. Calyx is 6–7 mm long, with linear pubescent lobes. Corolla is 17–20 mm in diameter, dove-blue. Nuts are elliptical, white, smooth. This vanishing species is found along the rocks in the mountains of southern Italy.

The genus *Lithodora* includes dwarf shrubs with flowers arranged in apical leafy umbel. Calyx is quinquelobate. Corolla is blue, purple, or white, funnel-shaped. Tube has no scales in the throat.

Oleiferous lithodora [*L. oleifolia* (Lapeyr.) Griseb.] is a sparsely branched dwarf shrub with weak, drooping, 10–45 cm long stems. Leaves are obovate or ovate-oblongate, short petiolate, dull green; upper surface has sparse bristles, lower surface is whitish silky, grouped in apical short vegetative shoots or surrounding the bases of peduncles. Umbels have 3–7 flowers. Calyx is 5–8 mm long, whitish silky with linear lobes. Corolla is initially pale pink, thereafter dove-blue, silky on the outer side, glabrous on inner side. Nuts have a short beak, are grayish-white, smooth. The plant grows on rocks in the eastern Pyrenees, approximately 24 km northwest of Figueras. A few locations are known on the French slopes at the border.

Glossy lithodora [*L. nitida* (H. Ern) R. Fernandes] is similar to the previous species, but its leaves are smaller, elliptical-obovate, short-pointed, sessile, both sides densely covered with white pubescence. Umbel has 2–5 flowers. Calyx lobes are lanceolate-oblongate, acute; corolla tube is 14 mm in length, its lobes ovate-oblongate, obtuse. A rare endemic of Spain, this plant grows in the southern regions (Sierra Magina) on stony places at the height of 1800–1900 m.

Doerfler's moltkia (*Moltkia doerfleri* Wettst.) is a 30–50 cm tall perennial herb. Stems are erect, simple, compressed-strigose; leaves are lanceolate, acute. Calyx is 10 mm in length; corolla is 19–25 mm in diameter, dark purple. The plant is endemic to the hilly northeastern part of Albania and is endangered.

The forget-me-not genus (*Myosotis*) includes herbs usually with soft setose pubescence. Flowers are regular, in helicoids. Calyx is campanulate, enlarged on fruiting. Corolla is usually dove-blue, rotuliform with yellow bristles in the throat.

Ruscinonian forget-me-not (*M. ruscinonensis* Rouy) is a short annual with a rosette of small, obtuse, lanceolate leaves up to 4 cm long with soft, scattered hairs. Flowers are axillary. Corolla is up to 3 mm in diameter, white to bright dove-blue, with a short tube and bent petioles. Distribution of this disappearing plant is in the eastern part of the Alber mountains between the Pyrenees and the sea where it grows on coastal sandy dune together with other annuals. As a result of construction and development of beaches its habitats have been destroyed. The seeds of this species of forget-me-not have been conserved, with the help of which it can be multiplied in botanical gardens. The species has been included in the Red Data Book of IUCN.

Soleirol's forget-me-not (*M. soleirolii* Gren. et Godron) is a biennial or perennial with densely branched and leafy stems up to 50 cm long. Leaves are elliptical. Peduncle is up to 2.5 mm long; calyx is densely pubescent. Corolla is white to pale dove-blue; diameter of limb is not more than 4 mm.

Corsica forget-me-not [*M. corsicana* (Fiori) Grau] is up to 25 cm tall, almost glabrous. Basal leaves are elliptical, often crenate at the apex, glabrous on the lower surface, with a few bristles on the upper surface; cauline leaves are ovate to lanceolate. Calyx, especially at the base, has dense, forked bristles; limb is up to 6 mm in diameter. Both species of forget-me-not are disappearing on Corsica.

Coastal navelwort (*Omphalodes littoralis* Lehm.) is an annual 4-15 cm tall; basal leaves are long-petiolate and cauline leaves are lanceolate, sessile. Inflorescence is sparse, with a few white flowers. An endemic of France, the plant is found along the coastal sands in the Bretagne-Loire valley and from Bretagne to Jironda. It is endangered as a result of change in its habitat.

Tornenian onosma (*Onosma tornensis* S. Jávorka) is a pubescent herbaceous perennial with a height of 15-30 cm. Leaves are lanceolate, 50 × 2-3 mm, densely pubescent with 2-2.5 mm long hairs, originating from tubers surrounded with short, scattered hairs. Peduncles are very short; flowers are yellow, pollinated by bees. Seeds have sticky hairs, which stick to wool and clothing. Its distribution is restricted to the limestone region, a larger part of which is located at the extreme east of Slovakia (Czechoslovakia), also entering into Hungary (Tornen Karst), about 30 km toward southwest of Košice. All these sites are within a belt up to 10 km long, where onosma grows on dry stony slopes facing south, in the crevices of Triassic limestone deposits at the height of 230-280 m above m.s.l. together with other relic and endemic plants. The plant is also found on sunny slopes and in dry meadows. Earlier, the plant was suffering heavily from cattle grazing, now it suffers because of stone mining. The Slovakian karst region is protected as a reserve landscape, but almost all the onosma population here consists of old plants. The highly ornamental **bubanian onosma** (*O. bubanii* Stroh) is an endemic of Spain, and **vaudenian onosma** (*O. vaudensis* Gremli) is highly local endemic of Switzerland, found in the Rhone valley near Aigle.

FAMILY BRASSICACEAE (CRUCIFERAE)

The Brassicaceae family mainly consists of herbs, rarely shrubs or semishrubs, often pubescent with simple, branched, or glandular hairs. Leaves are simple, alternate, exstipulate. The inflorescence is a raceme. Flowers are regular, bisexual, tetramerous; sepals are free; corolla is polypetalous, white, yellow, rarely lilac or violet. Stamens number 6, are tetradynamous. The fruit is a siliqua or silicula, usually splitting into 2 halves. The family has 375 genera and about 3000 species, mainly distributed in the northern hemisphere. Many of the cruciferous plants are important food, medicinal, ornamental, and other economic plants, several of them cultivated.

Fastigiate alison (*Alyssum fastigiatum* Heywood) is an erect, green or almost white perennial with numerous equal-sized shoots. Basal leaves are oblongate to oblanceolate-spatulate; upper leaves are linear or linear-spatulate; petals are 3–4.5 mm long, deeply incised. The plant is an endemic of southeastern Spain and is endangered.

The buckler mustard genus (*Biscutella*) consists of herbs or shrubs with entire or pinnatifid leaves; fruit is a compressed silicula with wingless seeds. Among the rare and disappearing species of this genus, 8 are found in France and 5 in Spain.

Neustrian biscutella (*B. neustriaca* Bonnet) is 20–40 cm tall and has simple or branched stems. Basal leaves are 4–8 × 1–1.5 cm, straight, spatulate, denticulate, pubescent, with a long petiole. Cauline leaves are broad, entire, or serrated. Racemes are dense, compact at blooming, and elongate during fruiting. The plant grows on limestone rocks on the Seine, near Paris. The species is endangered.

Hard-fruited biscutella (*B. sclerocarpa* Revel) has long, woody rhizomes and rosettes of leaves. Flowering shoots are up to 30 cm long. Inflorescence is branched, loose; sepals are 1.5 mm long; petals are 3 mm long. The plant is an endemic of southern France and is endangered.

The cabbage genus (*Brassica*) includes more than 50 species of wild and cultivated plants. Many of them are cultivated as valuable vegetable or oil crops; therefore, it is particularly important to conserve the germplasm of wild cabbage.

Balearian cabbage (*B. balearica* Pers.) is a glabrous, small shrub with thick, long, woody rhizome. The leaves are mainly basal, long-petiolate, lyrate or pinnatilobate. Flowering shoots are 25–30 cm long. Petals are 10–15 mm long, yellow. The plant is found only on Majorca Island.

Macrocarpous cabbage (*B. macrocarpa* Guss.), an endemic of Sicily, is threatened with extinction. It is a glabrous, small shrub with stem up to 30–60 cm long. Basal leaves are 10–18 × 4–10 cm, lyrate-pinnatifid or pinnatilobate, lobes are crenate-dentate; cauline leaves are oblong or lanceolate, entire or serrulate. Petals are up to 20 mm long, yellow. The plant is found on the coastal rocks of the Egadi Islands (near Sicily).

Scurvy grass (*Cochlearia polonica* Fröhl.) has reniform or ovate-cordate leaves. Petioles are 5.5–9.5 mm long. It grows on moist sand along the streams in the environs of Olkusz (southern Poland). The plant is endangered.

Aragonian scurvy grass (*C. aragonensis* Coste et Soulié) is a biennial with poorly branched stem up to 25 cm long. Basal leaves are 4–8 mm long, ovate-cordate, entire or with a small tooth dent on each side; petals are 4–5 mm long, white. The plant is found on screes in the Sierra de Guara (northeastern Spain).

The genus *Degenia* is represented by a single species: **velebitian degenia** [*D. velebitica* (Degen) Hayek]. This is a perennial silvery-gray due to dense stellate hairs. Leaves are linear-lanceolate, in basal rosettes. Flowering shoots are up to 10 cm long, each bearing a drooping raceme of a few flowers. Sepals number 4, are 7–8 mm long, the inner 2 sepals slightly sacciform at the base. Petals are yellow, 15 mm long, with long claws. Fruits are elliptical silicula, 10–14 × 7–8 mm long. The plant is an endemic of Yugoslavia, known only from the Velebit range, where it grows at 2 places on open rock debris at the height of 1200–1300 m together with other rare plants. It suffers heavily due to cattle grazing and collection. The wild population is not protected at present; constant observations on it are necessary. It is cultivated in the botanical garden of Lublin, where it is easily multiplied by seeds. The plant is included in the Red Data Book of IUCN.

The whitlow grass genus (*Draba*) includes perennials, rarely annuals or biennials, mostly turfy. Leaves are oblong, ovate, or linear-lanceolate, forming rosettes. Flowering shoots are leafless; flowers are white or yellow. Fruit is oblong, oval, or an ovate silicula.

Haynald's whitlow grass (*D. haynaldii* Stur) is a very small perennial with stems up to 6 cm long and 3–8 flowers. Leaves are 5–7 × 1 mm, linear, acute; petals are 4–4.5 mm long. The plant is an endemic of the southern Carpathians (Rumania), close to extinction.

Ladin whitlow grass (*D. ladina* Braun-Blanquet) is the only flowering plant indigenous to Switzerland. It is a small turfy perennial with thin, branched rhizomes, bearing a few dense basal leaf rosettes. Leaves are lanceolate, up to 8 mm long, slightly fleshy; their lower surface and margins are covered with sparse stellate and numerous simple hairs. Flowering shoots originate from the center of the rosette, leafless, with 1–4 terminal flowers. Petals are 3.5 mm long, pale yellow. Fruit is a silicula, 5–8 mm long, with a few short hairs. Nine locations of this small alpine plant are known in the Dolomite mountains, north of the Ofenpass. It grows at the height of 2600–3040 m, forming dense turf on the rocks up to 8 cm in diameter, rarely on limestone screes in typical alpine communities. A large part of its habitat is within the territory of a national park. However, the population is very small, only a few pockets over an area of about 30 km^2 have been found. The plant is included in the Red Data Book of IUCN.

Loiseleur's whitlow grass (*D. loiseleurii* Boiss.) is a dwarf perennial with short, rarely pubescent stems. Petals are 5–9 mm long, yellow. Silicula is flat, with hairs less than 0.25 mm long. The plant is found along rocks at the height of 2300 m in the Corsica mountains.

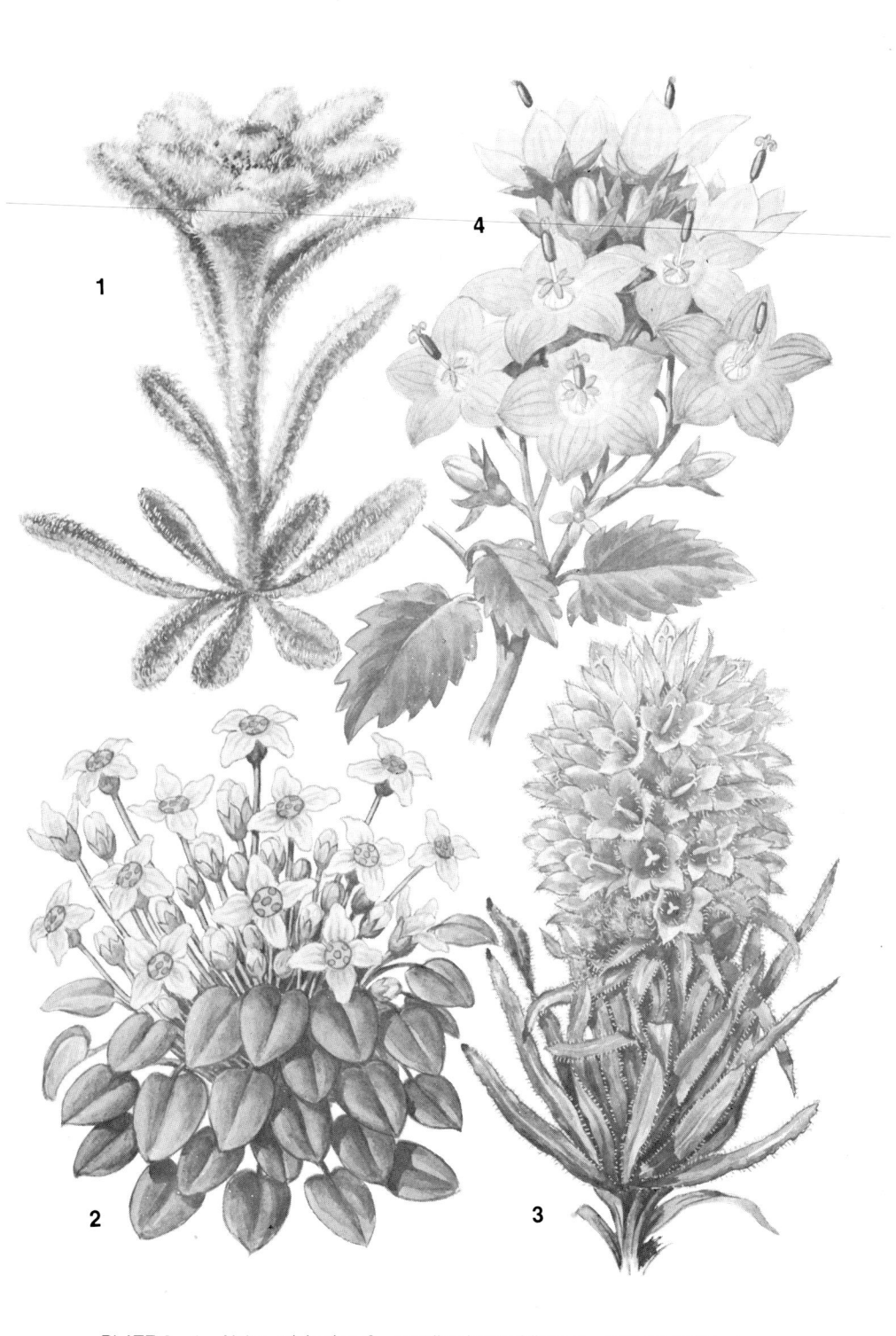

PLATE 3. 1—Alpine edelweiss; 2—acauline ionopsidium; 3—thyrsoid bellflower; 4—isophyllous bellflower.

PLATE 4. 1—Glacial pink; 2—glossy carnation; 3—Wulfen's houseleek; 4—bent oxytropis.

Marshy erucastrum [*Erucastrum palustre* (Pirona) Vis.] is a biennial or perennial, 20–80 cm tall, entirely glabrous. Basal leaves are lyrate-pinnatisect, lower cauline leaves are pinnate-pinnatisect, with 16–18 lobes on each side. It is a vulnerable species of northern Italy. The **pennian edge mustard** [*Erysimum pieninicum* (Zapal.) Pawl.] is a plant with lanceolate or oblong serrated leaves with 5–8 pairs of teeth; hairs on the lower leaf surface are arranged in 2 rows. Silicula is 45–65 × 0.9–1.3 mm. The plant is endemic of the eastern Carpathians (Poland).

Oblong-leaved dame's violet (*Hesperis oblongifolia* Schur) is covered with hairs. Leaves are serrated, semiamplexicaul. Petioles are purple. Silicula is glabrous. The plant is an endemic of the eastern Carpathians (Rumania).

The genus *Hutera* includes biennial or perennial herbs. Flowers are in corymbose racemes; petals are clawed, yellow, with violet veins. Fruit is a transversely segmented silicula. The genus belongs to the tribe Brassicae, which is characterized by maximum variability in the western Mediterranean region (24 endemic genera).

The **rocky hutera** (*H. rupestris* P. Porta) is included in the Red Data Book of IUCN. It is a biennial with a height of 60–100 cm and erect, branched stems bearing basal lyrate-pinnatifid, 10–20 cm long leaves (young leaves are silvery-silky with serrated margin). Flowers are in diffused panicles, which become oblongate at the time of fruiting, their length reaching 20 cm. Sepals are green; petals are pale yellow, with dark veins, cuneate, narrowed at the base. Fruit is a silicula, divided into 2 cup-shaped, 2–6 mm long lower segments and an upper obovate, 12–20 mm long segment, with an acute beak at the apex. Seeds are black, spherical, numbering 4–7 in each segment.

This endemic of Spain is found only in the Sierra de Alcaraz in the southwestern province Albacete, where it grows on a very limited area, on limestone cliffs along hilly canyon with a depth of 50–100 m, at the height of 1040–1120 m, above m.s.l. In 1973, its population consisted of 1050–1100 plants and in 1975, only 500–700 adult plants were found in a 2–3 km belt along the margin of a canyon and rocks. The reduction of rocky hutera population is, most probably, a result of intensive land use and collection. Construction of dams is also posing a danger to the population. The seeds of this plant are being preserved in the seed bank in Madrid.

Leptocarpous hutera (*H. leptocarpa* Gonz.-Albo) is biennial or perennial, densely pubescent. The silicula consists of 2 segments; the lower segment is 4–15 mm long, cylindrical, with 5–8 seeds, and the upper segment has 4–6 seeds and is somewhat constricted between the seeds. The plant grows along flinty rocks at the altitude of 800–1000 m in the eastern part of Cuedad-Real province. It is endangered.

The candytuft genus (*Iberis*) has 2 rare species.

Font-Quer candytuft (*I. fontqueri* Pau) is an annual with simple or branched puberulent stems up to 15 cm in height. Leaves are spatulate to linear-spatulate, with 1–2 pairs of teeth near the apex or almost entire; lower leaves are obtuse, narrowed into a fairly long petiole. Inflorescence is a short and dense corymb. Silicula is 5 mm long, broadly roundish, winged. The plant grows in the mountains of southern Spain (Malaga province).

Sampayo candytuft (*I. sampaiana* Franco et P. Silva) is a prostrate, pubescent annual with numerous shoots, up to 20 cm tall. Leaves are spatulate, regular pinnate-pinnatifid, with 3–4 pairs of oblongate obtuse segments. Inflorescence is a corymb, not very dense during fruiting. Siliqua is 5 × 4 mm, ovate-rectangular, glabrous or sparsely papillate-verrucose. The plant is found only in lime soils along the southwestern coast of Portugal.

The woad genus (*Isatis*) comprises 60 species. The broad-lobate isatis (*I. platyloba* Link ex Steudel) is an annual, 30–100 cm tall. Cauline leaves have acute auricles. Silicula is 8–10 mm long, elliptical-roundish. The plant grows along granite scarps in southeastern Portugal (in the environs of Miranda do Douro).

Acauline ionopsidium [*Ionopsidium acaule* (Desf.) Reichenb.] is a turfy acaulescent or short-stemmed annual. Basal leaves are roundish-ovate, entire or trilobate. Flowers are usually solitary, with long pedicels in the axils of basal leaves. Petals are lilac or purple, sometimes white. Silicula is obovate-roundish, with 2–5 seeds in each locule (Plate 3). The plant is an endemic of Portugal but widely cultivated in parks and naturalized at some places in southern Europe.

Sava's ionopsidium [*I. savianum* (Caruel) Ball ex Arcangeli] is a 7–15 cm tall annual with many shoots. Basal leaves are in a loose rosette, oblong-spatulate, short-petiolate, and irregularly serrated. Inflorescence is in a raceme. The plant grows on limestone and serpentine rocks in central Italy (Mount Calvi). It is threatened with extinction.

Pyrenees ptilotrichum [*Ptilotrichum pyrenaicum* (Lapeyr.) Boiss.] is a small turfy shrub, with a height up to 50 cm and thick, branched shoots bearing terminal clusters of obovate-elliptical, soft, silvery-haired, 1.5–4 cm long leaves. Flowering shoots are erect, 7.5–10 cm long, with dense corymbs of flowers. Petals are white, up to 6 mm long, ovate or roundish, but sharply narrowed at the base. Fruit is rhombic or ovate silicula, flat, 6–9 mm long. The plant grows on limestone cliffs in the eastern Pyrenees (France). In 1973, only 10 bushes were found. Possibly, it is being destroyed due to collection. The plant is ornamental and is cultivated in several botanical gardens (Geneva, Neuchâtel, Lausanne).

Lundy cabbage [*Rhynchosinapis wrightii* (O.E. Schulz) Dandy] is a perennial with a thin primary root. A rosette of sessile, pubescent, pinnatisect, 15–30 cm long leaves with coarsely serrated segments is followed by erect flowering shoots up to 1 m long, woody in the lower part and branched above, covered with squarrose hairs, and a few cauline leaves and scattered panicles, each bearing about 20 flowers. Sepals are erect, 8 mm long; petals are yellow, 16 mm long. Fruit is a linear siliqua, with a peduncle, terminating in a beak. Seeds are spherical, purple-black, finely reticulate. It is a species endemic to Lundy Island (at the northern Devon coast of England). The area of the island is about 4 km^2, and the plant is found only in its southeastern part over a distance of 0.5 km in the coastal cliffs and slopes facing east and south, on Devonian shales, sandstone rocks, and even on granite, usually at places protected from Atlantic storms, where humidity is high during winter but summer is hot and sunny.

The plant suffers due to goats, sheep, and deer; it may be replaced by bracken and introduced pontic rhododendron. This species is close to the cultivated cabbage and has great importance in breeding. At present, its population is being regulated. It is included in the Red Data Book of IUCN.

Madrid hedge-mustard (*Sisymbrium matritense* P.W. Ball et Heywood) is an annual or biennial with basal branching, 50–90 cm tall. Leaves are pinnatifid or serrated, narrow-oblong, forming a rosette. Racemes have 50 small, pale yellow flowers arising from the center. This plant either has disappeared or is close to extinction. At one time it was found on sandy soils, known only from 2 locations, one of which is at present a square in Madrid, and the other, about 15 km away from the first, planted with pine. It has not been found during recent years. It is included in the Red Data Book of IUCN.

The cress-rocket genus (*Vella*) includes 2 species and both are indigenous to Spain.

False cytisus (*V. pseudocytisus* L.) belongs to the vulnerable species, included in the Red Data Book of the IUCN. This is a small, highly branched and densely leafy shrub, 60–70 cm tall; shoots are covered with short and hard bristles and long hairs. Leaves are sessile, with lustrous hairs on both sides. Flowers are arranged in long panicles, yellow. Fruit is a silicula. This plant is collected from several locations. A larger part of its distribution is in Aranjuez province (south of Madrid), the rest in Teruel province (east of Madrid). It was also observed in southern Spain, but a search for it there during recent years did not yield results. The number of plants was estimated to be 2000–3000 over an area of 3 km^2 in Aranjuez province in 1976. The number of cytisus plants is declining because of destruction of its habitats in the process of artificial plantation of pine.

In 1976, only 100 plants were found in Teruel province at 2 places with the area of about 1000 m². A part of its habitat is now being protected. The seeds are being stored in the seed bank in Madrid, and experiments on reintroduction are in progress. The plant is ornamental and is easily reproduced through seeds.

FAMILY CAMPANULACEAE

The family Campanulaceae comprises 50 genera and about 800 species distributed throughout the world. Three centers rich in bellflowers can be distinguished: Middle Europe and Mediterranean; East Asia; and South Africa and Madagascar.

The members of this family are herbaceous or woody plants with entire, alternate, rarely opposite, or verticillate leaves and bisexual flowers. Perianth is pentamerous, corolla is gamopetalous, usually campanulate. Fruit is a multiseeded capsule, rarely a berry.

The bellflower genus (*Campanula*) is widespread in meadows and forests. Bellflowers are perennial or annual herbs, many of them being distinguished for their ornamental value and introduced into cultivation. The plant is disappearing in Europe due to collection for bouquets as well as destruction of its habitats. Twelve species are rare or threatened with extinction, half of which are indigenous to Italy.

Isophyllous bellflower (*C. isophylla* Moretti) is a perennial plant with prostrate or almost ascending shoots up to 10–15 cm high and roundish crenate-dentate leaves. Corolla is broadly campanulate or funnel-shaped (Plate 3). The plant grows along the crevices in limestone rocks in north-western Italy.

Transylvanian bellflower (*C. transsilvanica* Schur ex Andrae) is a biennial with erect, simple stem, 20–40 cm tall. Basal leaves are oblong-spatulate, dentate; cauline leaves are lanceolate, sessile; upper leaves are cordate. Inflorescence is terminal, capitate, multiflorate. Calyx lobes are lanceolate, ciliate; corolla is 15–25 mm long, violet. The plant is found in the alpine meadows and pasture in the Carpathians and mountains of southwestern Bulgaria. It is endangered.

Thyrsoid bellflower (*C. thyrsoides* L.) is 30–50 cm tall, with simple, erect, grooved shoots. Leaves are entire; basal leaves are oblongate-lanceolate, cuneate; cauline leaves are linear-lanceolate to ligulate, acute. Inflorescence is dense and compact, ovate or oblongate. Calyx lobes are setose, linear. Corolla is 17–22 mm long, villous, yellowish-white (Plate 3). The plant grows in meadows in the mountains of the Balkan peninsula, the Jura and the Alps.

FAMILY CARYOPHYLLACEAE

The Caryophyllaceae includes herbaceous plants, rarely semishrubs or shrubs. Leaves are entire, opposite, often fused at the base with membranous stipules, or exstipulate. Inflorescence is a panicle or raceme. Flowers are regular, bisexual, rarely unisexual, pentamerous or tetramerous. Fruit is a capsule. The family comprises about 80 genera and 2100 species, distributed mainly in the northern temperate zone. The largest number of genera and species are concentrated in the Mediterranean region and western and central Asia. Several species are used as medicinal, technical, and ornamental plants.

The carnation genus (*Dianthus*) includes perennial or annual herbs, rarely semishrubs with linear or linear-lanceolate leaves. Flowers are arranged in dense heads or clusters at the end of shoots and branches. Calyx is gamosepalous; petals have long claws and horizontal dentate limbs. The genus consists of about 200 species, distributed in Eurasia, particularly in large numbers in the Mediterranean region. Many species are very ornamental and cultivated since ancient times.

Urumoff's pink (*D. urumoffii* Stoy. et Acht.) is a pubescent perennial with a few erect, 30–60 cm long shoots. Leaves are linear, pale green, 7-veined, 8–16 mm long, paired. Flowers are in groups of 2–7 in almost spherical terminal heads, with linear-lanceolate, green or purple, serrated bracts almost of the same length as calyx; petals have a limb, 5–7 mm long, fimbriate at the apex, bright red with a yellow spot at the base (on the inner side), often glandular. The plant is indigenous to Bulgaria, found only in the mountains near Sophia up to the height of 1250 m in mixed forests. It is an ornamental plant, endangered, and is included in the Red Data Book of the IUCN.

Glossy pink (*D. nitidus* Waldst. et Kit.) is a glabrous, somewhat turfy perennial with branched stems. Leaves are oblong, obtuse. Flowers are arranged in pairs; bracts, numbering 2–4, are ovate with pointed tip. Calyx is 10–12 mm; corolla limb is 8–10 mm long, purple-red with white spots (Plate 4). The plant is indigenous to the western Carpathians (Czechoslovakia, Poland). It is endangered due to collection.

Glacial pink (*D. glacialis* Haenke) is an almost glabrous, turfy perennial with a height up to 5 cm and linear, obtuse leaves. Flowers are solitary, purple-red (Plate 4). The plant is indigenous to the Carpathians and eastern Alps and is protected in several European countries.

Callous pink (*D. callizonus* Schott ex Kotschy) is 5–20 cm tall, with glabrous stems, each bearing a single flower. Leaves are linear-lanceolate; upper leaves are acute, lower leaves are obtuse. Calyx is violet, tubular, broader toward the upper side; petals have large, up to 15 mm long, divergent, pouch-shaped limbs, with white and purple spots at the base and fimbriate apex. Only one habitat of the plant known in the mountains toward the southwest of Brasov (Rumania), where it is found on limestone rocks at 1650–2200 m. Its biology and ecology have been poorly studied; taxonomically it is an isolated species, possibly a relict. It is mostly destroyed by tourists. The plant is protected in Rumania and included in the Red Data Book of IUCN.

A total of 9 carnation (pink) species require special protection measures in Europe.

Chalk plant genus (*Gypsophila*) has 120 species. **Papillate chalk plant** (*G. papillosa* P. Porta) is a perennial with woody rhizome. Stems are 5–10 cm tall, erect, often branched, glabrous; internodes are shorter than leaves. Leaves are linear-lanceolate, long-pointed. Calyx lobes are sharp-pointed. Flowers are numerous, arranged in a fairly dense inflorescence; petals are incised, white to pale red. The plant grows on limestone rocky hillock near Garda (Italy). It is endangered.

Bachelor's buttons (*Lychnis nivalis* Kit.) is a turfy perennial with ciliate or dentate leaves. Stems are simple, 5–20 cm long. Basal leaves are oblong, lanceolate to almost spatulate; cauline leaves are in 1–2 pairs. Flowers number 1–3; corolla is pale red or white. The plant grows at humid rocky places at 1820–2290 m in the Rodnem mountains (eastern Carpathians, Rumania). It is endangered.

The genus *Minuartia* comprises about 100 species growing mainly in the northern hemisphere.

The **olon minuartia** [*M. olonensis* (Bonnier) P. Fourn.] is a perennial with 3–10 cm long stems. Leaves are clustered at the base; flowers number 1–3. It is an extinct species. Earlier it was found along the coastal rocks and sands in western France.

Waterfall minuartia (*M. cataractarum* Janka) is threatened with extinction. It is a sparse, turfy, usually glandular, pubescent perennial. Flowering shoots are erect, up to 20 cm long. Semiumbellate inflorescences are sparse, with 3 or more flowers. Petals are ovate-lanceolate, with 5–7 veins. The plant is endemic to Rumania.

Grisbach's meringia (*M. grisebachii* Janka) is a perennial with erect and fairly rough stems. Leaves are sometimes fleshy, linear. Inflorescence consists of 3–6 flowers. Distribution is restricted to eastern Bulgaria and eastern Rumania.

Of the 400 species of the catchfly genus (*Silene*), distributed mainly in the temperate and subtropical regions of the northern hemisphere, 5 species are endangered.

Diclinous catchfly [*S. diclinis* (Lag.) M. Lainz] is a dioecious rhizomatous perennial with prostrate, branched stems, up to 20 cm long, and lanceolate leaves. The plant is covered with long and soft white hairs. Flowers are solitary, in leaf axils. Corolla is dark pink; in female flowers divided up to a quarter of its length, in male flowers incised. The plant is found only on the rocks in mountains in the surroundings of Hativa (eastern Spain, Valencia province). Three populations and isolated scattered plants of this species are known. One population has already disappeared, and in the second only 500 adult plants were found in 1974 over an area of only 150 ha (seedlings were not found, although female flowers did produce seeds). The plant grows on dry terraces in olive groves. Possibly it reproduces vegetatively under conditions of irregular cultivation. It may disappear due to frequent ploughing of the land, use of herbicides, and fire. The male and female plants have different degree of resistance to various treatments; after treatments, the population structure may also change. For conservation, it is essential to control the population structure. The plant is included in the Red Data Book of IUCN.

Holtzmann's catchfly (*S. holzmannii* Heldr. ex Boiss.) is an annual up to 20 cm tall. Leaves are lanceolate, opposite, borne on a strong solitary stem. The axis terminates in an inflorescence. Calyx is green, sometimes gibbous during flowering, later (at the fruiting stage) papery and compressed. Petals number 5, are small, dull-colored, and bilobate. This is the only species in the genus with large fruits capable of floating in brackish water (they do not lose viability even after remaining in sea water for 40 days). Scattered habitats of Holtzmann's catchfly are known in the islands of the Aegean Sea between Attica and Crete (Greece). The best known population is located on the reef of 2 rocks with a length of 150 m near the coast of Crete; here the catchfly grows in a herbaceous community typical of small islands. It is very sensitive to cattle grazing. Tourism also has a harmful effect on it, as the islands are located near the most popular beaches of Crete. The plant is included in the Red Data Book of IUCN.

Stickseed catchfly (*S. viscariopsis* Bornm.) is a biennial or short-lived perennial with a height of 30 cm and perennating rosette, mainly flowering shoot and one or a few lateral potential flowering shoots. The rosette has leaves 1–4 cm long, linear-spatulate, pubescent. Inflorescence is in the form of slender panicles with sticky sheath and 3–20 flowers on horizontal pedicels; calyx is 4–6 mm long, gibbous, with 10 prominent greenish veins; petals number 5, are deeply bifid, have bright purple, 5 mm long limb and whitish claw. Ripe pod is 7 mm long, conical, with divergent denticles. The species is endangered. Two habitats of this species are known in Yugoslavia (Macedonia) in the extreme south: near the city of Prilep on volcanic rocks and mountain slopes, and 12 km east of the city, on limestone deposits. The plant grows at an altitude of 800–900 m, on the slopes facing north and northeast, in dry meadows, where 25 percent of the area is bare with rock outcrops. One pocket with a small number of plants occupies an area less than 1 ha. It is an extremely isolated species, which does not cross with other species. Seed production is high; viability (germination 50–100 percent) is maintained up to 2 years. The plant is cultivated easily but becomes biennial under cultivation. A mild sheep grazing stimulates branching, and many of the lateral branches produce flowers and seeds, but the plant may easily disappear under changed conditions due to small population size. The species has been included in the Red Data Book of IUCN.

Velvety catchfly (*S. velutina* Pourret ex Loisel.) is a perennial with lignified stem. Lower leaves are lanceolate or spatulate. Inflorescence is drooping, with opposite branches. Sepals are 18–25 mm long, with glandular and nonglandular hairs. Petals are whitish or reddish. The plant is indigenous to Corsica.

FAMILY CHENOPODIACEAE

The family Chenopodiaceae includes perennial and annual herbs or semi-shrubs, rarely shrubs and small trees. Leaves are usually entire, often fleshy or poorly developed. Flowers are small, with unpleasant smell, arranged in most species in spicate or paniculate inflorescences. Fruit is single-seeded, mostly dry.

The beet genus (*Beta*) includes (usually biennial plants characterized by pentamerous flowers, in which the stamens are fused at the base with a glandular disk. Fruits have a stony pericarp, fused in groups of 2–3 in a typical aggregate fruit. The genus includes valuable fodder plants.

Dwarfbeet (*B. nana* Boiss. et Heldr.) is a small, nonpubescent perennial with strong, cylindrical fleshy roots. Leaves are clustered at the base, 15–20 mm long and 8–10 mm broad, ovate-oblong, entire. Flowering shoots are prostrate to ascending, leafless. Inflorescence has small bracts; thalamus is hemispherical, crestate, segments bent in the presence of fruit. The plant is indigenous to Greece, known from several mountains (including Parnassus, Olympus, Tagetes). It grows above the forest boundary in low grassy meadows. On mount Parnassus, it is preserved in the crevices of rocks, suffering due to cattle grazing and tourism. It is preserved in the national park of Olympus, but conservation of the species on mount Parnassus is also essential. Other rare indigenous species grow on the mountain which are known only there. Dwarfbeet is included in the Red Data Book of IUCN.

It is the only mountainous species of the genus, which is taxonomically isolated.

FAMILY CISTACEAE

About 200 species of the family Cistaceae are distributed mainly in the Mediterranean region and North America.

The rock rose genus (*Cistus*) includes small, evergreen shrubs with aromatic entire leaves and large flowers, similar to the flowers of briar. About 20 species of this genus are distributed in the Mediterranean. They are typical representatives of maquis, drought-resistant; many species are used as ornamental plants or for extraction of laudanum.

Albanian rock rose (*C. albanicus* E.F. Warb.) is a highly indigenous plant of central Albania, threatened with extinction. Stems are up to 20 cm tall; leaves are elliptical, green, their dorsal surface pubescent with long, simple, and stellate glandular hairs, almost sessile. Inflorescence is in the form of semiumbels, 4-flowered, one-sided. Sepals number 5, have long, white hairs; corolla is 3–4 cm in diameter, white.

Palhin's rock rose (*C. palhinae* Ingram) is a short, evergreen shrub. Stems are up to 50 cm long, erect. Leaves are 20–60 cm long, oblanceolate to spatulate, green and glabrous on the upper surface, with dense white pubescence on the lower surface, almost sessile. Sepals number 3; flowers are solitary, white. It is a rare endemic of southwestern Portugal, and can disappear easily with the destruction of habitats.

FAMILY CRASSULACEAE

The family Crassulaceae includes annual and perennial herbs, rarely semishrubs or shrubs. Leaves are simple, usually thick, succulent, fleshy, sometimes in dense basal rosette. Flowers are regular, bisexual, arranged in various inflorescences. Sepals number 3–20, are free or fused. Petals also number 3–20, are usually free, rarely fused. Fruit is multiple, very rarely a capsule. Seeds are small. The family has about 1500 species distributed almost all over the world. The dry and hot regions, especially in South Africa, are particularly rich in Crassulaceae.

The stonecrop genus (*Sedum*) is represented by annual or perennial herbs with fleshy leaves and usually pentamerous, rarely tetra- to hexa- or nonameric flowers of different colors.

Kostov's stonecrop (*S. kostovii* Stef.) is a small plant with linear-oblong or oblanceolate leaves with a length of 5–6 mm and short, dense, terminal umbels of pale yellow flowers. **Stefco's stonecrop** (*S. stefco* Stef.) is a small, glabrous perennial with flowering shoots up to 7 cm long. Flowers are tetramerous, with acute, pale pink petals, forming a fairly dense, compound umbel. **Zollikofer's stonecrop** (*S. zollikoferi* F. Hermann et Stef.) is a perennial with glabrous, nonfruiting, very short branches and 5–8 cm long flowering shoots. Flowers are yellow, arranged in small, dense, compound umbels. All these species of stonecrop are rare endemics of southwestern Bulgaria.

Pruinose stonecrop (*S. pruinatum* Link ex Brot.) is a perennial with nonfruiting, prostrate stem, often rooting. Flowering shoots are 7–20 cm long, erect. Flowers are straw-colored, arranged in scattered, nonequilateral, compound umbel.

This species and **Willcomm's stonecrop** (*S. willkommianum* R. Fernandes) are rare endemics of northern and central Portugal.

The houseleek genus (*Sempervivum*) is represented by perennial plants. Leaves are fleshy, borne on nonfruiting branches, in spherical or widely open basal rosettes. Inflorescence is corymbose-paniculate. Petals are free. **Pitton's houseleek** (*S. pittonii* Schott, Nyman et Kotschy) is a very rare indigenous plant of the eastern Alps (Austria). Its rosettes are dense; leaves are densely pubescent, grey-green with red glands; flowering shoots are 12–15 cm long; flowers are 9–12-merous. **Wulfen's houseleek** (*S. wulfenii* Hoppe ex Mert.) is a monocarpic perennial with a rosette of 4–5 cm diameter. Leaves are 30 × 12 mm, oblong, cuneate, glabrous, bluish-green. Stolons are long, strong, woody. Flowers are borne on 15–25 cm long shoots; petals are lemon-yellow with a purple spot at the base (Plate 4). The plant grows at an altitude of 1700–2700 m in the eastern Alps, but is reported locally at some places even in the central parts. It is an ornamental plant and suffers from the effects of tourism.

FAMILY DIPSACACEAE

The family Dipsacaceae includes herbs or semishrubs with opposite or verticillate leaves. Flowers are irregular, usually arranged in dense heads, surrounded by an involucre. Corolla is funnel-shaped, consisting of fused petals and 2 lobes often fused. The fruit is an achene. The family comprises about 300 species, predominantly distributed in the countries of the ancient Mediterranean. Some plants are cultivated as honey producers and medicinal and ornamental crops.

The genus *Cephalaria* consists of tall perennial or annual herbs with lyrate-pinnatifid cauline and entire basal leaves. The **radiate cephalaria** (*C. radiata* Griseb. et Schenk) is a rare endemic of the mountains in Rumania. It is a perennial, 60–120 cm tall. Leaves are 12–40 × 4–10 cm, lyrate or regular pinnate, with 7 pairs of ovate-lanceolate lateral lobes and a larger terminal lobe, pubescent; cauline leaves are often linear, long-petiolate. Bracts are 4–7 × 3–5 mm, ovate, obtuse; corolla is 12–17 mm in diameter, yellow.

The scabious genus (*Knautia*) is represented by perennials, biennials, or annuals. Leaves are opposite, entire or pinnatisect. Flowers are arranged in heads; corolla has 4 unequal lobes. The genus consists of about 50 species distributed in Europe and the Mediterranean region, predominantly in mountains.

Basalt scabious (*K. basaltica* Chass. et Szabó) is a perennial with sympodially branched rhizome; flower-bearing shoots are 45–75 cm tall, with lateral leaf rosettes. Lower internodes are short and glabrous, upper internodes are longer and pubescent. Leaves are almost leathery, elliptical-lanceolate, dentate, glabrous, ciliate along the margin; leaves in the rosette have a few yellowish awns. Heads number less than 5, are 3–5 cm in diameter; corolla is light bluish-lilac. The plant grows in meadows. It is indigenous to the mountains of southern and central France. It is endangered. The **Forez scabious** (*K. foreziensis* Chass. et Szabó) is similar to *K. basaltica,* but taller, 50–100 cm tall, with long lower internodes and rosette leaves without awns. It is indigenous to central France and is a rare plant.

Godet's scabious (*K. godetii* Reuter) grows in humid meadows and marshes in the mountains of central and eastern France and in northwestern Switzerland. It is externally similar to *K. basaltica.* The plant is endangered.

Nevada scabious [*K. nevadensis* (M. Winkler ex Szabó) Szabó] is a perennial with sympodially branched rhizome. Leaf-rosette is absent at the time of flowering. Shoots are 35–60 cm. Leaves are green, crenate to deeply serrate; upper leaves are ovate to almost cordate; upper surface of leaves is glabrous, lower surface has soft pubescence. Heads are 3–4 cm in diameter, sheaths have 8 awns; corolla is lilac-colored. The plant is found in the outgrowth of shrubs in the mountains of southern Spain and, possibly, northwestern Portugal. It is indigenous to Spain and is a rare species.

Rupicolous (rocky) scabious [*K. rupicola* (Willk.) Szabó] is a dense turfy perennial, woody at the base, with 10–20 cm tall, weak stems. Leaves are almost coriaceous, finely verrucose, ciliate along the midrib on the lower surface; lower leaves are oblong, upper leaves are lyrate, with 2–6 lateral lobes and an ovate terminal lobe. The plant is indigenous to Spain and is a rare species.

FAMILY FABACEAE

The family Fabaceae contains 600 genera and about 1200* species, distributed all over the world, but mainly in countries with temperate climate, where predominantly herbaceous and shrubby forms grow. The tropical

---

*Error in the Russian original, should read 12,000 (see Willis: *Dictionary of Flowering Plants and Ferns*).

species are fewer in number; these are mainly trees. Leaves are alternate, stipulate, simple or compound. Flowers are bisexual, pentamerous, zygomorphic; upper petal is called 'standard', the 2 laterals are called 'wings', and the 2 lower petals are fused into a keel, enclosing the pistil and 10 (rarely 5–9) stamens. The fruits are dry pods. A great majority of the plants are pollinated by insects because of the complex structure of the flower. Economically, this is one of the most important families of plants. A peculiarity of most of its species is formation of root nodules with bacteria assimilating atmospheric nitrogen. Therefore, the plants are very rich in proteins. Some of the most important food, fodder, dye, medicinal, and industrial plants belong to this family. Some of its members are highly ornamental.

The milk-vetch genus (*Astragalus*) is one of the most numerous genera of the family. It consists of annual or perennial herbs, semishrubs, rarely shrubs, usually pubescent. Leaves are imparipinnate or paripinnate with a terminal spine. Flowers are grouped in axillary racemes or heads. The genus comprises more than 1600 species. Some milk-vetches are a source of gum.

Arnacanthian milk-vetch (*A. arnacantha* M. Bieb.) is a shrub, forming large hemispherical pads up to 50 cm tall, with densely pubescent short branches, bearing 2–4 cm long pinnate leaves consisting of 4–5 pairs of softly pubescent leaflets. Leaflets are 10–16 mm long and 2–3 mm broad, with spiny apices. Bracts are lanceolate, membranous, up to 18 mm long. Flowers are minute, 14–18 mm long. The plant is found in Bulgaria and the USSR. It is rare in Bulgaria, a few habitats are found on hills in the eastern part of Old Planina, northwest of Burgas, near Aitos, over an area of about 1500 ha.

It grows at an altitude of 90–550 m in degraded oak forests with shrubby jasmine, Christ's thorn, and broad-leaved phillirea. Under natural conditions, regeneration through seeds is fairly good; up to 100 seedlings per 1 m^2. This plant is interesting for the study of taxonomy and geography of plants, as it belongs to the group of West Asian species. It is protected by law in Bulgaria; a part of its habitat is included in the Aitos city park. The plant is included in the Red Data Book of IUCN.

Swollen-calyx milk-vetch (*A. physocalyx* Fischer) is a turfy plant with condensed stems and 15–20 cm long leaves. Leaflets are in 15–22 pairs, ovate or elliptical, acute, sparsely pubescent. Racemes have 1–5 flowers. Calyx is 15–17 mm long (at anthesis) and 25 × 15 mm in overall size, tumid (swollen) at fruiting stage. Corolla is yellow. Standard is 35–45 mm long. Pod is 12–15 mm long, oblongate, glabrous. Found at one time in southwestern Bulgaria, near Plovdiv, it has disappeared from this habitat, as also from the other earlier known habitats in Bulgaria. Recently, it was found in Macedonia (Yugoslavia), but its present status is unknown. The species is included in the Red Data Book of IUCN.

In Europe, 11 more milk-vetch species need protection. Three indigenous species of milk-vetch are close to extinction in Rumania.

Pseudopurple milk-vetch (*A. pseudopurpureus* Gusul.) has weak, climbing, 10-40 cm tall stems. Leaves are 4-8 cm long, with 7-15 pairs of leaflets. Leaflets are oblong-elliptical, pubescent on the lower surface, almost glabrous on upper surface. Racemes are cylindrical, with numerous flowers. Corolla is light blue-violet, with incised standard and pointed keel. The plant is found only in the eastern part of the Carpathians, between Tyrgu-Muresh and Bakeu. **Roemer's milk-vetch** (*A. roemeri* Simonkai) is a plant with strong, erect, 50-70 cm tall stem and 7-12 cm long leaves. Leaflets are in 4-6 pairs, elliptical or lanceolate, sparsely pubescent with appressed hairs on the lower side and almost glabrous on upper side; stipules are absent. Corolla is whitish-lilac; standard is 20-25 mm long. Pod is 15-20 × 5-8 mm, oblong-lanceolate, densely pubescent, with a short, bent beak. The plant is indigenous to the eastern Carpathians. **Peterf's milk-vetch** (*A. peterfii* S. Jávorka) is up to 20 cm high. Leaflets are linear or linear-lanceolate. Raceme is sometimes up to 15 cm long; corolla is yellowish. Pod is 15-25 mm long with dense, procumbent hairs. The plant grows in steppes and grassy slopes in southern Rumania.

The broom genus (*Cytisus*) has more than 50 species, mostly Mediterranean. These are mostly shrubs, rarely small trees with ternate leaves and yellow, rarely white or red, flowers. Some species are ornamental and melliferous and have beautiful wood.

Emeriflorous broom (*C. emeriflorus* Reichenb.) is a prostrate or erect shrub with ridged, angular, 30-60 cm, rarely 100 cm stems. Young branches are tomentose. Leaves consist of 3 obovate-lanceolate 10-20 mm long leaflets, more or less glabrous on the upper surface and pubescent with short silvery hairs on the lower surface. Flowers are arranged in groups of 3-4 in racemes, forming dense, leafy terminal panicles; calyx is bilobed; corolla is yellow; pod is linear or oblong, glabrous.

Its distribution is restricted to Italy and Switzerland, in the south by the Bergamo Alps, and in the west by the Como and Lugano lakes. Seventeen habitats of the plant are known in the mountains, where it is still abundant at some places. It grows on limestone slopes at an altitude of 1000-1800 m in beach forests, heaths, covered empty areas, and alpine pastures. It belongs to the group of relict endemics of this region. It is an ornamental plant. The populations are still not endangered, but may suffer heavily due to goat grazing and herbarium collections. Its reserves should be marked out to protect the species. It has been included in the Red Data Book of IUCN.

Aeolian broom (*C. aeolicus* Guss. ex Lindl.) is an erect shrub, 100–200 cm, with ascending, pentagonal, pubescent branches. Leaves are 15–30 × 5–15 mm, oblong to elliptical; central leaf is longer than the lateral ones, glabrous on the upper surface, glandular on lower surface. Flowers are solitary or arranged in groups of 2–3. Corolla is yellow; standard is 15–18 mm long, dark red at the base. Pod is 20–45 × 4–7 mm, pubescent or glandular. It is a rare species, indigenous to Sicily.

The genus *Oxytropis* is typical for the temperate and cold belts of the northern hemisphere. The genus has more than 300 species, mostly shrubs or perennial herbs, acaulescent or with a well-developed stem. Leaves are imparipinnate or whorled. Flowers are arranged in heads or oblong racemes, multiflorate or with a few flowers. Corolla is variously colored; keel almost always has a characteristic cusp at the apex (from which the generic name has been derived). Pods are oblong to spherical.

Bent oxytropis, the Norwegian subspecies [*O. deflexa* (Pallas) DC. ssp. *norvegica* Nordh.] is up to 10 cm tall. Leaflets are in 11–16 pairs, lanceolate or oblong-lanceolate, appressed-pubescent on both surfaces. Stipules are free, racemes are oblong, elongating after flowering. Flowers are drooping, whitish. Pods are 8–15 mm long, pendulous, with short appressed hairs (Plate 4.) The plant is indigenous to the arctic part of Norway and is endangered.

The furze genus (*Ulex*) includes up to 20 species, distributed in the Atlantic part of Western Europe. These are prickly shrubs with flowers solitary or in pairs in the axil of scaly leaves or spines at the tips of reduced branches. Flowers are golden yellow. Pods are pubescent, with 2–4 seeds.

Prickly furze (*U. densus* Welw. ex Webb) is a compact, dense shrub with a height of 20–50 cm and numerous shoots. Young branches, phyllodes, and thorns are pubescent with white or whitish-brown hairs. Primary phyllodes are 5–8 mm long, narrow-triangular, leafy, with a short spine at the tip. Thorn is terminal, 8–20 mm long. Flowers are grouped in terminal semiumbellate inflorescences. Bracts are ovate-elliptical; calyx is 11–16 mm long, with compressed pubescence or almost glabrous. Standard is equal to or slightly longer than calyx; wings are 2 mm long. The plant grows on dry limestone soils. It is a rare endemic of western and central Portugal.

FAMILY FUMARIACEAE

The family Fumariaceae is close to Papaveraceae and is represented by herbs with multifid alternate leaves. Flowers are in terminal inflorescences, bisexual, zygomorphic. Calyx consists of 2 deciduous sepals; corolla of 2 saccate or spurred outer petals and 2 inner petals, there are 2 stamens; ovary is unilocular. The fruit is a nut or pod-shaped unilocular capsule. The family comprises 16 genera and about 400 species, mainly distributed in the temperate regions of the northern hemisphere.

The fumitory genus (*Fumaria*) includes annual herbs, grayish-green due to waxy coating, with bipinnatisect or quadripinnatisect leaves. Flowers are arranged in racemes; zygomorphic bilabiate. Fruit is a nut.

Jank's fumitory (*F. jankae* Hausskn.) is 8–40 cm tall, with tripinnatisect leaves. Sepals are 1.5–0.75 mm and corolla is 4–5 mm long, dirty crimson red. Fruit is ovoid, sharp-pointed. The plant is indigenous to northwestern Rumania. **Occidental fumitory** (*F. occidentalis* Pugsley) is an annual with 12–20 pink-white flowers borne in a raceme. Calyx is 4–5.5 × 2–3 mm, dentate; corolla is 12–14 mm long, with tips of inner petals dark purple; lower petals have curly or slightly wavy margin. Fruit is tuberculate-wrinkled. The plant is indigenous to southwestern England (Cornwall). It is endangered.

FAMILY GENTIANACEAE

The family Gentianaceae includes perennial or annual herbs, very rarely shrubs or small trees with entire, opposite or verticillate, exstipulate leaves. Flowers are bisexual, regular, solitary, or arranged in inflorescences. Calyx is fused; corolla has fused petal, funnel-shaped, campanulate, rarely tubular or trochoid. Ovary is superior. Fruit is a capsule, rarely a berry. The family has about 80 genera and more than 900 species distributed all over the world, but found in particularly large numbers in hilly regions.

The centaury genus (*Centaurium*) is represented by herbs with opposite leaves and tetramerous-pentamerous flowers arranged in multiflorate inflorescences. Corolla is tubular-trochoid. Fruit is a bivalved capsule. The genus includes 50 species.

Three Spanish endemics belong to rare centauries. **Trifoliate centaury** [*C. triphyllum* (W.L.E. Schmidt) Melderis] is a biennial with dense rosette of linear-oblong leaves. Stem is often branched in the lower part; cauline leaves are linear. Flowers are large, almost sessile, arranged in a more or less dense semiumbellate inflorescence. The plant grows on gypsum soils in the central part of Spain.

Sealed centaury (*C. rigualii* Esteve Chueca) is an annual with a height of 40–50 cm and without leaf rosette. Stem is branched from the base; cauline leaves are linear, with 1–3 veins. Flowers have long pedicels in a very loose semiumbel. The plant is found in dry meadows in southern Spain. The **enclusion centaury** (*C. enclusensis* O. Bolos, Molinier et Montserat) is a plant with a 30 cm tall stem, branched from the base and with a large flower. The species is still poorly studied. It grows only on Minorca Island (Baleares).

The gentian genus (*Gentiana*) comprises about 400 species of herbaceous perennials and annuals with opposite leaves and 4–5-merous rarely 6–8-merous flowers.

Ligustian gentian (*G. ligustica* R. de Vilm. et Chopinet) has oblong-obovate leaves, broad ovate calyx lobes, and corolla with green spots on the throat. It is found in the coastal Alps and central Apennines and is endangered.

Purple gentian (*G. purpurea* L.) is a perennial with a height of 20–60 cm and simple, straight stems. Leaves are lanceolate to broadly ovate, acute, petiolate, with 5–7 veins. Flowers are sessile, grouped in a small terminal bunch, sometimes in axillary clusters with a few flowers; calyx is membranous; corolla is reddish-purple with dark purple spots. The plant is found in the mountains of central Europe and southern Norway and is protected in many countries.

FAMILY GERANIACEAE

The family Geraniceae includes semishrubs and perennial or annual herbs, usually covered with simple or glandular hairs. Leaves are alternate, simple, palmatisect or pinnatisect. Flowers are usually arranged in inflorescences, bisexual. Calyx consists of 4–5 free or basally fused sepals; corolla consists of 4–5 free petals, alternating with nectar glands. Fruits are dehiscent into single-seeded units with or without awn-like projections.

The stork's bill genus (*Erodium*) is represented by annual or perennial herbs with palmatipartite or pinnatipartite leaves. Flowers are sometimes unisexual, pentamerous. Fruitlets have awn-like deciduous or persistent appendages which are spirally twisted at the base. Five species of stork's bill are proposed for protection, 3 of them indigenous to Spain. All of them are found only in the southern part of the country, in the Sierra Nevada. **Astragaloid stork's bill** (*E. astragaloides* Boiss. et Reuter) is an acauline perennial. Leaves are up to 4 cm long, whitish-pubescent on both surfaces, oblong-lanceolate, pinnate, with ovate lobes. Stipules are lanceolate, whitish pubescent. Calyx is 8–10 mm long, densely glandular-viscid; petals are 10–12 mm long, purple. Appendage of fruit is 25–45 mm long. The plant grows on sandy soils.

FAMILY GESNERIACEAE

The large family of Gesneriaceae is mainly distributed in the tropics and subtropics, represented by only 3 genera in Europe. The European

PLATE 5. 1—Myconian ramonde; 2—Natal ramonde; 3—coin-leaf St. John's wort; 4—Heuffel's crocus.

PLATE 6. 1—Candelabrian sage; 2—arenaceous crocus; 3—involucrate fritillary; 4—bulb-bearing (orange) lily.

PLATE 7. 1—Carniola lily; 2—saffron lily; 3—liliaster; 4—Urumoff's tulip.

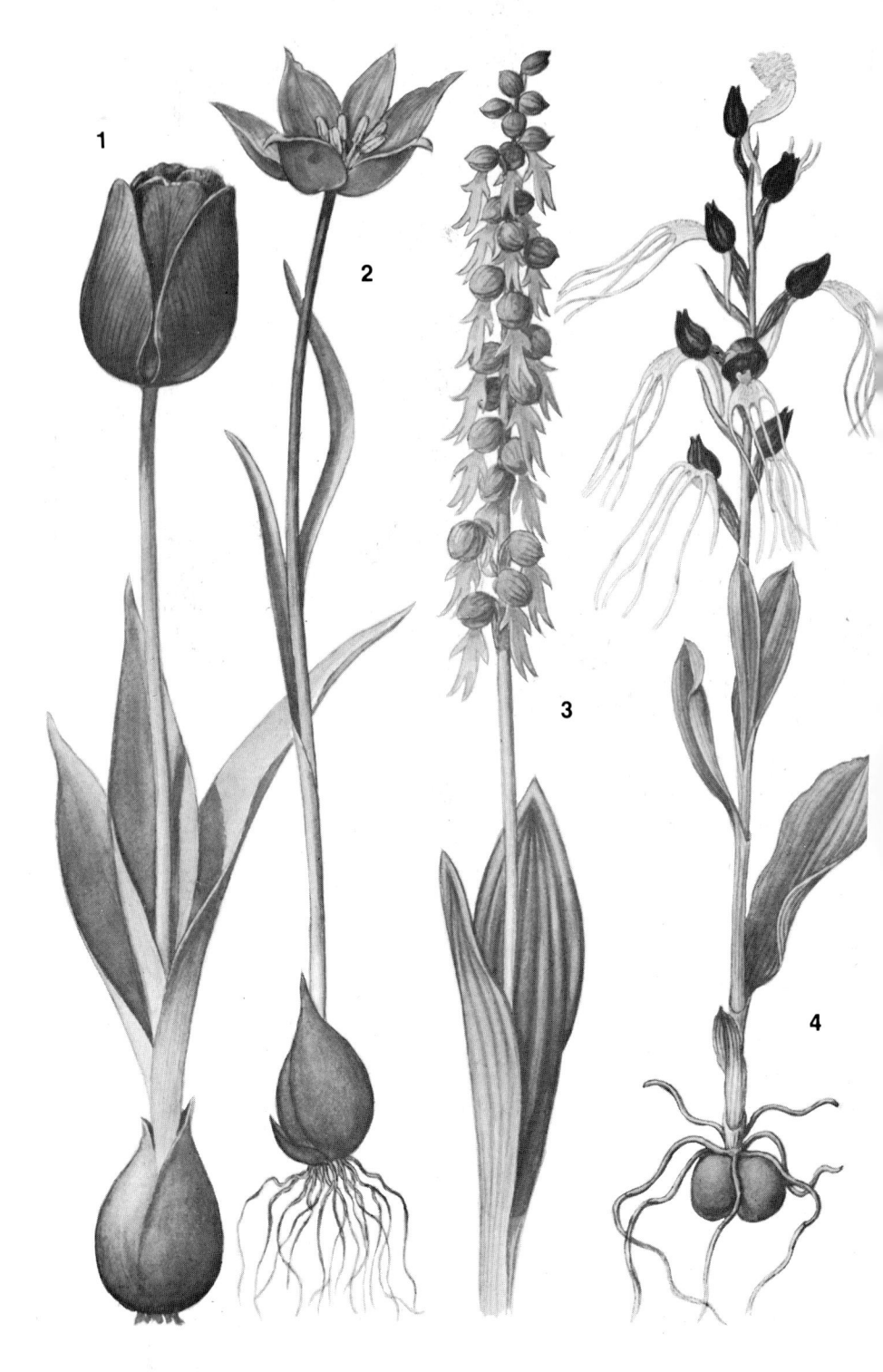

PLATE 8. 1—Rhodopean tulip; 2—forest tulip; 3—green man orchid; 4—Crimean comperia.

species are considered to be the relicts of the Tertiary period. These are perennial acauline herbs with a rosette of simple leaves. Flowers are solitary or in small axillary umbels, bisexual, with trochoid or tubular, sometimes bilobed corolla. Stamens are equal in number to the corolla lobes, sometimes one of them modified into a staminode or absent. Ovary is superior, unilocular; fruit is a 7-locular capsule.

The ramonde genus (*Ramonda*) consists of plants with wrinkled leaves; upper surface is densely pubescent with short white hairs, and the lower surface is sticky. Flowers are usually pentamerous. All stamens are fertile. Three species of this genus are found in Europe, all of them endangered.

Myconian ramonde [*R. myconi* (L.) Reichenb.] has rhombic-roundish (up to ovate), obtuse, crenate-serrated leaves, narrowing into a short petiole. Flowers are pentamerous; corolla is 3–4 cm in diameter, trochoid, dark violet with yellow centre (Plate 5). The plant grows in shady crevices of limestone rocks in the central and eastern Pyrenees (France) and adjoining areas of northeastern Spain.

The **Natal ramonde** (*R. nathaliae* Pančič et Petrovič) is found in shady places, along the rock crevices in southern Yugoslavia, northern Greece, and, possibly, Albania. Leaves are ovate, almost roundish, serrated at the base, with entire margin or crenate. One to three flowers are borne on glandular-pubescent peduncle. Corolla is 3–3.5 cm in diameter, trochoid, with 4–5 spreading lobes, violet or lilac, with orange-yellow centre (Plate 5).

Serbian ramonde (*R. serbica* Pančič) is similar to the Natal ramonde, but differs in having narrow, obovate leaves, cuneate at the base, unevenly and sometimes deeply crenate or serrated, with short petioles. Flowers are pentamerous; corolla is 2.5–3 cm in diameter. The plant is found in shady rock crevices in northwestern Bulgaria, southern Yugoslavia, Albania, and northwestern Greece.

Rhodopian haberlea (*Haberlea rhodopensis* Frivald.) has obovate (up to ovate-oblong), obtuse, crenate leaves, covered with short and soft pubescence. Peduncles are 6–10 cm long with 1–5 flowers. Calyx is 5-lobed, almost equal to the corolla tube; corolla is 15–25 mm in diameter, cylindrically tubular, with irregularly bilabiate limbs, whitish-light bluish-violet, pubescent on the inner side; stamens number 4, staminode one. The plant is found along the crevices of rocks in the southern and central mountainous parts of Bulgaria and in northeastern Greece.

FAMILY GLOBULARIACEAE

The family Globulariaceae includes perennial herbs or shrubs with alternate, entire leaves. Flowers are small, usually arranged in globular involucrate heads, bisexual, irregular. Calyx is tubular, 5-lobed; corolla is 4- or 5-lobed, bilabiate; lower lip is often highly reduced and bilobed. Stamens number 2–4. Ovary is superior, unilocular; it has one style. Nut-shaped fruit is enclosed in the persistent calyx. The family has 3 genera and 27 species, distributed in the northern temperate and subtropical zones of the Old World, mainly in the Mediterranean region, West Asia and Europe (*Globularia*), as well as on Socotra island (*Cockburnia*) and in Macronesia (*Lytanthus*).

The globe-daisy genus (*Globularia*) includes 2 rare species. The **white globe-daisy** (*G. incanescens* Viv.) is a 3–10 cm tall plant with slender rhizomes. Leaves are deciduous, not forming rosettes, roundish to lanceolate, incised, often dentate; lower leaves are long-petioled, cauline leaves have shorter petioles, uppermost leaves are almost sessile; all leaves are grayish-green, with calcium deposition. Head is 1.5 cm in diameter; bracts usually number 5, are linear-lanceolate, ciliate; calyx apices are awned; upper part of corolla tube is usually entire. The plant is found in the northern Apennines and the Apennine Alps (Italy). The Cambessedas globe-daisy (*G. cambessedesii* Willk.) is a plant with obovate or mucronate, usually serrated leaves with short petioles. Flowering shoots are 20–30 cm long; head is 3.5 cm in diameter. The plant grows only near the northern coast of Majorca Island (Baleares).

FAMILY GROSSULARIACEAE

The family Grossulariaceae consists of shrubs with alternate, simple, palmately lobed, exstipulate leaves. Flowers are solitary or in racemes, bisexual, rarely unisexual, regular, pentamerous. Ovary is inferior, unilocular. Stamens number 5. Fruit is a berry. The family includes 2 genera and about 150 species, distributed in the temperate zone of the northern hemisphere, in the mountains of Central America and along the Andes up to the Magellan Strait.

The currant genus (*Ribes*) comprises about 100 species, mainly shrubs with glabrous branches. Leaves are alternate, fairly large, palmately lobed. Flowers are arranged in racemes. The **Sardinian currant** (*R. sardoum* Martelli) is a small, nonprickly shrub, 1–1.5 m tall. The leaves on the young branches have numerous semisessile glands. Leaves are 10–13 mm broad, deeply lobed. Possibly, the plant is dioecious, as only female specimens are known. It is indigenous to Sardinia. The plant is included in Appendix 1 of CITES and is endangered.

FAMILY HYPERICACEAE

The family Hypericaceae includes trees, shrubs, herbs, or creepers with opposite or verticillate, entire and exstipulate leaves. Glands form dark or transparent dots on the leaves. Flowers are bisexual, regular, borne in inflorescences. Calyx and corolla are tetra- or pentamerous, free; stamens are numerous, in 3–5 bundles; ovary is superior, styles number 1–5. The fruit is a capsule or berry-like. The family has more than 40 genera and about 850 species, mainly distributed in the temperate regions of the northern hemisphere.

In Europe, 3 species of St. John's wort (*Hypericum*) need protection.

Caprifolious St. John's wort (*H. caprifolium* Boiss.) has erect or ascending stems, which are curly-pubescent and 20–100 cm tall. Leaves are 30–40 mm long, ovate to oblong, highly perforate, curly-pubescent. Inflorescence is corymbose to broad pyramidal, usually dense and glabrous. Calyx is lanceolate, awned, black glandular-ciliate; petals have black marginal glands. The plant grows in humid, shady places in southeastern Spain and is very rare.

Single-leaf St. John's wort (*H. haplophylloides* Halácsy et Baldacci) has 25 cm tall stems, branched, producing adventitious basal roots. Leaves are 12–28 mm long, oblong-linear to elliptical-linear, with black glandular hairs. Semiumbellate inflorescences consist of 3–5 flowers. Sepals are black glandular-ciliate; petals have minute, sessile black glands along margins. The plant grows in hilly forests and slopes in Albania and is endangered.

Coin-leaf St. John's wort (*H. nummularium* L.) has 8–30 cm tall branched stems, rooting at the base. Leaves are 5–18 mm long, broadly ovate to roundish, green on the upper surface, glaucous on lower surface, glabrous, with distinct pale glands along the margin and 2 black apical glands. Inflorescence consists of 1–8 flowers. Sepals are black glandular-dented. Petals sometimes have red veins (Plate 5). The plant grows in the rock crevices of stony slopes in the Pyrenees, in northern Spain, and in the southwestern Alps. The population of this species is declining.

FAMILY IRIDACEAE

The family Iridaceae includes about 60 genera and 1500 species, mainly distributed in the tropics and subtropics. The cape regions in Africa and tropical America are particularly rich in its species. Usually, the plants are perennial with rhizomes, bulbs, tubers, corms, or tuber-like rhizomes, rarely semishrubs. Flowers are regular or zygomorphic, usually in inflorescences, bisexual; perianth is corolloid; stamens number 3; styles are 3- or 6-lobed; ovary is inferior, trilocular. Fruit is a multiseeded, oblong or roundish loculicidal capsule. Many species are highly ornamental, widely used in ornamental gardening (i.e., gladioli, irises, saffron); some species contain essential oils and alkaloids.

The saffron genus (*Crocus*) comprises about 80 species, distributed mainly in the Mediterranean region and central Europe.

Multicolored crocus (*C. versicolor* Ker.-Gawl.) is a perennial corm plant. Inner membranes of corm are papery, outer membranes are hairy. Aerial stem is unbranched. Leaves number 3–6, are gray-green, 2–3 mm broad. Flowers are white to lilac or purple with pale yellow or white throat. The plant flowers from February to April. Its distribution is restricted to southeastern France and northwestern Italy. The species is close to extinction.

Endangered in Czechoslovakia are *C. albiflorus* Kit. and *C. heuffelianus* Herb. (Plate 5). They are known from not more than 5 locations; *C. olivieri* is protected by law in Bulgaria, and it is proposed to protect *C. maesiacus* Ker.-Gawl. in Rumania.

The genus *Iris* includes plants with annual flowering shoots and perennial reduced vegetative shoots, with rhizomes. Leaves are sword-like. Flowers are solitary or in inflorescences. Perianth is regular. About 250 species are distributed in the temperate and subtropical latitudes of all the continents of the northern hemisphere. The generic name is of ancient Greek origin and means 'rainbow'. Most species are highly ornamental.

Late iris (*I. serotina* Willk.) is indigenous to the mountains of southeastern Spain. It produces ovate bulbs and leaf bases covered with scales, stem 40–60 cm tall. Lower leaves are 30–60 × 2.6 mm, upper leaves are shorter and broader. Flowers number 2–3, are light blue-violet with dark violet veins and yellowish centre. Petals are 60–90 mm long, herbaceous, with a narrow, rough margin. Seeds are hemispherical, compressed, yellowish. The plant flowers in the second half of summer. It is found in pine forests and dry meadows.

FAMILY ISOETACEAE

The family Isoetaceae has only 2 genera. It comprises perennial heterosporous herbaceous plants with entire and ligulate leaves. Stems are simple, rarely forked or bifurcate. Rhizomorph is usually bilobed, rarely simple. Spirally arranged leaves are linear-subulate, with a broad and flat base and narrow subulate upper part. Sporangia are large, borne on the upper surface near the leaf base.

The quillwort genus (*Isoetes*) includes aquatic or amphibious plants, a few species growing in moist soil. Stem is simple, unbranched, located right at the soil surface. Leaves are straight or bent, divided into fertile and sterile. A well-developed ligule, up to 15 mm long, is borne at the leaf base. The genus consists of about 70 species, which are widely distributed throughout the world. The aquatic quillworts usually grow in oligotrophic lakes with mildly acidic and very transparent water, where they do not face competition from other plants. Earlier the quillworts were widely distributed in the glacial and alpine lakes of Europe, but they have already disappeared or continue to disappear at many places due to industrial and domestic pollution. Four European species are immediately threatened with extinction.

Bory's quillwort (*I. boryana* Durieu) is a fascicular, submerged or partly submerged plant with narrow subulate leaves, up to 20 cm long, almost round in cross-section. This is a plant of clean waters with very limited distribution in France and Spain. It is included in the Red Data Book of IUCN.

Delil's quillwort (*I. delilei* Rothm.) is an aquatic plant with triangular stems. Leaves are 12–40 × 1–2 mm, often numerous; ligule is oval. The plant is found in ponds and small lakes in southern France, northern Spain, and southern Portugal.

Slender quillwort (*I. tenuissima* Bor.) is a semi-aquatic plant with triangular stem and 7–12 cm long leaves. Ligule is small, triangular. The plant grows along small coastal parts of lakes near Montmorillon (central France).

Malinvernian quillwort (*I. malinverniana* Ces. et De Not.) is an aquatic plant with triquetrous stem. Leaves are 30–100 cm long, numerous, their bases having a broad membranous margin; ligule is triangular. The plant is found in fast-flowing water of irrigation canals in northwestern Italy, in the Vercelli and Novara regions.

FAMILY LAMIACEAE (LABIATAE)

The family Lamiaceae consists of herbs, semishrubs or shrubs with squarish stems. Leaves are opposite, rarely verticillate, exstipulate. Flowers are bisexual, brightly colored, zygomorphic, with very short pedicels, borne in condensed inflorescences in the axils of upper leaves, forming false whorls which themselves form raceme-like, spicate, corymbose, or paniculate inflorescences. Calyx fused, 5-dentate, sometimes bilabiate, persistent with the fruit. Corolla is pentamerous, fused, consisting of a tube and bilabiate limb. Stamens number 4, rarely 2, usually are didynamous. Style has a superior quadripartite or 4-lobed ovary. Fruit is dry, dehiscing into 4 single-seeded nutlets. The family consists of more than 180 genera and about 3500 species, distributed almost all over the world, especially found in large numbers in the Mediterranean region. Many of them are rich in essential oils, which are widely used in medicine, perfumery, and cooking. The plant is ornamental or a source of medicinal raw material.

Pulegian micromeria [*Micromeria pulegium* (Rochel) Benth.] is a 20–50 (90) cm tall plant with erect, usually simple, pubescent stems. Leaf is 10–30 × 5–20 mm, ovate, obtuse or acute, crenate-dentate, pubescent with short compressed hairs, and densely dotted on the lower surface. Flowers are in loose verticils of 10–70; lower verticils are shorter and upper ones longer than bracts. Corolla is white or lilac. The plant grows on rocks in very dry places. It is indigenous to Rumania and is a rare species.

The catmint genus (*Nepeta*) is represented by perennial, rarely annual herbs or semishrubs with entire serrated leaves. The genus contains up to 250 species, mainly distributed in the dry places in Eurasia and North Africa. Most species are concentrated in the Mediterranean region, 8 of which are included under rare plants in Europe.

Sphaciotic catmint (*N. sphaciotica* P.H. Davis) is a herbaceous perennial with a few 10–20 cm long shoots and woody rhizomes. Leaves are opposite, pale-green, oblong or ovate, with soft pubescence, 12–15 mm long, with crenate margin. Flowers are arranged in racemes modified into a spike at the apex of a stem. Calyx is 7 mm long, covered with gray hairs, 5-lobed. Corolla is pale, 12 mm long, bilabiate; upper lip is galeate, lower lip broadened and dentate. The plant is found only on the island of Crete. Its location is restricted to the White Mountains in the western part of the island. It may disappear as a result of sheep grazing, as its regeneration under such conditions is difficult. In 1966, only 40–50 plants were found. Seedlings were not found. Possibly, this is the only population. The sphaciotic catmint grows among limestone rocks at an altitude of 2300 m on the north-facing slopes. It is essential to control sheep population for protection of this species and also to introduce it into cultivation and organize reserves. This species is of great scientific interest for studying plant geography and taxonomy. It is included in the Red Data Book of IUCN.

The sage genus (*Salvia*) comprises perennial herbs or semishrubs with entire, rarely pinnatisect leaves and apical spicate or paniculate inflorescences. The genus contains about 700 species, distributed all over the world. All sage species yield essential oils; some of them have long been under cultivation. Six sage species have been listed as rare in Europe.

Candelabrian sage (*S. candelabrum* Boiss.) is a herbaceous perennial up to 80 cm tall, with woody base. Stem is erect, with a few branches, pubescent in the lower part, leafless above. Leaves are up to 9 cm long, simple and with a pair of small stipules at the base, oblong-elliptical, wrinkled. Inflorescence is loose, a complex umbel of 3–5 flowers. Corolla is 30–40 mm long; upper lip white with violet spots, lower lip violet-blue (Plate 6). It is a rare species, indigenous to southern Spain.

FAMILY LILIACEAE

The family Liliaceae is one of the largest families in the world flora in the number of genera (about 170) and species (above 2500), including many ornamental flowering plants. The species of this family are found in all parts of the world from tropics to tundra, from the plains to the alpine belt. They are perennial herbs with various types of underground organs (rhizomes, corms, bulbs), rarely annuals or shrubs. Stem is usually erect, leafy or leafless; leaves are entire, usually narrow. Inflorescence is a raceme, spike or panicle, rarely an umbel; sometimes flowers are solitary. Flowers are actinomorphic or slightly zygomorphic, bisexual, rarely dioecious. Perianth usually consists of 6 free or fused tepals. Ovary is superior, trilocular (sometimes uni- or tetralocular). The fruit is a capsule or berry. The family has many food and medicinal species.

The crocus genus (*Colchicum*) comprises cormous perennial ephemerals (the aerial organs die in summer). The corm is covered with membranous or coriaceous dry covering scales. Leaves appear in spring, together with flowers, and in the species flowering during autumn they appear in the next spring. Flowers are large, bisexual. About 65 species of the genus are distributed mainly in the Mediterranean countries. These are ornamental, alkaloid-producing plants. The **arenaceous (sandy) crocus** (*C. arenarium* Waldst. et Kit.) has 2–2.5 × 1–1.5 cm ovate bulbs, scale reddish to dark brown, thin. Leaves number 3–5, develop after flowering, are ligulate to linear-lanceolate, glabrous, obtuse, 8–20 × 4–17 mm in size. Flowers number 1–4; perianth leaves are 25–40 × 3–10 mm, purple-pink, oblong to narrow-elliptical (Plate 6). The plant flowers in September–October. It is found in sandy meadows in Czechoslovakia, Hungary, and Rumania, where it is protected by law.

The fritillary genus (*Fritillaria*) includes plants with narrow cauline leaves and one or a few drooping flowers. Perianth is campanulate, whitish-yellow to dark chocolate-brown, sometimes with chessboard-like markings (from which the generic name, meaning chessboard, has been derived). About 100 species are distributed in the temperate belt of the northern hemisphere. All species are ornamental, many plants are used in Tibetan and Chinese medicines.

Involucrate fritillary (*F. involucrata* All.) has 15–25 cm long stems. Leaves number 7–10, are linear-lanceolate or linear, sometimes glaucous; lower leaves are 0.5–1 cm long and 6–10 mm broad, opposite; middle leaves are opposite or alternate. Perianth is broad, campanulate; perianth leaves are green, with chocolate-brown or purple spots, obtuse; outer perianth leaves are 25–40 × 7–16 mm, inner ones sometimes broader (Plate 6). The plant grows in sparse forests, among shrubs, and along stony places in France and Italy. A highly ornamental plant, its population is decreasing in many countries due to collection for bouquets.

In some European countries, even the more widespread species, for example, checkered lily or common snake's head (*F. meleagris* L.) have been declared protected species (or protection is contemplated).

The lily genus (*Lilium*) includes perennial, bulbous plants with erect or slightly drooping leafy stems. Leaves are mostly cauline, in some species aerial bulbs develop in their axils. Flowers are usually borne in racemes or umbellate inflorescence, rarely solitary, often aromatic. The fruit is a trilocular capsule. The genus comprises about 100 species, distributed in the temperate belt of the northern hemisphere. The lilies are under wide cultivation since ancient times.

Rhodopean lily (*L. rhodopaeum* Delip.) forms bulbs with a diameter of 3.5–4 cm with numerous white or pale yellow scales. Stems are 80–100 cm tall with linear-oblong, 6–14 cm long leaves, slightly silvery-pubescent along the veins and margins. Flowers are grouped in terminal racemes, each with 5 flowers, on divergent pedicels. The perianth consists of 6 lanceolate divergent, bright lemon-yellow tepals, 8–12 cm long and 2–3 cm broad. Anthers are red. Distribution is restricted to the central part of the Rhodope Mountains in the Smolyan region of Bulgaria (a few locations known) and the adjoining region of Greece (isolated locations). The plant grows at the altitude of 1300 m in alpine meadows together with sheep's fescue and molinia. The plant is very ornamental and also of great scientific interest because of its relationship with the Caucasian and Turkish species. Its population is declining due to collection for herbaria and by tourists. It is included in the Red Data Book of IUCN. It is protected by law in Bulgaria. One of its locations has been earmarked to be maintained as a reserve. It is cultivated in several Bulgarian botanical gardens and can be cultivated relatively easily.

Saffron lily (*L. croceum* Chaix) is 80 cm tall. Inflorescence has 20 or more flowers. Flowers are glomerulate, erect, bright golden-yellow, up to 14 cm in diameter. Stem is rigid, glabrous, in the upper part tomentose-pubescent. Cauline leaves are alternate only at the base, lanceolate, glabrous. Bulb is broadly ovate, up to 2 cm in diameter, with broadly lanceolate white scales (Plate 7). The plant grows among small shrubs in mountains, on precipitous slopes in Mediterranean Europe (Switzerland, northern Italy, southern France). It is highly ornamental, therefore, protected and is widely cultivated.

The **Yank's lily** (*L. jankae* Kerno) is protected by law in Bulgaria. The bulb-bearing (or orange) lily (*L. bulbiferum* L.) with an inflorescence of red flowers (Plate 6) and the **Carniola lily** (*L. carniolicum* Bernh. ex C. Koch) with solitary bright red flowers (Plate 7) are protected in Yugoslavia. The **Albanian lily** (*L. albanicum* Griseb) with yellow flowers is protected in Greece.

The spiderwort genus (*Paradisea*) includes ornamental plants, whose population is declining. These include **liliaster** (paradise lily) [*P. liliastrum* (L.) Bertol.], a plant with 30–50 cm tall stem and flat leaves 12–25 cm long and 3–5 mm broad. Inflorescence is sparse, with 4–10 (20) flowers. Flowers are campanulate (Plate 7). The plant grows in mountainous meadows in the Pyrenees, Alps, Jura, and northern and central Apennines.

The tulip genus (*Tulipa*) is one of the most ornamental genera. The wild species have served as the initial material for obtaining numerous cultivars. The plant consists of 2–5 fleshy fertile scales and a membranous sterile scale. Leaves usually number 2–4, flowers 1–2 (rarely 3–15), tepals 6, falling at the end of flowering. Fruit is a dry capsule. The tulips are ephemeroid; their vegetative growth period continues for not more than 3 months. The plant is protected from incidental self-pollination by protandry (earlier maturation of anthers than pistils). Distribution of the genus covers Europe, Asia and North Africa. Central and northwest Asia, Kazakhstan, and the Mediterranean regions are richest in species.

Rhodopean tulip (*T. rhodopea* Velen.) has a height of 20–25 cm and ovate bulb, 2–3 cm in diameter. Covering scale is thin, papery; stem is green; leaves are broad, glaucous, lower leaves up to 16 cm long and 4 cm broad. Flowers number 1–2, are wide open, campanulate, dull red, with olive dark brown to black base and yellow border (Plate 8). Its distribution is confined to the Rhodopean mountains. The plant is endemic to Bulgaria, placed under protection in that country. The species *T. celsiana* DC., *T. sylvestris* L. (forest tulip) (Plate 8), *T. thracica* Davidov, and *T. urumoffii* Haek (Urumoff's tulips) (Plate 7) are also protected in Bulgaria.

FAMILY OLEACEAE

The family Oleaceae includes shrubs or trees with opposite or rarely alternate, simple or imparipinnate exstipulate leaves. Flowers are small, usually bisexual (rarely the plant is dioecious), regular, tetramerous, arranged in inflorescences. Calyx is small or absent. Corolla is gamopetalous; stamens number 2, rarely 4; ovary is superior, bilocular. Fruit is a capsule, winged nut, berry, or stone fruit. The family includes 29 genera and about 600 species, widespread in the temperate, subtropical, and tropical regions of the northern hemisphere, their largest number concentrated in Asia.

The golden-ball tree genus (*Forsythia*) comprises shrubs with opposite leaves, flowering profusely in early spring, usually much before the appearance of leaves. Flowers are golden-yellow. Five species of this genus are found in East Asia and one species in Europe.

European golden-ball tree (*F. europaea* Degen et Bald.) is a deciduous shrub up to 2 m tall, forming abundant growth from roots. The swaying branches have opposite, ovate or lanceolate leaves, 4–7 mm long. Flowers are 15–20 mm long, appearing in March and April; calyx is tetralobate, small; corolla is bright yellow, with a short tube and 4 narrow divergent lobes; stamens number 2. Fruit is a smooth ovoid capsule. This rare plant is found in Albania and Yugoslavia (the condition of the Albanian populations is unknown). Three pockets of distribution are known in Yugoslavia: the mountain slopes near Pec (the most visited and the smallest population); the hills between Pec and Gacko (detailed distribution unknown, but abundant at isolated spots); and the region west of the Gacko road (the largest population over an area larger than 4 km^2).

The European golden-ball tree grows only on serpentine rocks at an altitude of 300–900 m among shrubs and isolated trees. The submediterranean species generally predominant in the communities are: hornbeam-leaved ostrya, flowering ash, and rarely Austrian pine. At present almost all the trees have disappeared due to grazing and burning, and the golden-ball tree is included in the secondary plantations. It can be easily multiplied by seeds and vegetatively. Therefore, the plant is not threatened with extinction if grazing, burning, and collections are prohibited. The European golden-ball tree is the only European species of this ornamental genus; however, it is known only from special collections of the botanical gardens and not yet introduced into large-scale cultivation.

FAMILY ORCHIDACEAE

The family Orchidaceae includes perennial terrestrial or epiphytic herbs with rhizomes or root and stem tubers. Leaves are alternate, entire, and scaly in saprophytes; flowers are zygomorphic. All orchids are mycorrhizal plants. Their seeds germinate only when the hyphae of some fungus enter their embryonic tissues. During the first few years, the plants feed only at the cost of these fungi, after which they become independent, but the hyphae remain associated throughout their life. Seed propagation is difficult, but vegetative reproduction is well-developed in many species.

Orchidaceae is one of the largest families, consisting of 20,000–30,000 species (about 7 percent of all the flowering plants), widespread all over the global surface, except the Arctic regions. The tropics, especially Central and South America and southeast Asia, are particularly rich in orchids. The orchids are very ornamental, and several species are medicinal plants. They are very sensitive to environmental changes. At present, about 17,000 species of orchids are on their way to extinction. One of the most important causes for this is their collection and sale. The other reason is effect of human activity (cutting and clearing of forests, reclamation, air and water pollution, and application of chemical fertilizers).

At present, orchids are protected in all the countries or measures for their conservation are being evolved.

All the endangered orchids are included in Appendix 2 of CITES.

In Europe, the need for conservation of orchids was raised for the first time at the end of the 19th century, when lady's slipper was placed under protection in Switzerland (at the beginning of the 20th century, its protection was started in Denmark and France, followed by other countries). Wild orchids have been protected in East Germany since 1955. Seventeen orchid species (2 of them have disappeared) have been included in the Red Data Book of Great Britain. In France, 50 species need protection (4 species have disappeared). In Czechoslovakia, 3 species of orchids have already disappeared, one is suspected to have disappeared, 17 species are threatened with extinction, and 27 require protection, of which 18 need strict protection. In Yugoslavia, 7 species need to be protected, in Greece 4, in Liechtenstein 14, in Luxemburg 8, in Bulgaria 4, and in Iceland one. Thirty-five species have been included in the Red Data Book of the USSR. It is not possible to deal in this book with all the orchids that need to be protected; we will confine ourselves to the characterization of genera or, as an example, individual species.

The man orchid genus (*Aceras*) is represented by a single species in Europe. This is **green man orchid** [*A. anthropophorum* (L.) R. Br.], a 25–35 cm tall plant with entire oval root tubers. Leaves are glaucous-green, all leaves arranged in a basal rosette. Inflorescence is multiflorate, flowers are yellowish-green (Plate 8). The plant grows in broad-leaved forests and in the Mediterranean region of Europe. It is found rarely.

The phantom orchid genus (*Cephalanthera*) includes plants with 5–9 leaves and reduced horizontal or ascending rhizomes. Inflorescence is in the form of a sparse spike with flowers directed upward. The genus contains about 14 species, distributed in the temperate belt of Eurasia; one of them grows in North America. All phantom orchids are ornamental. Europe has 5 species of phantom orchids.

The genus *Comperia* is distributed in Crimea and Asia Minor.

The **Crimean comperia** [*C. comperana* (Stev.) Aschers. et Graebn.] is a herbaceous perennial with entire ovate or oval tubers. Stem is 25–50 cm tall, with 2–4 lanceolate or oblong leaves. Flowers have chocolate-brown to yellow-purple labellum and whitish pink lip aggregated in groups of 3–10, rarely up to 25 in a loose inflorescence. Lip is long, trilobed; lateral lobes are triangular, continued into long, up to 5–7 cm, filamentous lobes; central lobe is bifurcated, also having long filamentous lobes (Plate 8).

The distribution of this species is limited to the southern coasts of Crimea and Turkey. This particular comperia species grows in light deciduous or mixed forests in the lower mountain belt, on the fringes and in glades, and as an exception on limestone deposits. It is found rarely as isolated specimens or in small groups. Its population is declining due to collection for bouquets and destruction of its habitats.

The lady's slipper genus (*Cypripedium*) includes about 50 species growing in the temperate and cold belts of the northern hemisphere. Three species are found in Europe. These plants have horizontal creeping rhizomes, with 1–2 (rarely 3–5) lanceolate or elliptical leaves. Flowers are large, solitary or 2–3. Lip is sacciform. The plant flowers in 7–12 years.

True lady's slipper (*C. calceolus* L.) is the first orchid to be protected in Europe at the end of the 19th century. At present protected in all countries, it is included in the Red Data Book of IUCN. The plant has 3–5 leaves and up to 8 cm long flowers; the lip of flowers is yellow, with reddish spots on the inside. The plant grows in deciduous and mixed as well as humid and moss-covered coniferous forests, along shrubs and forest meadows. The **scattered lady's slipper** (*C. guttatum* Sw.) produces a stem with 2 leaves. Flowers are white with violet-pink or purple spots. It grows in deciduous, mixed, and coniferous forests. The **large-flowered lady's slipper** (*C. macranthon* Sw.) has a purple or violet-pink, spotless lip. It grows in deciduous forests.

All the 3 species mentioned above have a wild Eurasian distribution, but are found sporadically. During the last few decades, they have disappeared from many areas of Europe as well as Siberia which were located near large inhabitations.

The helleborine genus (*Epipactis*) includes plants with creeping or condensed rhizomes. Their leaves are alternate, and inflorescence is a sparse, long, straight raceme. Flowers are fairly large, drooping, on twisted pedicels, purple or greenish. The genus has about 25 species, distributed in the temperate belt of Eurasia, Africa and North America. Several species are found in Europe, many of which are protected.

The **dune helleborine** [*E. dunensis* (T. et T.A. Stephenson) Godfrey] has been included in the Red Data Book of England. It is a plant with yellowish-green leaves. The inflorescence consists of 5–25 yellowish-green flowers; perianth leaves are ovate-lanceolate (Plate 9). It grows on the coastal sandy dunes in the northern regions of England and in northern Wales. Only 11 locations are known. The populations of the species are very small and continue to decrease.

The genus *Nigritella* includes plants 10–30 cm in height, with slender stem and 2–3-lobed root tubers. Leaves are linear, long, forming a basal rosette. Inflorescence is a short, very dense raceme. Flowers are small, with vanilla flavor. Lip is triangular, entire, sometimes with a pair of short projections, directed upward. The genus comprises 2 species, both ornamental.

The **black nigritella** [*N. nigra* (L.) Reichb.] has a raceme that is conical in the beginning of flowering and later becomes spherical, of black-purple, rarely pink-red, pure white or light-yellow flowers (Plate 9). It grows on varied grass alpine meadows and unfertilized pastures at an altitude of 1700–2500 m in western, central and southern Europe.

The bilfoil genus (*Ophrys*) includes perennials with almost spherical or oblong bulbs. Leaves in the lower part of stem number 2–5, are oblong or oblong-lanceolate. Inflorescence is a sparse spike with 2–10, rarely 15–20, fairly large, sessile flowers. Perianth leaves are divergent, outer ones oblong or oblong-ovate, obtuse, glabrous; 2 perianth leaves of the inner whorl are smaller, linear, and densely pubescent. Lip has no spur, points downward, is flat or convex, entire or trilobed, brown to black-purple, with velvety pubescence (in some species, its shape and color resemble a bee or bumblebee), with 2 basal tubercles or horns (Plate 9). Most species are pollinated only by some Hymenoptera males, with whose females the flowers have amazing similarity (only bee orchid *O. apifera* is an autogamous species). The genus contains about 80 species, distributed in Europe, Asia, northern Africa, North America, and the Canary Islands. *Ophrys* is a typical Mediterranean genus. In Europe, 20 species of this genus are found, predominantly in the Mediterranean regions, and all of them are placed under protection in most European countries. The plants of this genus are disappearing as a result of destruction or pollution of their habitats as well as disappearance of the pollinator insects due to insecticides.

The genus *Dactylorhiza* includes plants with root tubers and flattened, leafy stems. Leaves are spatulate. Inflorescence is usually a dense, multiflorate spike. Flowers are small, with variable color. Thirty-one species of the genus are distributed in Eurasia, North America, and northwestern Africa.

The orchid genus (*Orchis*) is represented by plants with entire, ovoid, obovoid, or spherical root tubers. Leaves are usually uniformly distributed over the stem or in a basal rosette. Inflorescence is spicate (Plate 10). The genus comprises about 35 species of plants widely distributed throughout the world.

The dactylorhiza and orchid genera are close to each other and ornamental. Some of their species are collected as medicinal raw material.

The ladies tresses or dwarf orchid genus (*Spiranthes*) comprises plants up to 40 cm in height and a fascicle of tuberous roots, often with a rosette of leaves and cauline, leafy sheaths. Inflorescence is long, slender, multiflorate and spirally twisted, the spur being absent. The genus includes 25 species, distributed in Europe and Asia.

Summer ladies tresses [*S. aestivalis* (Poiret) L.C. Rich.] has small, pure white or greenish white, scented flowers (Plate 10). It grows in marshy meadows, on hills and sands in temperate and subtropical zones of western Europe. It is a rare species.

Romantsev's ladies tresses (*S. romanzoffiana* Cham.) has white flowers covered with green or cream-colored stripes (Plate 10). It grows in humid, marshy meadows and lake banks in Ireland, western Scotland, southwestern England, and North America. It is a rare species.

FAMILY PAEONIACEAE

The family Paeoniaceae includes perennial herbs with fleshy, thick roots, rarely small shrubs. Leaves are alternate, without stipules, ternate, biternate, or triternate. Flowers are large, bisexual, regular, solitary. Sepals and petals number 5 each, stamens are numerous, pistils number 1–15. Fruit is multiple with large, usually black seeds. The family has only one genus, comprising about 40 species, distributed in the temperate regions of Europe, Asia, and North America.

The peony genus (*Paeonia*) is represented by **Kambesedes peony** [*P. cambessedesii* (Willk.) Willk.]. This is a herbaceous perennial with lower leaves up to 25 cm long, divided into 9 lobes. Lobes are leathery, nonpubescent, lanceolate to ovate, pointed, up to 10 cm long, dark green with a purple lower surface. Peduncles are up to 30–45 (60) cm long, each terminating in a large cup-shaped flower, 6–10 cm in diameter. Petals are dark pink, obovate; stamens are numerous, golden-yellow. Fruit is aggregated, consisting of 5–8 pink, up to 6 cm long follicles with red or violet seeds (Plate 11).

The plant is found on the Balearic Islands, possibly, its distribution is now confined to the mountains in northwestern Majorca; it was known from Minorca (reports about its existence in Corsica are erroneous). It grows on limestone screes and on limestone cliffs below in association with *Smilax aspera*. It is becoming a rare plant. It suffers mainly due to grazing by goats as well as land reclamation and formation of tracking paths. Its regeneration is difficult as the goats eat the maturing fruits. Wild populations are not protected, but a national park is proposed at Cape Formentor at the northern tip of Majorca where, besides peony, many other indigenous plants grow. The flowers of peony have the fragrance of rose and are pollinated by bees and other insects. The roots of this plant are used in the popular medicines in Majorca. It is a highly ornamental plant and can be used in ornamental floriculture; it flowers early.

FAMILY PALMAE

The family Palmae includes 2400 species belonging to 270 genera. Its members are trees or shrubs with simple, rarely branched trunks from 1 to 100 cm in diameter, often reaching considerable height. Their leaves form a crown at the top of the stem, with the young leaves pointed upward, and the more adult leaves gradually spread in all directions; the older leaves are suspended along the trunk. Phyllotaxy is spiral; leaves are usually very large with the lamina entire, pinnatifid, palmatifid, or divided (compound). Flowers are borne in simple or complex inflorescences. Perianth is in 2 whorls, persistent. Fruit is a berry or drupe.

Most of its species are distributed in the tropics, and a few in the subtropics. The palms are found in most variable habitats: along the sea coasts, in mangrove plantations, oases, deserts and open savannas, marshy leylands, in mountainous terrains, and even in deciduous forests or moderately hot regions. Many species are extremely ornamental and some have food or commercial value. The present condition of a large number of palms is critical: conditions of individual species are not yet determined, but some species have already disappeared and others are on the verge of extinction. This is mainly due to destruction of their habitats. At present, the tropical forests with palms are being intensively cut and the IUCN has prepared a special programme for conservation of natural vegetation of the tropics. A subcommittee on palms has been appointed under the Commission on Threatened Plants of the IUCN. Only 2 palm genera are distributed in Europe (in the Mediterranean region).

The **date palm** (*Phoenix theophrasti* Greuter) is included in the Red Data Book of IUCN. It grows up to 10 m in height, each of its stems has a few sharp lateral branches and dense crown of thin imparipinnate leaves, 3–5 m long, which are vertical at the beginning but horizontal and drooping subsequently. The middle and upper pinnae are more than 20–50 cm long, each with a sharp tip, lower pinnae modified into hard spines. Male and female flowers are borne on different trees (dioecious), in large and highly branched panicles, which are upright even while fruiting and surrounded by leaf base. Fruits are elliptical, 14–16 mm long, inedible, almost nonfleshy, yellowish-brown. Its main distinction from the cultivated date palms is the presence of protruding racemes of small inedible fruits. This palm is known only from 5 points on the island of Crete, where it grows on sandy alluvial soil adjoining the sea. It reproduces by seeds and vegetatively. The main location of this palm is near Vai and at the northeastern tip of the island, where it forms a grove at the bottom of a large valley, about 1 km long, leading to sandy and turfy coasts. At this place, the date palm suffers due to tourism (all its seedlings and root sprouts are trampled). It is proposed to organize a reserve there. The 3 other habitats are located along the southern coast west of Iraklion. One habitat was almost completely destroyed, only a few bushes remained in 1967, but their regeneration started thereafter. Isolated specimens are also known at the northern coast. This palm has a great scientific value as being a close relative of the cultivated date palm, it can be used for raising hybrids resistant to cold, pests, and diseases. It was more widely distributed earlier on Crete. Cutting, burning, and trampling are the main causes of degradation of its thickets.

FAMILY PAPAVERACEAE

The family Papaveraceae contains herbaceous plants with milky white or colorless juice, rarely semishrubs or shrubs. Their leaves are usually alternate, entire, or divided. Flowers are dichlamydeous, regular, bisexual, often large. Sepals number 2–3, rarely 4, are caducous. Petals number 4–6, are free; stamens are many; ovary is superior, uni- or bilocular, with sessile stigma. Fruit is a capsule, opening through pores, or a pod. Several members of this family contain alkaloids, and some are poisonous. The family has 26 genera and about 450 species, mainly distributed in the temperate and subtropical regions of the northern hemisphere.

Lestadian poppy [*Papaver laestadianum* (Nordh.) Nordh.] is a turfy, short perennial with dense rosette of crenate leaves, covered with yellowish hairs. Leaflets are broadly lanceolate, pointed, in groups of 3–5. Juice is usually yellowish. The leaf bases are persistent, forming a dense cover. Flowers are solitary, on more or less straight 10 cm long branches, with squarrose hairs. Petals are densely covered with squarrose, mostly black hairs. This poppy is a self-pollinating plant. The seeds germinate very slowly. The distribution of this indigenous plant is limited to a small mountainous region in Scandinavia; 3 habitats are known in the north, in the border areas of Norway, and a few locations covering an area of about 12 km^2 in Sweden. This poppy grows in the middle-mountain alpine belt, at an altitude of 900–1300 m, in relatively humid places, on siliceous or shale soils, on the slopes facing north. It is a very rare species, included in the Red Data Book of IUCN and protected in Sweden. Its cultivation is extremely difficult.

PLATE 9. 1—Dune helleborine; 2—black nigritella; 3, 4—bilfoil (left—spider bilfoil).

PLATE 10. 1, 2—Orchid purple; 3—summer ladies tresses; 4—Romantsev's ladies tresses.

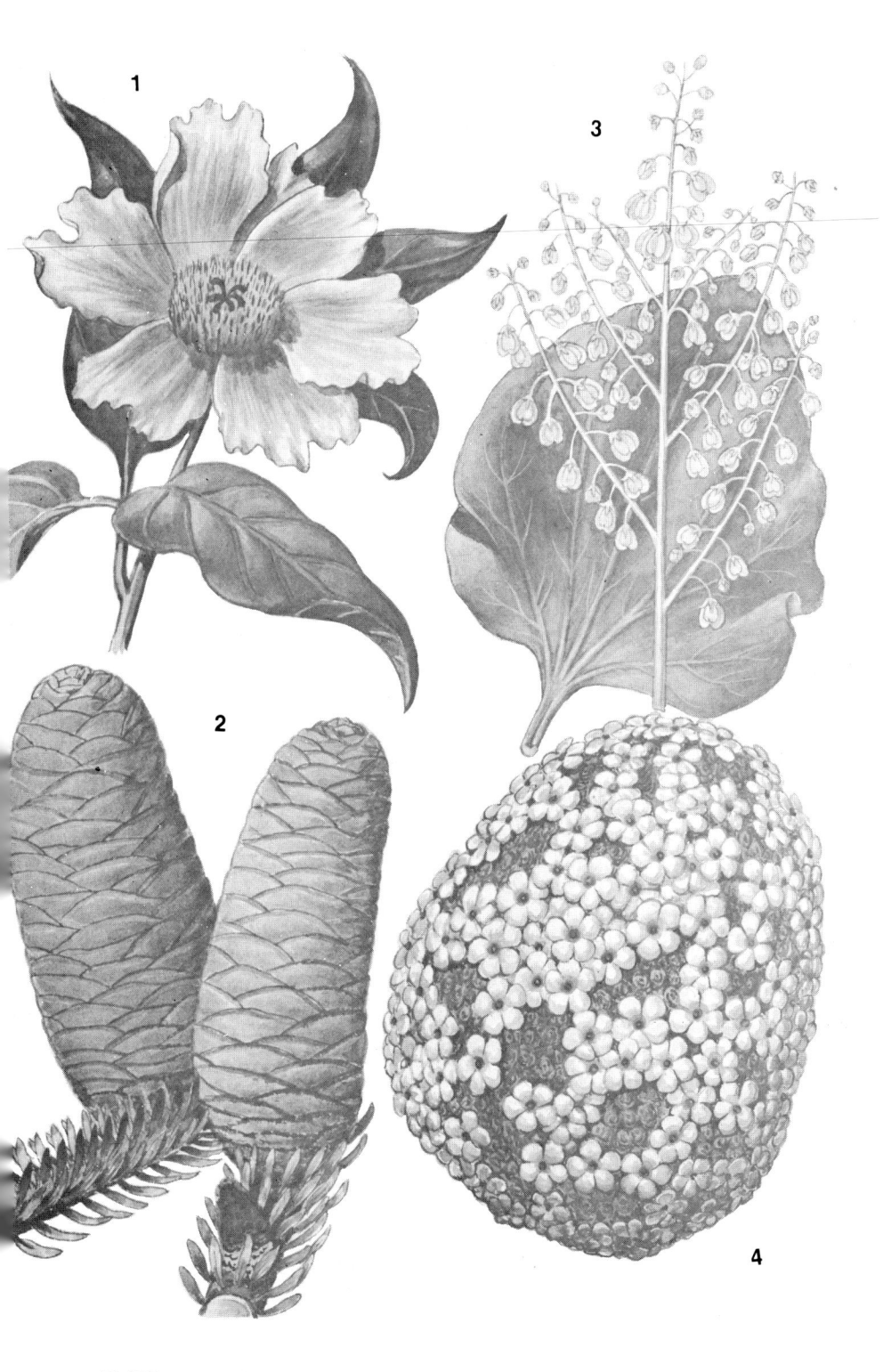

PLATE 11. 1—Kambesedes peony; 2—Spanish fir; 3—rhaponticoid rhubarb; 4—Swiss androsace.

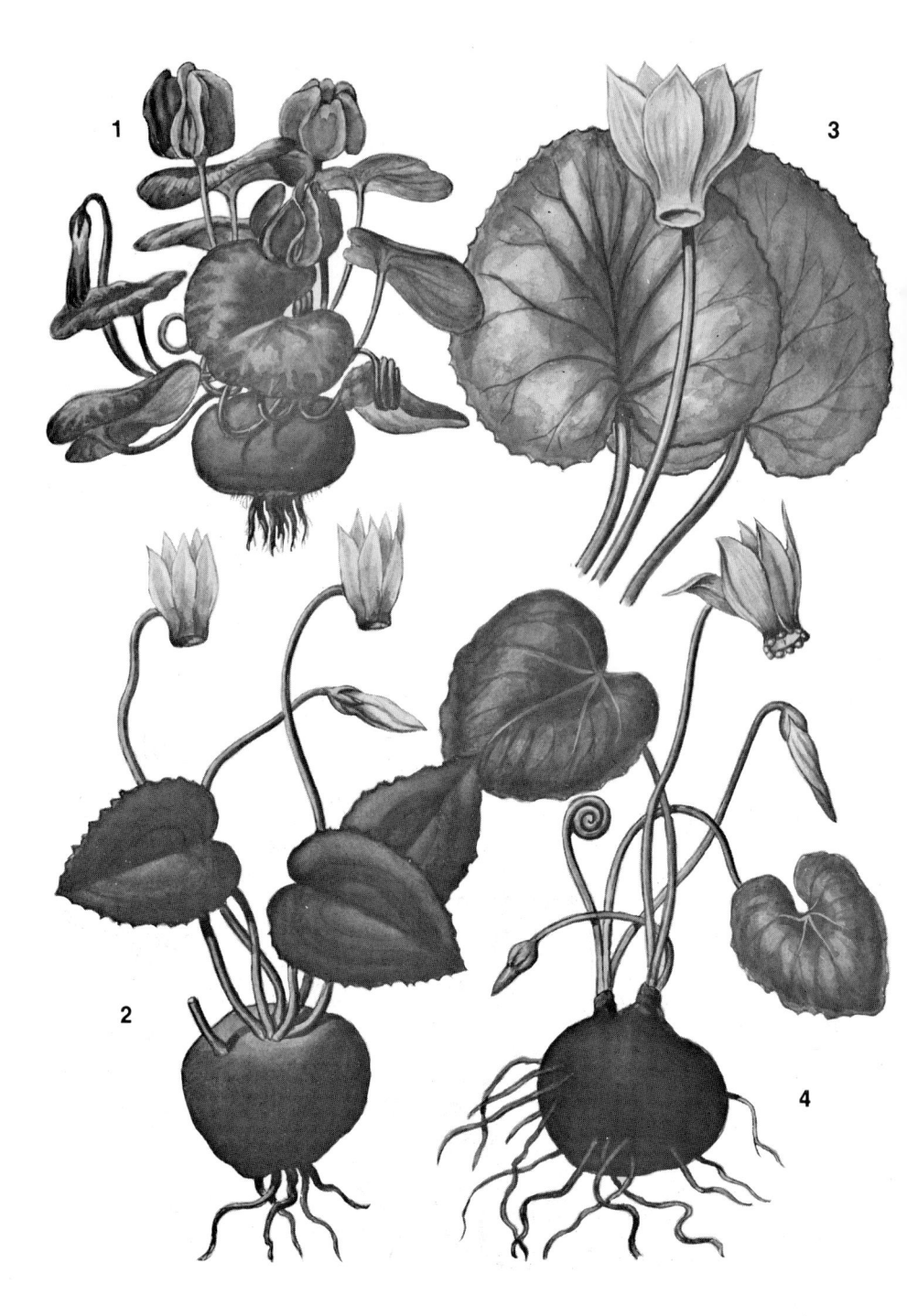

PLATE 12. 1—Coum sowbread; 2—Persian sowbread; 3—ivy-leaved sowbread; 4—purple sowbread.

FAMILY PINACEAE

The family Pinaceae comprises about 250 species of evergreen, rarely deciduous trees, shrubs, and dwarf birches, distributed almost exclusively in the northern hemisphere. Leaves are narrow-linear, needle-shaped, or scaly. Male and female cones are borne on the same tree. The species of this family are widely used as a source of timber, raw material for resins, essential oils, tannins, and vitamins.

The fir genus (*Abies*) consists of about 40 species, distributed in the temperate latitudes of the northern hemisphere. These are large evergreen trees with indistinctly verticillate conical crown. The needle is flat, soft, perennial, with 2 whitish stripes on the lower side. Cones are usually cylindrical, erect, and open in the first year. The plants are shade-tolerant, wind-resistant, but very sensitive to air pollution.

Sicilian fir [*A. nebrodensis* (Lojac.) Mattei] earlier formed a dense forest in the mountains of Sicily, but now not more than 20 young wild trees are found in the north of the island. It grows up to a height of 15 m, with broad, sometimes flat crown. Trunk is strong; needles are dense, thick, hard, obtuse; cones are 20×5 cm. Plantations are being raised on the island to restore this indigenous species.

Spanish fir (*A. pinsapo* Boiss.) is a tree up to 30 mm tall with a pyramidal crown and strong trunk with more than 1 m diameter. Needles are thick, acute, bristly, borne on the twigs, which differentiates this fir clearly from other species. Cone is $10-16 \times 3-4$ cm (Plate 11). It grows along the slopes facing north in limestone mountains at an altitude of 1000–2000 m in southwestern Spain, near Ronda. It is a very ornamental and drought-resistant plant.

FAMILY PLUMBAGINACEAE

The family Plumbaginaceae includes herbaceous perennials or shrubs, rarely semishrubs and climbers with alternate exstipulate leaves. Leaves are broad or narrow, rarely pinnatipartite. Flowers are regular, bisexual, variously colored, in uniflorate or multiflorate spikes, aggregated into inflorescences of various types. The family comprises about 15 genera and 500 species, widely distributed all over the earth's surface, but mainly in the Mediterranean region, west, central and middle Asia, in arid regions. Some species contain large quantities of tannins; many are ornamental.

The thrift genus (*Armeria*) comprises perennial herbs with numerous linear leaves in a basal rosette and violet-pink flowers arranged in dense spherical heads on leafless peduncles. Inflorescence has an involucre of a few membranous leaves on the lower side, fused into a tubular sheath, surrounding the upper part of peduncle. Calyx has a herbaceous tube and membranous limb; corolla consists of 5 petals fused only at the base. A large section of thrifts, which require special protection, are indigenous species of Portugal.

Arcuate thrift (*A. arcuata* Welw. ex Boiss. et Reuter) is a plant with reduced stem and no branches. Leaves are 24–30 × 0.2–0.7 mm, filiform, pointed. Outer involucral leaves are pointed, inner awnless. Calyx is dark chocolate-brown, with a broad membranous margin, velvety-membranous on the upper side. The plant was found on humid pastures in Vila Nova de Milfontes (southwestern Portugal). At present, it has possibly disappeared.

Pseudoarmerian thrift [*A. pseudarmeria* (Murray) Mansf.] is a compact dwarf shrub with strong, multibranched stems. Leaves are 100–230 × 14–22 mm, obovate to lanceolate-spatulate, pointed, awnless, flat, glabrous, peduncle is 25–50 cm long, glabrous, straight, or arcuate. Spathe is loose; heads are 30–40 mm across; bracts have a broad membranous margin. Spikelets are sessile, calyx is 10–11 mm long with the tube hairy along the ridges; awns are 1.2–1.7 mm long. Corolla is white. The plant grows on pastures and thickets on granite at Cape Rock. It is endangered.

Rouy's thrift (*A. rouyana* Daveau) is a pubescent herbaceous plant with fairly strong, short stem. Leaves are up to 2.5 mm broad, 90–160 mm long, linear-lanceolate. Bracts are leathery, with broad membranous margin; stipules of spikelets are longer than calyx, leathery, and broadly membranous in the upper part. Calyx has very short awns; corolla is pink or white. The plant is found in thickets of shrubs on sandy soils in southwestern Portugal. It is endangered.

Soleirol's thrift [*A. soleirolii* (Duby) Godron] is a dwarf shrub with long branches. Leaves are acute, dimorphic, verrucose; outer leaves are 40–50 × 2–4 mm, linear-spatulate, flat, glabrous; inner leaves are 70–80 × 2 mm, linear with longitudinal furrows. Bracts have narrow membranous margin; outer bracts are shorter than inner ones, obtuse, awnless. Spikelets are usually almost sessile. Calyx is 6 mm long, its tube hairy along the ridges. Corolla is pink or white. This plant, indigenous to Corsica, is found on the rocks and sands near Calvi. It is endangered.

The sea-lavender genus (*Limonium*) includes perennial herbs or semi-shrubs. Leaves are arranged in rosettes; flowers are small, in uniflorate or multiflorate spikelets, aggregated in loose or dense spikes which, in turn, form paniculate or corymbose, often capitate inflorescences. The genus comprises about 300 species, among which are valuable tannin-producing, dye, and ornamental plants. The largest number of plants that need protection are found in Spain (13 species) and Italy (12 species). Several plants of this genus distributed in England and Ireland also require protection.

Recurved sea-lavender (*L. recurvum* C.E. Salmon) is included in the Red Data Book of IUCN. This is a perennial herb with a stem bearing a rosette of narrow, obovate, obtuse, 2–4 cm long, tough leaves. Flowering shoots are rough, 5–20 cm long, bearing short, dense, bent or divergent, sometimes horizontal, 1–2 cm long spikes with 2 rows of closely compressed spikelets. Spikelets have 1–3 flowers. Calyx is funnel-shaped; corolla is up to 6 mm in diameter, with very short tube and 5 divergent lobes, violet or dove-blue. The plant is indigenous to England, growing only on cliffs of Portland Island (Dorset). It was first found in 1832, but its habitat was determined during stone quarrying. In 1977, there were 1100–1600 plants in 4 populations at a short distance from each other. One of the populations is endangered (it may be already destroyed during quarrying of stones for construction purposes). Two other populations are located near stone quarries, and although their number is still not declining, there is a potential danger of their extinction. This particular species of sea-lavender has been included in the Red Data Book of England. It is cultivated in the botanical garden of Cambridge University.

The **Transwell sea-lavender** [*L. transwallianum* (Pugsley) Pugsley] and **paradoxical sea-lavender** (*L. paradoxum* Pugsley) are found in the coastal cliffs in England and Ireland. Both species are similar to the recurved sea-lavender and are very rare. A rare indigenous species, **stellate hairy sea-lavender** [*L. asterotrichum* (C.E. Salmon) C.E. Salmon)], grows in the dry saline meadows of central and southeastern Bulgaria. It is a densely pubescent plant with a height of 40–65 cm. Leaves are 100–180 × 10–13 mm, linear-spatulate, acute, pinnately veined, with stellate hairs. Spikes are up to 1.5 cm long, dense; spikelets have 1–2 florets. Bracts have a narrow transparent margin, inner bracts are 2.2–2.8 mm long. Corolla is pale violet.

FAMILY POACEAE

The family Poaceae includes herbaceous, rarely turfy plants with fibrous root system or underground rhizomes or stolons. Their stem is usually cylindrical, jointed, mostly with hollow internodes and swollen nodes, usually branched only in the lower part. Leaves are distichous (in 2 rows), consisting of a long, narrow sheath and a narrow lamina, and a membranous ligule or bristles at their junction. Flowers are minute, wind-pollinated, arranged in spikelets which, in turn, form terminal panicles, spikes, or racemes. Each spikelet consists of one or a few bisexual florets, with lemma and palea; stamens mostly number 3. The fruit is a caryopsis, which usually drops together with the floral glumes. The family has about 700 genera and 10,000 species, which are widely distributed almost throughout the world and constitute the major grass cover of many plant communities (e.g., steppes, meadows). The cereals are the most important food and fodder plants. They are the source of flour, sugar, and construction material and the main fodder for cattle is obtained from them.

Many cereal species were brought under cultivation in very ancient times. The wild cereals are valuable material for breeding.

Interrupted brome [*Bromus interruptus* (Hackel) Druce], possibly, has already disappeared from nature. This is an annual, sometimes biennial cereal, 20–100 cm in height with straight, unbranched, loosely clustered or isolated stems, with 2–4 finely pubescent nodes. Leaves are green; leaf sheaths have soft pubescence on the lower side; ligule is membranous, dentate, 1–2 mm long; lamina is acute, linear, with soft pubescence, 6–10 cm long and 2–6 mm broad. Panicle is straight, dense, oblong, usually interrupted, grayish-green, 2–9 cm long and 2 cm broad. Spikelets are in dense bundles, broadly ovate or broadly oblong, 10–15 mm long and 5–8 mm across, consisting of 5–11 florets; caryopsis (fruit) is 7–9 mm long.

The interrupted brome was first found in England in 1849. It was widespread and abundant at some places till the 1920s, but became rare after the 1930s, and is known only in 6 locations during the last 25 years (bromes were last collected in 1972 in Cambridge). It was also found in the Netherlands, where it was considered an introduced species (it was found in the fields together with sainfoin, ryegrass, and clover; its seeds possibly spread together with the seeds of these plants). The species is included in the Red Data Book of IUCN.

The feathergrass genus (*Stipa*) includes perennials forming dense turf. The feathergrasses are common and often dominant plants in steppes and semideserts. A large number of species of this genus are typical for the Russian feathergrass steppes, North American prairies, South American pampas, and other areas. However, this type of vegetation has already disappeared over large areas due to cultivation of steppes, and many species of feathergrass, which were found on a large scale earlier, are now endangered or have become rare.

Bavarian feathergrass (*S. bavarica* Martinovsky et H. Scholz) is a dense turfy grass up to 70 cm tall, with grayish-green, fairly long leaves. Sheaths of lower leaves are densely pubescent, upper leaves are glabrous. Ligule is 1–3 mm long, obtuse or serrated, pubescent. Lamina is acute, with numerous ridges; upper surface is pubescent, lower surface rough or hairy. Panicles consist of 6–9 spikelets, glumes unequal, 3–8 cm long, converted into a thin network; lemma and palea are 23–25 mm long, with 7 rows of hairs and a 33–41 cm long awn. The only habitat of Bavarian feathergrass is preserved in Bavaria and Danube region, near Neuberg (West Germany). It grows along the upper margin of stony outgrowths of Jurassic limestone deposits in a narrow valley covered with vegetation similar to the vegetation of steppes, in association with variable moor-grass, creeping sedges, and elecampane. It forms a few turfs on a stony area, often visited by tourists. The population strength is critically low, although measures have been taken for its salvation. The plant is protected by law, a reserve has been created, and it is being reintroduced into the earlier habitats. This species has been included in the Red Data Book of IUCN.

Danube feathergrass (*S. danubialis* Dihori et Roman) is indigenous to Rumania, found only on stony slopes along the Danube river above Turnu-Severin. It is endangered.

FAMILY POLYGONACEAE

The family Polygonaceae consists of herbs, semishrubs, shrubs, creepers, and in the tropics, even trees with alternate ochreate leaves. Flowers are small, usually bisexual, regular. Perianth leaves number 3–6, are uniform, in 1–2 whorls; stamens number 5–7. Ovary is superior, unilocular; fruit is a nut, enclosed in enlarged perianth leaves in many species. The family has about 40 genera and 900 species, universally widespread in the tropics, subtropics, and temperate regions; rarely in cold regions.

The rhubarb genus (*Rheum*) includes perennial rosette-forming herbaceous plants with a strong primary root. Leaves are large with long petioles, cauline leaves are smaller. Inflorescence is spicate or paniculate. Flowers are small, fruit is a triquetrous nut. The genus contains about 40 species. The **rhaponticoid rhubarb** (*R. rhaponticum* L.) is included in the Red Data Book of IUCN, which is a strong herbaceous perennial with woody rhizome and large, glabrous, palmate, ovate or orbicular leaves up to 50 cm across, on long and thick petioles. Flowering shoots are up to 2 m long, bearing a few cauline leaves and a loose panicle of small, yellowish flowers; perianth consists of 6 segments. Fruit is a small, hard nut with 3 membranous wings (Plate 11). The habitats of the species are known in Bulgaria and Norway. In Bulgaria, they are all located at an altitude of 1800–2100 m in the western part of Rila mountains south of Sofia. In Norway, it has been found 120 km northwest of Bergen, on a scrap facing north (height 480 m), in a patch with scattered grass cover. Rhaponticoid rhubarb is one of the parents of cultivated rhubarb. It was first introduced into cultivation as a medicinal plant and thereafter used as food. The reserves of rhubarb in Bulgaria are extremely small and the plant is protected. In Norway, the plant may disappear with a change in water regime. The species has been included in the Red Data Book of IUCN.

FAMILY PRIMULACEAE

The family Primulaceae includes perennial or annual herbs, rarely semi-shrubs. Their leaves are simple, alternate, opposite or whorled, in a basal rosette or all over the stem, entire, spatulate or crenulate; shoots are often leafless. Flowers are regular, bisexual, solitary or in panicles, racemes, or umbellate inflorescences. Calyx is gamosepalous, with 4–9 teeth or lobes, and is persistent with the fruits. Corolla is gamopetalous, usually with a long tube and 5, rarely 4–9 lobes; sometimes corolla is trochoid. Stamens number 5; ovary is superior or semi-inferior, unilocular. Fruit is a capsule with a large number of seeds. More than 30 genera and 800 species of this family are mainly distributed in the northern hemisphere. Many of them are ornamental.

The genus *Androsace* consists of annual or perennial herbs, sometimes forming a carpet. Leaves are in a basal rosette. Flowers are numerous, bracteate, in an umbel on a floral scape. Calyx is 5-lobed from the base or midpoint, campanulate or almost spherical. Corolla is quinquepartite, with a short tube and saucer or funnel-shaped limb, white, pink, or pinkish-violet.

Dwarf androsace [*A. brevis* (Hegetschw.) Ces.] is a very small perennial, forming dense, thick carpets, with small spatulate leaves, surrounding the stem at the base and forming dense rosettes. Flowers are on 4–23 mm long pedicels; corolla is 5–7 mm in diameter, its tube short, narrowed in the throat, with slightly incised lobes at the apex, pink with a yellow throat. Fruit is a small, 5-loculed capsule. The plant is found in the mountains of Italy and Switzerland on both sides north of Lake Como. Its distribution is restricted to siliceous region in the northwest of the Bergmo Alps and adjacent mountains on the west. It grows on fixed screes and along stony crevices at an altitude of 1950–2600 m. The sunny places, mostly free of snow during winter, are most favorable for this plant. This alpine species has been included in the Red Data Book of IUCN and is protected by law in Switzerland.

Ciliated androsace (*A. ciliata* DC.) is a turfy perennial with loose carpets. Leaves are 6–15 × 2–6 mm, ovate-spatulate or oblong-ovate, ciliate, pubescent. Corolla is pink or violet with orange or yellow throat. The plant is found at greater heights in the central Pyrenees (Spain, France). It is endangered.

Swiss androsace [*A. helvetica* (L.) All.] is a dense, turfy perennial. Stems are regularly branched, forming dense hemispherical pads. Leaves are 2–6 × 0.5–1.5 mm, lanceolate, oblong or spatulate, densely pubescent with simple hairs. Flowers are solitary; calyx is 2–3.5 mm long, pubescent; corolla is 4–6 mm in diameter, white, with yellow throat (Plate 11). This is a very rare plant, found on rocks in the Alps and, possibly, the Pyrenees.

The sowbread genus (*Cyclamen*) includes endangered plants whose trade is regulated in accordance with CITES. These are perennial plants with spherical or slightly flattened tubers. Leaves are basal, long-petiolate, broadly ovate, with cordate base, sometimes fleshy, usually dotted or with marbled pattern and purple lower surface. Flowers are solitary, on long basal pedicel, spirally twisted with fruits. Calyx is quinquepartite. Corolla has 5 recurved bent lobes. Stamens number 5; ovary is unilocular; fruit is a spherical or broadly ovate capsule; tubers are poisonous. Almost all sowbreads are ornamental, many flowering in early spring. Due to collection for bouquets, the populations of these plants are declining. Some species have been under cultivation as ornamental plants, others are used as medicinal plants. About 8 species are found in Europe, all of them growing in forests or outgrowth of shrubs or on shaded rocky slopes, usually on limestone deposits.

Ivy-leaved sowbread (*C. hederifolium* Aiton) has almost spherical or ovoid tubers. Leaves are 3–14 × 2–10 cm, cordate, with indistinct and obtusely dentate margins, often angular or lobate, appearing in autumn after flowering. Corolla is pale pink or white with dark purple and furcate spots in throat (Plate 12). The plant flowers from August to November. It is found in southern Europe (from France to Bulgaria).

Purple sowbread (*C. purpurascens* Miller) has tubers 2–3 cm in diameter, irregularly spherical or ovoid, often with bulges, with roots over the entire surface. Leaves are 2.5–8 × 2–6 cm, evergreen, broadly cordate or kidney-shaped. Flowers are very aromatic; corolla is reddish-pink or purple with a dark spot in the throat; lobes are 15–20 mm long (Plate 12). The plant flowers from June to October. It is found in the southeastern regions of France to the western Carpathians and central regions of Yugoslavia.

Coum sowbread (*C. coum* Miller) has tubers 2–3.5 cm in diameter, ovoid, pubescent, rooting from the base. Leaves have mosaic pattern, kidney-shaped or almost orbicular, margins entire or indistinctly serrated. Corolla is fuchsin-colored, a dark violet spot prominent at the base of each petal, a pale spot in the center (Plate 12). The plant flowers from January to April. Its area of distribution is southeastern Europe.

Persian sowbread (*C. persicum* Miller) produces a tuber with 4–15 cm diameter or more, almost spherical or ovoid, rooting from the base. Leaves are 3–14 × 3–11 cm, their margins sometimes thickened. Corolla is white or pinkish with a darker purple portion around the throat (Plate 12). The plant flowers from January to May. It is found on the islands of the Aegean Sea (Athos, Karpathos) and in southwest Asia.

The loosestrife genus (*Lysimachia*) includes erect or prostrate perennial herbs with leafy stems and variously colored flowers arranged in spikes, racemes, or corymbose-paniculate inflorescences.

Minorca loosestrife (*L. minoricensis* Rodriguez) is indigenous to the Balearic Islands (Minorca). It is possibly already extinct. This is a glabrous rhizomatous perennial. Stems are 30–60 cm tall, rarely branched from the base. Leaves are 4–8.5 × 1.5–2.5 cm, alternate to almost opposite, lanceolate to ovate, entire. Flowers are solitary, in the axils of upper leaves. Calyx is 3.5–4.5 mm long, its lobes linear-lanceolate. Corolla is 4–5.5 mm long, almost campanulate, pinkish at the base and greenish yellow above. The Minorca loosestrife was found in a forest-covered ravine (Sa-Val).

The primrose genus (*Primula*) consists of perennial, rhizomatous herbs with a basal rosette of leaves. Leaves are entire or lobate, petiolate or sessile. Flowers are arranged on a leafless floral scape, in umbels or almost forming heads. Calyx is tubular, campanulate, or funnel-shaped, 5-lobed. Corolla has a tube and flat, funnel-shaped limbs. Ovary is superior. Capsule is spherical to cylindrical.

Palinuro primrose (*P. palinuri* Petagna) has been included in the Red Data Book of IUCN. It is a rhizomatous perennial with a rosette of leaves. Leaves are obovate, about 4–16 cm long, fleshy, sticky, pale green, and serrated in the upper half. The leafless stem, 8–20 cm tall, bears an umbel of 5–25 drooping, yellow flowers. Calyx is 5–8 mm long, densely covered with white, powdery coating (similar to the pedicels and bracts). Corolla is golden yellow, funnel-shaped, with dark yellow circle inside. Capsule is brown (Plate 13). The plant is pollinated by large bees, butterflies, and flies (some specimens have a short style, others have a long style; therefore, self-pollination does not occur).

This Italian endemic is known from a few locations at the coast of the Tyrrhenian Sea in Campania, Lucania, and Calabria between capes Palinuro and Skalea. A large portion of its habitats are located in the neighborhood of Cape Palinuro. The Palinuro primrose grows on hanging limestone and sandstone rocks facing north, northwest, and west near the sea. It does not tolerate shade. On Cape Palinuro, it is protected by special law, and is included in the list of protected plants of Campania. So far, it is not affected by tourism. This is a very beautiful plant, easily multiplied from seeds.

Allionian primrose (*P. allionii* Loisel.) is a very rare plant. Leaves are 1.5–4.5 × 0.7–1.2 cm, oblanceolate to almost orbicular, entire, sometimes crenate or with fine serration, fleshy, sticky, covered with dense, glandular pubescence. The dry leaves are retained for a few years. Glandular hairs have 3–4 cells, with a light-colored tip. Stem is short, up to 8 mm at fruiting, with 1–5 flowers. Corolla is pale pink to reddish-purple (Plate 13). The plant is found on shaded cliffs of the coastal Alps in Italy and France.

Hirsute primrose (*P. hirsuta* All.) is a rare plant without powdery coating. Leaves are ovate, obovate, or almost orbicular, fairly sharply narrowed into a winged petiole, serrated, fleshy, sticky, and covered with glandular hairs. Calyx is 3.5–9 mm long. Corolla is pale lilac to dark purple-red, usually with a white center (Plate 13). The plant is found on the rocks and stony alpine pastures in the Alps and central Pyrenees.

Scottish primrose (*P. scotica* Hooker) is covered with powdery coating. Leaves are 1–5 × 0.4–1.5 cm, elliptical, oblong or spatulate, entire or sparsely crenate-dentate, usually highly powdery on the lower surface. Calyx is 4–6 mm long, its lobes obtuse, powdery. Corolla is 5–8 mm in diameter, dark purple with yellow throat, rarely white (Plate 13). This rare primrose is found only in humid pastures near the sea in northern Scotland.

FAMILY RANUNCULACEAE

The family Ranunculaceae mainly consists of herbaceous plants, rarely shrubs or creepers. Their leaves are mostly alternate, entire, divided or compound, without stipules. Flowers are regular or irregular, bisexual, arranged in inflorescences, rarely solitary. Perianth is simple or double. Stamens and pistils are numerous, sometimes pistil is single. Fruit is multiple, multiple-nut, rarely single, sometimes juicy, berry-shaped. The family has about 50 genera and 2000 species, predominantly distributed in the temperate belt of the northern hemisphere. Many of them are ornamental, some of them are valuable as a source of medicinal raw material.

The genus *Adonis* includes annual or perennial herbs with alternate pinnatisect or palmatisect leaves divided into narrow segments. Flowers are solitary, large, yellow or red. The genus includes about 30 species, distributed in the temperate and subtropical regions of Eurasia. The **twisted adonis** (*A. distorta* Ten.) is a vulnerable indigenous species of Italy (central Apennines). This is a perennial plant with bent stems up to 15 cm long. Basal leaves are long-petiolate. Petals are pubescent, yellow.

The columbine genus (*Aquilegia*) includes perennial plants with twice or thrice ternate-dissected or compound leaves and regular, large, solitary flowers. Calyx consists of variously colored sepals; corolla consists of 5 oblique, funnel-shaped petals with spurs pointing downward between the sepals. Fruit is polycarpous.

Meadow-rue columbine (*A. thalictrifolia* Schott et Kotschy) is a rare indigenous plant of northern Italy. It is 10–45 cm in height and has glandular pubescent stems. Basal leaves are biternate; leaflets are narrow, bicuneate or tricuneate, lower surface is glabrous, cauline leaves are entire, glandular-pubescent. Flowers are blue-violet, diameter of perianth is 20 mm (Plate 14).

FAMILY RUTACEAE

The family Rutaceae consists of trees, shrubs, and herbs, often with aromatic glands. The flowers are tetramerous or pentamerous, regular or irregular, usually bisexual. The family has about 1600 species, distributed in the tropical, subtropical, and temperate regions. It includes citrus plants, cultivated for their fruits, and many medicinal plants.

Corsican rue (*Ruta corsica* DC.) is a perennial herb with 18–45 cm tall stems woody below, branching almost from the base. Lower leaves have petioles up to 8 cm long, upper leaves are almost sessile, alternate, bipinnate or tripinnate, terminal leaflets are obovate to cuneate-orbicular. Inflorescence is sparse: sepals are deltoid-ovate, obtuse; petals are broadly ovate, pale yellow, serrated (Plate 14). The Corsican rue is a rare species, found in the mountains of Corsica and Sardinia.

FAMILY SAXIFRAGACEAE

The family Saxifragaceae includes herbs, semishrubs, shrubs, or trees, usually with alternate, rarely with opposite or whorled leaves without stipules. Flowers are arranged in various types of inflorescences or are solitary, regular or irregular, sometimes unisexual. Hypanthium is often well developed. Perianth is double or simple. Sepals and petals usually number 4–5 each, rarely up to 10. Fruit is a capsule or berry. The family has more than 100 species distributed primarily in the subtropical, temperate, and cold regions of the northern hemisphere. Many species are ornamental.

In Europe, it is required to protect 21 saxifrage (*Saxifraga*) species, 8 of which are endemic to Spain and 7 to Italy.

The **biternata saxifrage** (*S. biternata* Boiss.) has been included in the Red Data Book of IUCN. This is a small, loosely tufted perennial, fairly densely pubescent, bearing a rosette of fern-shaped, bright green leaves, with the upper surface velvety and lower surface glossy, mostly 2.5–8 cm long, twice divided into several lobes or dentate segments. Numerous axillary buds are located in the axils of lower leaves, producing new branches in autumn. Stems are often woody, prostrate or ascending, up to 20 cm long, bearing relatively large campanulate or funnel-shaped flowers. Sepals are small, oblong; petals are white with green veins. Capsule is spherical, 8 mm in diameter. This plant sharply differs from other saxifrages in its external appearance (no rosette, large flowers, and when not in bloom resemblance to a small fern).

Most probably, the biternate saxifrage is a relict. It is known from only 2 distant places. One of them is in the mountains north of Malaga (Spain), where it grows on vertical rocks facing the sun at an altitude of 1000–1100 m. In one of the habitats it is threatened with extinction by tourists attracted by the strange looking rocks (recently a road was constructed to that place). The plant also grows in some botanical gardens and in personal collections where it is easily reproduced both vegetatively and by seeds.

Vandell's saxifrage (*S. vandellii* Sternb.) has numerous branched shoots, forming a dense and rough carpet. Leaves are 8–11 × 1.5–2 mm, almost flat, keel-shaped below. Petals are white (Plate 15). The plant is found on limestone rocks in the Italian Alps.

Facchin's saxifrage (*S. facchinii* Koch) is found in the dolomite Alps, along stony screes and on the rocks at an altitude of 2200 m. It resembles small pads of short and straight branches; flowering shoots are very short, often uniflorate. Petals are oblong-cuneate with yellow veins (Plate 14).

Spoon-shaped saxifrage (*S. cochlearis* Reichenb.) is a monocarpic plant. Its leaves are highly variable in size, crenate, grayish, often dark red at the base, mostly basal, forming rosettes. The plant reproduces vegetatively through stolons bearing new rosettes. Peduncles are 4–4.5 cm long, inflorescence is glabrous or sparsely pubescent with glandular hairs. Petals are white with numerous minute red spots. The plant grows on limestone rocks in the coastal Alps (France, Italy). The species is endangered.

Florulent saxifrage (*S. florulenta* Moretti) is a slow-growing, long-lived perennial with dense, flat rosette of numerous leathery, dull green, narrow-spatulate leaves. Dry leaves persist, so that the rosette has the appearance of a short cylinder. Flowers are arranged in narrow, long, glandular-pubescent, dense panicles; corolla is flesh-red. This is a very rare species, close to extinction, found only in the southern tip of the coastal Alps, in the border zone between France and Italy. It grows on overhanging siliceous rocks at an altitude of 2000–2500 m, usually on rocks facing north or northwest. As a rule, it is found in isolated groups, numbering a few dozens, with some populations having a few hundred plants. This extremely beautiful alpine plant is disappearing for natural reasons as well as because of collection. A monocarpic, relict species, grows very slowly and flowers in the 10th to 12th year. In Italy (in the Cuneo and Piedmont provinces), it has been protected by law since 1975 (collection of flowers and seeds is prohibited). The species is included in the Red Data Book of IUCN.

Mutated saxifrage, a lowland subspecies [*S. mutata* L. ssp. *demissa* (Schott et Kotschy) D.A. Webb], bears loose tufts of several rosettes. Leaves are 10 × 7–12 mm, ligulate-obovate, dark green. Flowering shoots are strong, branched from the middle or below the midpoint. Flowers are numerous, borne in loose, densely glandular-pubescent racemes. Petals are 5–8 mm long, linear-lanceolate, acute, orange (Plate 14). It is a rare indigenous plant of the southern Carpathians (Rumania).

FAMILY SCROPHULARIACEAE

The family Scrophulariaceae includes herbs, sometimes semiparasitic or parasitic, rarely shrubs or trees. Leaves are alternate, opposite or whorled, simple or pinnatisect, without stipules. Flowers are bisexual, irregular, rarely regular, solitary, or arranged in inflorescence. Calyx consists of 4–5 fused sepals. Corolla is gamopetalous, trochoid to bilobed. Pistil has superior bilocular ovary and entire or bilobed stigma. Fruit is a capsule, rarely a berry, usually with numerous seeds. The family has about 220 genera and 3000 species, widely distributed over the entire global surface, but mainly concentrated in the temperate region.

The (eyebright) genus (*Euphrasia*) comprises annuals and perennials. Almost all the rare or disappearing species of eyebrights are found in northern Europe.

Dune eyebright (*E. dunensis* Wiinst) is an annual semiparasitic herb with 12 cm tall stem, and up to 3 pairs of short branches. Internodes are 1.5–3 times longer than the corresponding leaves, the first flower being borne on the fifth to the ninth node. Leaves are ovate, up to 11 mm long, with 5 pairs of serrate, densely gray pubescent lobes. Calyx is tubular, tetramerous. Corolla is 6–7 mm long, white, with violet stripes, bilobed, upper lip galeate, lower lip flat. Capsule is 4.5–5 mm long, oblong or oblong-elliptical.

The dune eyebright has been included in the Red Data Book of IUCN. This is the only indigenous species of Denmark, known from only 2 coastal points at the northwestern coast of Jutland. It grows in the coastal meadows, at some places on low grass eroded peat beds and in other places, on shifting dunes. Both types of habitats are vulnerable. One of them is on the territory of Bulberg and Lilstrand reserves, and the other is 20 km east of the first and is not protected. The grassy cover is usually destroyed rapidly when places are visited by people.

Hot spring eyebright (*E. calida* Yeo) is a 25 cm tall annual, erect, with 1–4 pairs of long arcuately bent branches. Internodes are not more than 3 times, and lower flowering shoots not more than 2.5 times as long as the leaves. Leaves are thin; lower bracts are broadly ovate or broadly rhombic. Calyx is whitish and membranous, except the teeth and prominent veins. Capsule is usually shorter than calyx, broadly obovoid to oblong. The plant is found near hot springs. It is indigenous to Iceland and is endangered.

The genus *Lafuentea* comprises perennial herbs with stems woody at the base. The **round-leaf lafuentea** (*L. rotundifolia* Lag.) is densely grayish pubescent to hairy, glandular, aromatic. Leaves are kidney-shaped to roundish-deltoid, irregularly crenate. Spikes have numerous flowers. Bracts are linear-oblong, shorter than flowers. Calyx is 4–6 mm long, divided almost up to the base into linear-lanceolate lobes. Corolla is 7–8 mm long, white with purple tinge. Capsule is ellipsoidal. It is a rare plant, found along the crevices and on the projecting rocks in the valleys of southern Spain, in the Malaga region.

The toadflax genus (*Linaria*) has more than 150 species, distributed in the temperate latitudes. The list of rare European plants includes 18 toadflax species, including 5 species from Portugal and 8 from Spain.

Algarve toadflax (*L. algarviana* Chav.) is an annual, glabrous below and glandular pubescent above. Stems are 10–30 cm long, prostrate or ascending, simple. Leaves are few, linear-oblong, almost acute, alternate. Inflorescence is sparse, with 1–8 flowers. Calyx is 3–4.5 mm long, its lobes oblong-lanceolate, acute, slightly uneven, with white or violet membranous margin. Corolla is 20–25 mm long, violet, with white or yellow spots, spur 11–12 mm long. The plant grows on dry sandy soils in southwestern Portugal. The species is close to extinction.

Hellena toadflax (*L. hellenica* Turrill) is indigenous to Greece and is endangered. It is included in the Red Data Book of IUCN. It is an annual with weak straight or bent branched stems up to 60 cm tall. Leaves are linear or linear-oblong, somewhat thick, obtuse, 5–45 × 1–2.5 mm. Inflorescences consist of 5–20 flowers in terminal racemes. Calyx is 4.5–5 mm long with oblong, almost equal-sized lobes. Corolla is 13–16 mm long, yellow, bilobed, upper lip bifurcated, lower lip trilobed; spur is 6–7 mm long. The species is known from only 6 places on the Malea peninsula (southeast of Peloponnisos), in the Neapolis Bay and on the neighboring Elaphonisos Island, on an area of about 20 km. It grows on coastal sands and sandy soils, near the sea, and on flat, open places. It has become very rare due to economic development. At present, there are only 4 places where this plant is found; only 2–6 plants are found in 3 of these places, and about 100 plants at the fourth. Other locations are already destroyed, and the remaining ones are not protected. Reproduction through seeds is often suppressed due to change in habitat, but this toadflax is easily regenerated vegetatively.

The mullein genus (*Verbascum*) consists of 360 species distributed in Europe and West Asia. Ten of its species are rare, of which 5 grow in Bulgaria and 3 in Spain.

Unequal-leaved mullein (*V. anisophyllum* Murb.) is a biennial 40–100 cm tall, with basal leaves 5–10 × 2.5–4 cm, oblong-lanceolate to ovate-elliptical or with crenate, densely grayish yellow-tomentose leaves. Inflorescence is sparse, branched, glandular-pubescent. Calyx lobes are oblong or linear-lanceolate. Diameter of corolla is up to 30 mm. The plant is indigenous to western Bulgaria, found northwest of Sofia. It is a vulnerable species.

FAMILY THYMELAEACEAE

The family Thymelaeaceae includes trees, shrubs, rarely perennial or annual herbs with alternate or opposite, always entire leaves. Inflorescences are terminal, rarely axillary. Flowers are arranged in heads, umbels, racemes, or spikes, rarely solitary; they are bisexual, rarely unisexual. Perianth is simple, petaloid, tabular or funnel-shaped. Perianth lobes number 4, rarely 5, stamens usually twice as many. There is one pistil. Ovary is free, unilocular. Fruit is a nondehiscent nut or juicy drupe. The family has 47 genera, mainly distributed in the subtropical regions.

The spurge-olive genus (*Daphne*) includes poorly branched, sometimes evergreen shrubs with simple, entire leaves. Flowers are regular, fruit is a single-seeded drupe. Many species are ornamental, some are good honey producers. The genus contains about 100 species, distributed in southeast Asia, the Himalayas, Iran, and the Mediterranean region.

Arboreal daphne (*D. arbuscula* Celak.) is a dwarf evergreen shrub 30–50 cm tall with prostrate or ascending coral-red branches, bearing compressed and deeply incised coriaceous (leathery) leaves at the apices, from linear to oblong. Four to ten aromatic flowers are grouped in terminal racemes. Corolla is pink, with a narrow, 12–20 mm long tube and 4 divergent, 5–7 mm long lobes. Fruits are grayish, wooly, maturing very rapidly (within 3 weeks) and immediately dropping. Biology of the species is poorly studied. Possibly, it reproduces by rooting of branches. The plant grows slowly: stems with a diameter of 35 mm are usual in 85-year-old plants. This is the most famous indigenous species of the western Carpathians, with a very limited area of 15 × 14 km. Its 14 sites are on the limestone deposits in the centre of the western Carpathians (Czechoslovakia). The arboreal daphne grows on northern and western slopes, usually in calciphyte meadows or along the margins of broom meadows together with other relict indigenous species (woody-fruited bellflowers, pointed-petal larkspur, and Valenberg's saxifrage) at an altitude of 750–1300 m often in the rock crevices. The reserves of this species are declining due to collection of its twigs and delicate beautiful flowers as well as digging of the bushes for transfer into parks. It is protected by law, and some of its habitats are situated within reserves. It is cultivated in the alpine mountains. The species is included in the Red Data Book of IUCN.

Rodrigues' daphne (*D. rodriguezii* Texidor) is a small, evergreen dwarf shrub up to 60 cm tall with numerous short, lateral branches and more or less glabrous, grayish bark. Leaves are alternate, oblong-oblanceolate, 1–2 cm long, dark green on the upper surface and lighter-colored on the lower surface, with short and fine pubescence.

Flowers are small, mildly aromatic, arranged in clusters of 1–5 flowers each. Corolla is 4-lobed, white, with a 5–8 mm long tube and 4 bent, creamish white, 3–5 mm long lobes; tube is greenish, sometimes purple. Fruit is small, greenish-brown.

The distribution of Rodrigues' daphne is confined to the Balearic Islands. Earlier, it was known from several coastal places on Minorca, but possibly because of development of tourism, it has already disappeared there. It is preserved only at a small island near Minorca with an area of about 70 ha. The island is stony and bare, and dry in summer. This daphne grows on dry stony places, among outgrowths of pistachio, rockrose, and heath. It is being cultivated in several botanical gardens. The species is endangered and, therefore, it is desirable to ban tourism and economic exploitation of the island. It is included in the Red Data Book of IUCN.

Rock daphne (*D. petraea* Leybold) is an evergreen dwarf shrub with numerous strong, wavy, prostrate, branched shoots forming a pad. The young branches are greenish-brown, with scattered pubescence; old shoots are covered with residues of petioles. Leaves are 8–12 × 2–3 mm, linear-oblanceolate, obtuse, glabrous, side keel-shaped below, clustered near the tips of branches. Flowers are aromatic, pink, in heads of 3–5 flowers, rarely more, with membranous bracts. The rock daphne grows in a very limited area of northern Italy (with the center around Lake Idro, northeast of Brescia) along the crevices in limestone rocks at an altitude of 700–2000 m. The species is close to extinction.

FAMILY VIOLACEAE

The family Violaceae includes herbs, shrubs, or trees with simple, usually entire, alternate, rarely opposite leaves with stipules. Flowers are bisexual, irregular, rarely regular, solitary or arranged in inflorescences. Calyx consists of 5 free sepals, usually persistent on fruits. Corolla comprises 5 free petals, of which the lower petal is spurred. Stamens number 5. Pistil has a superior, unilocular ovary. Fruit is a capsule, rarely a berry or nut. The family has about 900 species, which are spread throughout the world, especially in the tropics and subtropics. The violet genus (*Viola*) is the most widespread.

Jaubert's violet (*V. jaubertiana* Marés et Vigineix.) is a 5–15 cm tall perennial with a short rhizome, rosette of leaves, and long, strong stolons. Leaves survive through winter, are ovate or deltoid-ovate, cordate, dark green, and glabrous. Flowers are bright violet, more than 2 cm in diameter, aromatic. The plant is found in humid lowlands on Majorca (Balearic Islands). It is endangered.

Bristled violet (*V. hispida* Lam.) is a mixed-up hairy perennial with stems up to 25 cm long and leaves 1–3 cm long, on petioles of the same length. Lower leaves are orbicular or cordate, upper leaves more or less ovate, usually with crenate margin. Axils have noticeable palmatipartite, 1–2 cm long stipules. Flowers are 2 cm in diameter, violet or yellowish, with 5 petals (lower petals larger than the upper ones) and a 4 mm long, sacciform spur. The fruit is a capsule splitting into 3 parts.

This endangered plant is found in France in the Seine valley near Rouen; only 2 populations are known from steep, chalky slopes, usually facing west or south. The plant readily reproduces by stolons; its cultivation for seeds is difficult. The ornamental species is suitable for the Alps. The **Cryan violet** (*V. cryana* Gillot) is very similar to bristled violet, but is shorter and nonpubescent. At one time it was found on the chalky slopes southeast of Tonnere (central France); the species has now disappeared.

PLATE 13. 1—Palinuro primrose; 2—hirsute primrose; 3—allionian primrose; 4—Scottish primrose.

PLATE 14. 1—Meadow-rue columbine; 2—Corsican rue; 3—Facchin's saxifrage; 4—mutated saxifrage.

PLATE 15. 1—Vandell's saxifrage; 2—snake-wood; 3—Indian turnip; 4—Gardner's wax-flower.

PLATE 16. 1—Indian frerea; 2—Formosan cypress; 3—Dalhousie rhododendron; 4—Edgeworth's rhododendron.

2
Asia

The vegetation of Asia is exceptionally rich and diverse. Two phytogeographic kingdoms—Holarctic and Paleotropic—are represented in Asia. The vegetation changes with the changing climate from north to south, from tundras to humid equatorial forests. The distance from the ocean has a great influence on the vegetation. The maximum diversity of species (more than 25,000 species) is found in the equatorial tropical forests of Malaysia, which includes the Greater and Lesser Sunda Islands, the Philippines, the Malaccas, and numerous groups of smaller islands. The humid tropical forests are also distributed in India, Indochina, and Sri Lanka. The remaining part of Asia belongs to the Holarctic kingdom. The East Asian regions (eastern Himalayas and adjoining regions, a large part of continental China, Taiwan, Korea, the Japanese islands, southern Sakhalin and southern Kuril) have subtropical vegetation. Fourteen families and more than 300 genera are indigenous to this part of Asia. The southwestern regions of Asia are included in the Sahara-Arabian floristic regions. Asia Minor, Syria, Lebanon, a large part of Palestine, and the northern part of western Transcaucasia belong to the Mediterranean region. The large Iranian-Turkish region is located toward the east of this region and the Circumboreal region is in the north of the Iranian-Turkish region.

The Asian flora is still incompletely known, and the effect of economic activity is so great that some species may be lost before they are discovered. This mainly relates to the tropical flora, which is richest in species. A catastrophic reduction in the area of tropical forests is taking place in all the countries of this zone. This is indicated by the critical condition of some genera of tropical palms. The family Palmae is one of the leading families in the number of genera threatened with extinction. The situation of tropical orchids is no better; their number is declining, on the one hand, as a result of reduction in the area of forests and changing environmental conditions and, on the other, because of commercial trade.

Possibly, there is no pocket of land left on the large territory of Asia which has not experienced and is not experiencing the effect of human activity. This is true of the desert regions of the Arabian Peninsula and Middle Asia, northern tundras and mountain ranges of Tibet, Pamir-Alai, Tien-Shan, the widespread Siberian forests, and subtropical Mediterranean and East Asian vegetation.

The existing and developing national parks, forest reserves, and other protected territories cannot accommodate all that diversity of nature. The new repositories should be established mainly at the places of maximum concentration of vulnerable and endangered plant and animal species. However, in many countries, lists of such species and information on their habitats are not available. There is only scattered information about the status of individual species or taxonomic groups.

For instance, the list of orchids in India that need protection includes more than 50 names. The list of protected plants of Israel includes 34 species, and a few dozens of rare and endangered species have been included in the Red Data Book of the USSR. Lists of plants that require protection have been prepared in Sri Lanka and Japan. There are published reports about the extinction of some rare and endemic species of limestone deposits in Malaya.

FAMILY APOCYNACEAE

The family Apocynaceae includes more than 200 genera and 2000 species, mainly distributed in the tropical and subtropical regions, only a few being found in temperate regions. Most species are woody creepers, shrubs, and semishrubs, very rarely trees, with milky juice. The leaves are opposite, simple, and entire. Many plants of the family have found practical use.

Soft holarrhena [*Holarrhena mitis* (Vahl.) Roemer et Schultes] is a small tree, up to 10 m in height. Young branches and leaves are glandular; leaves are opposite, oblong-lanceolate. Flowers are white, arranged in short terminal cymes. This endemic plant of Sri Lanka is found extremely rarely in the foothills up to 800 m in humid as well as dry habitats. Its wood and bark are used as antiseptic preparations, the bark being known as 'kalinda'.

Ceylonese petchia [*Petchia ceylanica* (Wight) Livera] is a shrub. Leaves are grouped on the branches in threes. Flowers are creamy white borne in short inflorescences. The plant blooms round the year. It is a very poisonous plant. In Sri Lanka, it grows in dense rain forests up to 700 m above m.s.l. This endemic plant, the only member of genus *Petchia*, is disappearing as a result of destruction of the rain forests. At present, it is endangered and is being protected in small forest reserves.

Tendrilar willughbeia (*Willughbeia cirrhifera* Abeywickr.), endemic to Sri Lanka, is the only member of the genus on the island, and is distributed in the Malayan archipelago. It is a tall and strongly branched shrub. Leaves are elliptical, leathery, pinnately veined. Shoot apices are modified into tendrils. Flowers are small, yellow, borne in dense racemes. Fruits are yellow-red, with a diameter of 8–15 cm, sweet, relished by monkeys. The willughbeia grows in mountains up to an altitude of 1300 m. It is becoming rare as a result of destruction of tropical forests.

The snake-wood genus (*Rauwolfia*) includes about 50 species, growing in the tropics of southeast Asia, Africa, and South America. The **snake-wood** (*R. serpentina* Benth.) has a fairly wide area of distribution, including the tropical part of the Himalayas, the Indian peninsula, Sri Lanka, Burma, and Indonesia. The leaves of this evergreen shrub are borne in groups of 3–4 in whorls, dense, glabrous, oval, and slightly pointed. Flowers are in dense umbellate inflorescences; corolla is pink, tubular, pentamerous, with the petals overlapping each other. Fruit is red, consisting of 2 juicy drupes, fused up to their middle (Plate 15). Roots and rhizomes contain various alkaloids which are used in medicine. This plant is under cultivation in India, Sri Lanka, and Java: experiments on its cultivation are in progress in the United States and USSR. The biochemical composition of other *Rauwolfia* species is also being studied. In spite of all this, the natural reserves of this plant are declining, especially after reports of its medicinal properties appeared in literature.

FAMILY ARACEAE

The family Araceae includes about 100 genera and more than 500 species, mostly distributed in the tropical regions. These are perennial rhizomatous or tuberous terrestrial herbs or epiphytes, shrubs, or trees with large leaves. Flowers are bisexual or unisexual, arranged in simple spicate inflorescences or cobs, with a large, bright bract at the base. Perianth is well developed or completely reduced; stamens number 4–6; carpels 1–3. Fruit is a berry.

The Indian turnip genus (*Arisaema*) comprises about 150 species, growing in the tropical and temperate regions of East Asia; a few species are found in Africa and North America.

The **Indian turnip** (*A. ternatipartium* Makino) is disappearing from the Japanese flora. Tubers are hemispherical, producing 2 long-petioled, trifid, 7–20 cm long leaves; lobes are rhomboid, 7–15 cm long, 3–8 cm broad, pointed, densely papillose along the margins. Peduncles are 10–15 cm long; bract is dark purple, widening in the upper part of the tube; blade is oblong-deltoid (Plate 15). The plant flowers during April–May. It grows in shady places in humid areas. Distribution area of this species covers the Honshu, Shikoku, and Kyushu islands. In spite of such a wide distribution, the total population of Indian turnip is small, as it does not form large concentrations but is found as isolated specimens. It is highly ornamental in flower. The disappearance of this plant is related to changing environmental conditions and digging for transfer to flower beds.

FAMILY ARALIACEAE

The family Araliaceae consists of 60 genera and 450 species, predominantly distributed in the tropical countries of both hemispheres, but mainly in the Indo-Malayan flora and in tropical America. Some genera, including ginseng, are typical for the temperate region of East Asia and North America.

The **American ginseng** (*Panax quinquefolius* L.) was initially represented by 2 populations: Asian and North American. However, in 1847, C.A. Meyer Separated the Asian population under an independent species, called **Chinese ginseng** (*P. ginseng* C.A. Mey.). This is a relict endemic of the Manchurian flora, a perennial up to 80 cm tall, with rhizome and a thick primary root. Stem is solitary; peduncle arises from the center of quinquepalmate, long-petioled, whorled leaves. Inflorescence is an umbel. Flowers are white or pink, fruits are bright red. The distribution area of ginseng covers North Korea, northeastern China and Primorye territory (USSR). It grows in dense forests (cedar-broad-leaved, cedar-spruce-broad-leaved, and cedar forests) on well-drained, mildly acidic soils. Found very rarely, it grows in small groups and as scattered, isolated specimens at a distance of many kilometers.

Ginseng is multiplied exclusively by seed. The seeds fallen on the ground germinate under natural conditions only in the second spring, i.e. after 1.5–2 years. The fourth to fifth leaves form after dozens of years. Flowering occurs still later. According to some reports, ginseng lives up to 100 years. The population strength of plants and their age in natural habitats have been declining constantly since its search and use for medicinal preparations started. At present, rhizomes older than 20 years are found very rarely. The distribution area of the species is also declining. In less than 100 years, the northern and western borders of its area of distribution in the Primorye territory have shifted toward the south and east. The change in environmental conditions has a significant effect on the area and population strength of ginseng: cutting of forests, forest fires, reclamation of forests and vigorous growth of varied grass, and spring burning of forests are fatal for the young plants.

Ginseng is one of the most ancient medicinal plants. It is considered to be a universal raw material for the oriental medicines (of China, Korea, and Tibet). The failure in meeting the demand for this plant from natural resources served as a stimulus for its cultivation in Korea even at the end of the 19th century: ginseng cultivation started in China and Japan a little later. In the USSR, the first experiments on its cultivation were initiated in the 1930s in the V.L. Komarov Ussurian Reserve and the 'Kedrovaya Pad' reserve. The rhizome quality suffers under cultivation. The Chinese ginseng is more sensitive to symbiotic fungi than the American ginseng.

At present, the total ginseng production, including the American ginseng, which is also cultivated in Asia, is 3500 ton, and two-thirds of this quantity is produced by South Korea. Since the Chinese ginseng is considered to be of superior quality to the American ginseng, China and Korea are trying to maintain their monopoly in the trade of its raw material. Export of fresh rhizomes and seeds is prohibited in these countries. Till now, Chinese ginseng was not known to be cultivated in the United States. The absence of sharp difference between the 2 species, export and re-export between the countries, and need for conserving the natural germplasm necessitate international control of the export and import of Chinese and American ginseng and inclusion in CITES.

FAMILY ASCLEPIADACEAE

The family Asclepiadaceae is represented by a few genera in Asia, as Africa is its center of origin.

Gardener's wax-flower [*Ceropegia elegans* Wallich var. *gardneri* (Thwaites ex Hooker) Huber] is a herbaceous biennial with tuberous rhizomes. Leaves are oblong-lanceolate, slightly hairy. Flowers are greenish-white with pink spots, in umbels (Plate 15). The main habitats of this variety are the humid forests in central Sri Lanka. Possibly, the plant has disappeared from its natural habitats, as nobody has found it during the last 100 years.

Cuspidate bidaria [*Bidaria cuspidata* (Thunb.) Huber] is an endemic of Sri Lanka. Formerly it occurred in the Central province, where it grew under the canopy of mixed forests at an altitude of 700 m. Possibly, by now this plant has become extinct.

Indian frerea (*Frerea indica* Dalz.) is a short and almost glabrous shrub with short branches. Leaves are opposite, oblong. Flowers are purple, large, few in number, arranged in pairs (Plate 16). It is a highly ornamental plant. Its natural habitats are concentrated in Konkan, in the foothills near Hernai, at an altitude of 1500 m. The population of Indian frerea has decreased significantly as a result of their digging and sale by collectors within the country as well as abroad. Frerea and all wax-flower species have been included in Appendix 2 of CITES.

FAMILY ASTERACEAE

One of the largest genera of the family Asteraceae in Asia is the saussurea genus (*Saussurea*), which includes 150 species of biennial or perennial plants with heads, umbels, or raceme inflorescences, distributed in the hilly regions of Asia. A few species grow in Europe and North America.

Costus (*S. lappa* Clarke) is a perennial herb with thick stem, up to 2 m tall. Basal leaves are 5–7.5 cm long, broadened toward the apex; cauline leaves are 15–30 cm. Inflorescence is a head, rough, with numerous purple flowers and bracts. The plant grows in Kashmir at an altitude of 2400–3600 m. At one time, it was fairly common, but its population has sharply decreased within a few years as a result of export to China. Costus has been included in CITES to regulate the volume of its export.

FAMILY BALANOPHORACEAE

The members of the family Balanophoraceae do not have chlorophyll and are root parasites of woody plants. The plants have tuberous rhizome, located right on the host root without haustoria; leaves are absent or scaly. Inflorescence develops within the rhizome and then emerges on soil surface, flask-shaped or spherical with a large number of scaly bracts. Perianth is often absent. The family includes about 15 genera and 100 species, distributed in the Paleotropic kingdom.

The balanophore genus (*Balanophora*) has about 70 species.

Involucrate balanophore (*B. involucrata* Hook f.) is a glabrous, fleshy, mushroom-like, herbaceous plant, parasitic on the roots of trees. It was first collected more than 100 years ago in a small area between Narkanda and Kotgarh and after almost 100 years it was found in the Kulu valley. These 2 discoveries in the western Himalayas indicate not only the extremely rare occurrence of balanophore, but also poor study of the flora. One more representative of this genus, dioecious balanophore, was recently found in western Sikkim.

FAMILY BERBERIDACEAE

The family Berberidaceae includes 14 genera and about 650 species, predominantly distributed in the temperate and subtropical regions of the northern hemisphere. These include arboreal plants (shrubs or rarely small trees) as well as perennial herbs.

Japanese ranzania [*Ranzania japonica* (T. Itô) T. Itô] is the only species of the genus, rarely found in the mountains on Honshu Island. It is a herbaceous plant with branched, creeping rhizome, a simple stem up to 30–50 cm tall, 2 ternate leaves and an apical cluster of light purple, drooping flowers with a diameter of about 2.5 cm. Fruit is a berry. The plant flowers in June–July. It grows in the hilly forests in central and northern regions facing the Sea of Japan.

FAMILY BIGNONIACEAE

The family Bignoniaceae includes 120 genera and almost 850 species, representing different forms: trees, shrubs, climbers, and herbs. The members of Bignoniaceae are distributed in the tropical and subtropical zones of Asia and America. The distribution of incarville (12 species) is restricted to East Asia.

Semirech'e incarville [*Incarvillea semiretschenskia* (B. Fedtsch.) Grierson] has extended beyond the East Asian distribution area of the genus and is found in southern Kazakhstan, in the eastern part of the western slopes of the ancient Chu-Ili mountains. This is a short (up to 30 cm tall) semishrub with herbaceous, basally woody stems, pinnate or palmatisect leaves, and large, tubular, funnel-shaped pink-red flowers, borne in racemes. The fruits are winged capsules. The plant grows on clayey and pebbled slopes and terrains of desert foothills. It flowers during May–June. Old plants of incarville produce fruits abundantly, up to 350 capsules per plant. Young plants and seedlings have been found among adult plants.

The plants constitute 3 populations, with an area not exceeding 7 ha, and 25,000–28,000 plants of different ages. The Semirech'e incarville was discovered in 1915. Subsequently, it was collected in 1925–26, 1936, 1955, and 1960. Thereafter, it was not found in natural habitats for a long time and was believed to have become extinct. However, the workers of the Central Botanical Garden, Academy of Sciences of Kazakhstan, rediscovered it in the 1970s. The studies on this plant have established its relation with the East Asian members of family Bignoniaceae.

Semirech'e incarville is a relict of Paleozoic Central Asian subtropical savannas, and as a rare plant with limited distribution it needs to be protected in its natural habitats. It is being cultivated in the Tashkent and Alma-Ata botanical gardens. Under cultivation, incarville has more rapid growth and produces flowers and fruits within the very first year after sowing. It is recommended as an ornamental plant for city plantations. It is included in the Red Data Book of the USSR.

FAMILY BRASSICACEAE

The specimens of the monotypic genus of the family Brassicaceae, **Bardunov's megadenia** (*Megadenia bardunovii* M. pop.), were found and described in 1953. Its only habitat is near the village Turan in the Tunkinsk region of the Buryat ASSR. Megadenia grew near a spring and at a little distance along its flow on the left bank of the Irkut River where fairly large groups were formed. In its external morphology, it has little resemblance to the plants of Brassicaceae. It is a small, acaulescent annual with a rosette of roundish-cordate leaves on long petioles. Flowers are small, white, almost basal, on short pedicels. Siliculas are like spectacles, with a width about 5 mm, length 2 mm; they are constricted in the middle and highly compressed laterally. The plant flowers in June.

In the 1970s, a road was constructed through the distribution area of megadenia that buried it under gravel. Efforts to locate the plants near its habitat did not yield any result. It is hoped that megadenia will grow in the vicinity and can still be found. This endemic relict species of the ancient Paleogene flora has great scientific importance for studying the geological history of the northern hemisphere. The plant has been included in the Red Data Book of the USSR.

FAMILY CARYOPHYLLACEAE

The genus *Gymnocarpos* with 2 species from the family Caryophyllaceae belongs to the ancient elements of the old Mediterranean flora.

Przheval'skii's gymnocarpos (*G. przewalskii* Bunge ex Maxim.) is distributed exclusively in the Afro-Asian deserts. It is a 0.5–1 m tall shrub with nodose, almost cylindrical, light gray branches. Leaves are opposite, linear-lanceolate. Inflorescence is short, sparse; scaly bracts are large, membranous. Perianth is sepaloid, with naked tube. The plant grows in gravelly deserts on gravel-stony slopes and alluvial fans along the foothills and lower hills at an altitude of 1000–2000 m, especially along dry river belts of seasonal streams, rarely on sandy-pebbled foothill plains. It is sometimes included in the mixed scrub communities, often found in joint fir-saltwort deserts.

The distribution area of the species is located in Central Asia. It is known from several places in northern Kashgaria, lower hills and foothills of the southern slopes of eastern Tien-Shan, the northeastern part of Jungarian Gobi, the extreme east of Gobi Altai, Alashan Gobi, and the western part of the foothills of Nan-Shan (Mongolia, China). This species has great scientific value in the study of phytogeographic relationships in the desert flora of Central Asia and Africa. Besides, gymnocarpos belongs to the most primitive subfamily Paronychiadeae of the family Caryophyllaceae. At present, Przheval'skii's gymnocarpos is included in Appendix 1 of CITES.

FAMILY CUPRESSACEAE

The genus *Calocedrus* of the family Cupressaceae includes 3 species, of which one is distributed along the Pacific coast of North America and the other 2 in the subtropical regions of East Asia.

Formosan calocedrus [*C. formosana* (Florin) Florin], an endemic of Taiwan, grows in the northern and central parts of this island. This tree is up to 25 m tall and 3 m in diameter, with broad conical or pyramidal crown and light-colored scaly bark. The branches are dichotomous; leaves are scaly, dark green. It grows in the ravines and on the rocks in a community with camphor tree at an altitude of 300–1900 m. Its wood is resistant to rotting and possesses excellent mechanical properties for furniture.

Formosan cypress (*Chamaecyparis formosensis* Matsum.) is also a rare endemic of Taiwan, growing up to a height of 65 m and a diameter of 6.5 m. According to some reports, the trees are 3000 years old. Bark is reddish chocolate-brown; branches are spreading; twigs are compressed; leaves are scaly, triangular, green. The mature cones are oblong-ovate, triangular, green. Green (young) cones are oblong-ovate, 10–12 mm long and 8–9 mm across (Plate 16). It grows in the hilly regions of the northern and central parts of the island at an altitude of 1000–2900 m, forming pure or mixed plantations with other Cupressaceae plants (*C. obtusa* Sieb. et Zucc. var. *formosana*).

FAMILY CYCADACEAE

The sago palm genus (*Cycas*) is one of the 9 genera of the family Cycadaceae. It has a wide area of distribution in tropical and subtropical regions of Asia, Africa, and Australia, as well as the islands of the Indian and Pacific Oceans. Southeast Asia is the centre of diversity of this genus, where 15 species are concentrated.

Fern palm (*Cycas revoluta* Thunb.) is the rarest species of the genus. It is an evergreen small tree, 1.5–3 m tall and up to 1 m in diameter. The trunk is covered with woody remains of leaf petioles and scales. Leaves are up to 2 m long, forming a dense crown at the tip of the stem, pinnate, hard, dark green. Male inflorescence is cylindrical (cone), up to 60–80 cm long; female inflorescence is of sparse, leafy, bright red megasporophylls. The fern palm grows near the coast on the Japanese islands Kyushu and Ryukyu, forming large outgrowths, and is being widely exploited. The pith of trunk and roots contain starch (sago), which is eaten in Japan. The leaves are used for making fences and decorations. It is also cultivated in the USSR. It can be found in the gardens and parks of the Black Sea coast of Caucasus.

FAMILY DIOSCOREACEAE

The family Dioscoreaceae includes 10 genera and 650 species, mainly growing in the tropics, and a few species found in East Asia, Mediterranean regions, the Pyrenees, and North and South America. They are herbaceous climbers with alternate or opposite leaves with reticulate venation and underground rhizomes, tubers, or tuberous roots. Flowers are small, regular, mostly unisexual, in axillary spikes, panicles, or racemes. Fruits are winged capsules or berries. The yam genus (*Dioscorea*) with 600 species is the largest of the family. Many species of this genus are sources of valuable medicinal preparations, obtained from the rhizomes and roots. The tubers are eaten, as they contain starch, and in several species the fruits are also edible. Because of the wide economic use of yam in the tropical countries, some species are under cultivation.

The **wild yam** (*D. deltoides* Wall.) is found only in Pakistan. The roots of this plant are rich in diosgenin and are used in producing cortisones. Large quantities of raw material of wild yam are exported, with significant effect on its natural reserves. The biochemical studies of recent years have established that the wild yam contains steroid hormones, which has further increased the export of its raw material. That is why this plant has been included in Appendix 2 of CITES. The possibilities of organizing trade in the following species is also being explored: *D. elephantipes* Endl., *D. montana* Spreng., *D. sylvatica* Ecklon, and *D. hemicrypta* Burkill.

Caucasian yam (*D. caucasica* Lipsky) is a relict endemic to Caucasus. It is a dioecious perennial, herbaceous creeper with rhizome up to 2 cm thick, from which 1–2 shoots develop every year. Its distribution covers the western regions of Transcaucasia: Adler regions of the Krasnodar territory and Gagry, Gudauta, Sukhumi, and Gul'ripsh regions of the Abkhasian ASSR from the basin of Mzymta region in the northwest to the Kodor' river in the southeast. The Caucasian yam mainly grows on the southern slopes up to an altitude of 1600 m, and on the eastern and western slopes it occupies the most illuminated areas where oak, oak-hornbeam forests, and outgrowths of oriental hornbeam, hawthorn, Christ's thorn, and other plants are predominant. Besides Caucasian yam, the herbaceous-shrubby tiers also have other relict Mediterranean species; including pontic ruscus, Anatolian moor-grass, Kolkhida epimedium, and smoke tree. The vegetative growth of Caucasian yam begins in late May and early June, and the fruits mature in September. The fruit is a triquetrous capsule with membranous wings. This plant reproduces through seeds and vegetatively. The process of natural restoration is extremely slow; at least 15–20 years are required for restoration of the bushes.

The yam is a source of diosaponin, used widely in medicine. The collection of rhizomes and roots as a raw material for medicine has resulted in the decline of its area and population. The weak competitiveness of yam (it is excluded in the process of forest felling and cropping, where the growth of rhizomes and seed propagation decrease) plays a significant role in the disappearance of some of its populations. Experiments on yam cultivation did not yield positive results. The only source for raw material of diosaponin is the natural reserves. The Caucasian yam may completely disappear as a result of excessive exploitation. The plant has been included in the Red Data Book of the USSR.

FAMILY DIPTEROCARPACEAE

All members of the family Dipterocarpaceae are large trees. Their leaves are stipulate, flowers are bisexual, perianth is pentamerous, and the fruit is a nut. A characteristic feature of this family is the presence of resin canals as well as persistent calyx in almost all the plants, because of which the fruit size increases. Members of Dipterocarpaceae are distributed in tropical Asia, some species form forests, some produce resin, oil, or timber.

The camphor tree genus (*Dryobalanops*) with 9 species is endemic to Malaysia. The **camphor tree** (*D. aromatica* Gaertn.) is a relict species of the Borneo and Sumatra islands. The tree is up to 50 m tall with reddish chocolate-brown wood with large patterns. It is used for ship building and furniture making. The so-called Borneo camphor is collected from the resin-rich trunks. The natural populations have significantly declined because of exploitation of its plantation, and the plant has completely disappeared from the southeastern part of Sumatra.

FAMILY EBENACEAE

The family Ebenaceae includes 7 genera with 450 species, distributed in the tropics and subtropics of both hemispheres. They are deciduous trees and shrubs with alternate, simple, entire, leathery leaves. The fruit is a berry. Many species provide valuable timber, some produce edible fruits. The largest genus of the family, persimmon (*Diospyros*), includes about 200 species.

Opposite-leaved persimmon (*D. oppositifolia* Thw.) is the rarest species of the genus. Only 50 trees are growing in southwestern Sri Lanka at the Hinidumkand peak. The population of this endemic plant is declining, being the main source of wood.

FAMILY ERICACEAE

About 2500 species of the family Ericaceae are distributed among 80 genera. These are shrubs or small trees with simple leaves, persistent for many years. Flowers are solitary in racemes or panicles, regular; corolla is gamopetalous, rarely divided. The fruit is a capsule, berry, or drupe. The plants of this family grow in the temperate and cold zones of both hemispheres and in the hilly regions of tropical countries.

The ground laurel genus (*Epigaea*) has only 3 species, found at different places in the world. They are small, evergreen, prostrate semishrubs with hairy branches, alternate leaves, and few-flowered inflorescences.

Gaultherioid ground laurel [*E. gaultherioides* (Boiss. et Bal.) Takht.] has large, funnel-shaped, pink flowers, borne in groups of 1–2, rarely 3–4. Only one of its habitats is known in the USSR: the ravine of the Namtsvavistaskali River (Ajarian ASSR). A small patch has been preserved in a patch of beech forest at an altitude of 800–900 m on a stony slope where this ground laurel grows together with other relict plants, such as Medvedev's birch and Ungern's rhododendron. It flowers in early spring, but if the ravine has a lot of snow in spring, then flowering occurs in June. Propagation through seeds is poor, the plant mainly being propagated vegetatively. It is an ornamental plant. This declining relict of the Tertiary period has great scientific importance as the only member of the discontinuous subtropical genus in the flora of the USSR. It was cultivated in the Batumi botanical garden for many years. It is included in the Red Data Book of the USSR. The present status of the Lazistan part of its area of distribution (Turkey) is unknown.

Asian ground laurel (*E. asiatica* Maxim.) is a scattered shrub with drooping branches, 10–25 cm tall, covered with reddish chocolate-brown glandular hairs. Leaves are oblong-ovate, acute, ciliate, dark green. Small, cup-shaped, pale-pink flowers are aggregated in a very dense cluster. The plant flowers during April–May. It grows in sparse hilly forests in the southwestern regions of Hokkaido Island and on Honshu Island. It is very attractive while flowering; it is collected in large quantities for bouquets, which leads to population decline.

The *Rhododendron* genus includes more than 600 species growing in the cold and temperate zones of the northern hemisphere and in high mountains in the tropics. These are deciduous or evergreen shrubs, sometimes small trees with umbellate inflorescence in many colors. Most of the rhododendrons are ornamental plants. They suffer heavily due to collection for bouquets. The population of some species has reduced to such an extent that special measures are required for their conservation.

Thomson's rhododendron (*R. thomsonii* Hook. f.) is a shrub up to 4.5 m tall with the leaves about 8 cm long and large and wide open flowers

(Plate 17). It grows in eastern Nepal and in Sikkim (India) in the mountains at the altitude of 3300–3900 m.

Dalhousie rhododendron (*R. dalhousiae* Hook. f.) is a terrestrial or epiphytic, prostrate shrub with large, aromatic, yellow, green, or pink flowers (Plate 16). In Sikkim, it grows at an altitude of 1800–2700 m, and in Bhutan at 2100 m. In India, this plant uses, for support, the branches of red magnolia, another endangered endemic of India.

Narrow-leaved rhododendron (*R. stenophyllum* Makino) is an Indonesian plant on the verge of extinction. It is a shrub with slender branches, narrow, almost needle-shaped, sessile leaves in whorls, and bright red flowers, tubular at the base with a broad limb in the upper part (Plate 17). It grows on Kalimanthan Island in the mountains, in highly shaded places with a dense, mossy cover.

Edgeworth's rhododendron (*R. edgeworthii* Hook. f.) is an epiphytic plant. Leaves have deep, prominent veins. Flowers are creamish-white (Plate 16). The species is distributed in Sikkim and Bhutan, where it grows at an altitude of 2100–2700 m.

Van der Bilt's rhododendron (*R. vanderbiltianum* Merrill) grows on the Indonesian islands in the forest communities of pine and oak at an altitude of 2100–3000 m. It is distinguished by its pale yellow flowers. The taxoid rhododendron (*R. taxoides* J.J. Smith) with red flowers is found still higher (altitude 3250 m) on rocks.

FAMILY FABACEAE

About 50 thornless species of trees and shrubs with pinnate leaves and small flowers in heads belong to the silk tree (*Albizia*) genus of the family Fabaceae. Almost all of them grow in the tropics.

Glabrous silk tree [*A. glabrior* (Koidz.) Ohwi] is a deciduous tree with slightly pubescent, spreading branches. Leaves are 15–25 cm long, bipinnate (4–5 pairs), on 4–6 cm long petioles; leaflets are in 15–20 pairs. Flowers are numerous, pink, in heads. The plant is a rare endemic of Amakuza Island, near Kyushu Island (Japan), and Tokara Island.

The genus *Ammopiptanthus* has only 2 species, whose areas of distribution are in Central Asia. The **Mongolian ammopiptanthus** [*A. mongolicus* (Maxim.) Cheng f.] is an evergreen, highly branched shrub up to 1.5–2 m tall. The branches are thick, with yellowish bark, young branches being pubescent. Leaves are trifoliate, silvery due to dense pubescence. Yellow flowers are arranged in groups of 8–10 in racemes at the tip of branches; fruit is a pod (Plate 17). The plant flowers during April–May, fruits maturing in May–June. It grows on loose sand and barkhans in the Huan-hey valley, scattered and in groups in the southern part of the Alashan' range (Alashan' Gobi) on sands, stony slopes, scattered on hummocky terrain, sometimes forming small thickets (China, Mongolia).

Dwarf ammopiptanthus [*A. nanus* (M. Pop.) Cheng f.] is very close to Mongolian ammopiptanthus, but its leaves are simple, entire, almost ovate, 2–4 cm long and 1.5–2.4 cm broad, 3-veined, and silvery pubescent. It grows on stony and pebbled slopes and mountainous terrains in western Kashgaria and central Tien-Shan (USSR, China). Both species are used as a source of medicinal raw material and their reserves are declining; they may also become extinct with such limited distribution of the genus and small population.

The Mongolian ammopiptanthus is included in Appendix 1 of CITES and dwarf ammopiptanthus in the Red Data Book of the USSR.

For a long time, the podded carob, or horn tree, growing in the Mediterranean region was considered to be the only representative of the carob tree genus (*Ceratonia*). In 1945 a new carob species was found in the hilly regions of the northeastern and the northwestern districts of Somalia, and one more species of this genus was discovered 30 years later in Oman.

Oreothaum carob (*Ceratonia oreothauma* Hillcoat, Lewis et Verdc. sp. n.) is found in Oman. The only population of this species, occupying an area of 50 ha consists of about 100 trees, up to 8 m tall with dark gray bark, paripinnate leaves (up to 10 pairs) and spicate inflorescence. A dioecious plant, it grows in the area between the cities of Muscat and Sur, on rocky limestone slopes predominantly facing north at an altitude of 900–1800 m. It forms sparse plantations together with olive, acacias, and rheptonia. The oreothaum carob is located on the territory of the Vadi Serin natural reserve established in 1976. Unfortunately, goat grazing in the reserves has a fatal effect on the young plants (for this reason, careful search over many years for young plants did not produce any results); the goats graze the leaves and pods fallen from adult trees. The local population also causes serious damage to the plantation. In order to protect the natural plantations, it is planned to ban goat grazing in the natural reserves and introduce carob into cultivation. At present, the status of this species is such that it has been included in the Red Data Book of IUCN.

The area of distribution of the locust tree (*Gleditsia*) is in the warm regions of Asia, North America, and tropical Africa. The genus includes 12 species of deciduous trees with simple or branched needles, pinnate leaves, small, greenish flowers in racemes or spikes.

Caspian locust tree (*G. caspia* Desf.) is up to 5–10 m tall with dense, usually spheroid crown and branched spines on the trunks. Its flowers are greenish and have an unpleasant smell. In the USSR, it is found only in the Caucasus, and in the southern part of Talysh; it is also found in the adjoining regions of Iran. It grows in low-lying areas and in the foothills mixed in the girkan forests of chestnut-leaved oak, ironwood, smooth-leaved elm, and other trees; isolated trees and groves are found along river terraces, forest fringes, and glades. The locust tree has a valuable, compact, strong, beautiful wood, suitable for artistic carving. It is a very prickly tree, used for live fences, protection belt, and fixing ravines. The fruits are collected for cattle feeding. It reproduces by seed and also vegetatively. It is used widely for green plantations in the northern Caucasus, at the southern coast of Crimea, and in the south of the steppe zone. The use of lands for agricultural crops, mass grazing of cattle, collection of fruits for cattle feeding, and mining in sand quarries at the places of its growth has led to a sharp decline in its reserves during recent years; only isolated groves over an area of less than 50 ha have been preserved. This species has been included in the Red Data Book of the USSR.

FAMILY FAGACEAE

The genus *Trigonobalanus*, including 3 species, belongs to the family Fagaceae. Of these, 2 species are found in tropical Asia. This genus was discovered relatively recently. The morphology of female flowers of the genus displays several primitive characters, which permits us to consider it an ancient relict.

Verticillate trigonobalanus (*T. verticillata* Forman) has a crown similar to that of old pruned beech; however, it has undulate serrated leaves, up to 5–10 cm long and 3–4 cm broad, with 6–8 pairs of lateral veins; leaves are in a whorl of 3. The plant grows in the mid-mountain belt at an altitude of 850–1500 m in the south of the Malaccas peninsula, north of Kalimanthan Island, and on Sulavesi Island in a hilly rain forest. It is a very rare plant.

Doichang trigonobalanus [*T. doichangensis* (A. Camus) Forman] has alternate phyllotaxy. It grows only in the northern regions of Thailand (Chiang Mai province), in the middle-mountain zone at an altitude of 1200–1765 m where it is a common tree.

Cyclobalanopsis hypophaea (Hayata) Kudo is an endemic plant of Taiwan. This evergreen tree grows up to a height of 17–18 m and a diameter of 1.2 m. Its bark is gray to chocolate-brown, branches are yellow-gray, twigs are dark chocolate-brown. Leaves (11–14 pairs) are about 10 cm long and 2 cm broad with pinnate venation. The plant flowers from December to January. Its distribution is limited to the extreme southeastern part of the island, and it grows on the slopes or on water divides at an altitude of 300–700 m in broad-leaved mountain forests.

FAMILY GESNERIACEAE

The members of the genus *Didymocarpus* belonging to the family Gesneriaceae are herbs or semishrubs with opposite leaves and umbellate inflorescences. They are ornamental and cultivated in botanical gardens and private collections.

Woolly didymocarpus (*D. floccosus* Thwaites) is a perennial herb. Leaves are numerous, broad, basal. Flowers are arranged in umbels. It is an endemic of Sri Lanka, known from only one habitat, in the environs of Candy, where it grows in the rock crevices at an altitude of 600 m together with another rare species of this genus *D. zeylanicus* R. Br. and the common species *D. humboldtianus* Gardn. (Humboldt's didymocarpus, Plate 17).

Reticulate championia (*Championia reticulata* Gardn.) is the only member of the genus. It has been found only in Sri Lanka in the Central province, growing in the humid forests up to an altitude of 1000 m, including the Adam's Peak mountains. It is found in shaded and highly moist areas along river streams and rivulets in pebbled soils. It is a perennial herb, most probably self-pollinated. Flowering of individual plants in a single population does not occur synchronously. The flowers can, therefore, be found throughout the year.

Indian jerdonia (*Jerdonia indica* Wight.) is an endemic species of *Jerdonia*. It is a perennial, short-stemmed (length 5 cm) herb, covered with stiff hairs. Flowers are pale lilac with reddish veins. The plant grows in the Deccan peninsula, in the Nilgiri and Annamalai mountains.

FAMILY GINKGOACEAE

Maidenhair tree (*Ginkgo biloba* L.) is a 'living fossils'. This is the only member of the family and the ginkgo class, which was predominant in the Mesozoic Era, and is one of the most primitive gymnosperms of the present plant kingdom.

The discovery of this relict plant was made at the end of the 17th century by Kempfer, a Japanese doctor (he also provided its scientific description). In the Japanese language, ginkgo means 'silvery fruit' as the edible seeds of its fruit were sold in Japan by this name. Since ginkgo was cultivated near temples, it was considered a holy plant and was worshipped in China and Korea. In the beginning of the 17th century, ginkgo was introduced into western Europe, and a few years later into North America. Cultivation of this plant made it possible to conserve it till the present time.

Maidenhair tree reaches more than 30 m in height and up to 3 m in diameter. The crown of young plants is pyramidal, but becomes scat-

PLATE 17. 1—Narrow-leaved rhododendron; 2—Thomson's rhododendron; 3—Mongolian ammopiptanthus; 4—Humboldt's didymocarpus.

PLATE 18. 1—Maidenhair tree; 2—Nazareth iris; 3—pretty lily; 4—brilliant lily.

PLATE 19. 1—Red magnolia; 2—common nutmeg; 3—Raffles' pitcher plant; 4—curly air plant.

PLATE 20. 1—Spotted air plant; 2—aloe-leaved cymbidium; 3—dense-flowered dendrobium; 4—Fairey's paphiopedilum.

tered with age. Leaves are divided into symmetrical halves (Plate 18). Ginkgo is a dioecious plant and belongs to the group of a few deciduous gymnosperms. Every year during late autumn the trees acquire a golden-yellow color due to change in the color of leaves, after which the leaves fall. Under natural conditions, the plant has been conserved on a small territory in eastern China, in the Dyan-Mu-Shan mountains, along the border between Zhejiang and Anhoi provinces, where it forms forests together with conifers and broad-leaved trees.

In China, Japan, and Korea, ginkgo has been known since time immemorial; it has been mentioned in Chinese books of the 7th and 8th centuries. Its seeds are used in Chinese medicine. Moreover, it is a food in the Orient, and the soft wood of the tree is used in cottage industry.

At present, maidenhair tree grows in almost all the botanical gardens and in many parts of the subtropical and warm temperate zones of Europe and North America. In 1818, ginkgo cultivation was started in the Nikitskii Botanical Garden (Crimea), and now its cultivation has extended up to the latitude of Kiev. In spite of successful cultivation, it is necessary that this rare relict plant be preserved in its natural habitat.

FAMILY HAMAMELIDACEAE

The members of the family Hamamelidaceae had maximum distribution in the Tertiary period, according to paleobotanical evidences. At present this family comprises 30 genera and nearly 100 species. Half the genera are monotypic. The Hamamelidaceae are evergreen, sometimes deciduous shrubs or trees. The leaves of most of the plants are stipulate; flowers are unisexual or bisexual, each flower having 1–2 or more carpels.

The genus *Chunia* is one of the monotypic genera of this family. The **Buckland chunia** (*C. bucklandioides* H.T. Chang) is an evergreen tree with a height up to 20 m and leathery, trilobed, palmately veined, stipulate leaves. The inflorescence consists of a few spikes; flowers are borne spirally on a short, fleshy axis in the upper part of the spike, resembling a cob. The plant is an endemic of Hainan Island. The present status of this relict genus is not known.

FAMILY IRIDACEAE

The members of the family Iridaceae are distinguished for their ornamental value. Many species, especially those flowering in spring, are cultivated in gardens. Often the amateur floriculturists obtain rhizomes or bulbs as the plantation material, which makes cultivation easier but at the same time depletes the natural reserves.

Lortet's iris (*Iris lortetii* Barley) produces large, whitish-pink flowers. The rhizomes of this plant were in great demand by horticulturists; however, the rhizomes died after two seasons of plantation and fresh material was required for cultivation. The studies of the Dutch worker Van Thubergen established that the rhizomes of iris should be maintained from late summer and through autumn at 23° C, imitating the dry hot Mediterranean climate in which this plant grows in nature. With such maintenance, the weight of rhizomes decreases by 30–60 percent, but they do not lose moisture completely. This discovery has, to some extent, reduced the demand for fresh rhizomes for plantation. However, the species population has been reduced so much by commercial exploitation as well as destruction of the natural habitats that it is now endangered. At present, 4 small iris populations are known in the upper part of Upper Galilee and in the south of the Litani River (Israel), where iris grows in scrub communities at an altitude of 700–1000 m. The plant remains dormant during the hot period of the year. It is very sensitive to competition from other species and shade. However, grass cutting is helpful for the survival of plants, as it reduces competition. The flowers of iris are pollinated by bumblebees of genera *Bombus* and *Xylocopa*, and the seeds are dispersed by ants.

The Lortet's iris is protected by law for the conservation of nature in Israel. It is proposed to establish several reserves for conserving the remains of its populations. The plant is included in the Red Data Book of IUCN.

The **Nazareth iris** (*I. nazarena* Hort. ex Hasselbr.) is also threatened with extinction. It has been preserved in only 2 habitats, one of which is the Nazareth mountains (Israel). This is a tall and slender plant with cauline leaves and solitary flower. The tepals of the outer whorl of the perianth are whitish on the upper side with numerous purple veins, which are fused at the base into a dark, almost black spot, and greenish yellow on the lower side with pale purple veins. The tepals of the inner whorl are pale lilac white, with wavy margins, and with a distinct midrib (Plate 18).

Dark brown iris (*I. atrofusca* Bak.) belongs to the section of dark-flowered (black) irises, which are characterized by strict adaptation to certain ecological conditions and do not tolerate any change, being degraded under cultivation. This endemic of the Israel flora grows on elevations near the northern part of the Negev desert. It has a thick, compact rhizome, producing a shoot with a pair of broad leaves surrounding it; leaves reach almost up to the flower. The tepals of the outer whorl are white with dark veins, inner tepals are also white with numerous dark brown, almost black veins and dark spots along the margin.

Grant-duffi iris (*I. grant-duffi* Bak.) grows in marshy areas in the river valleys in northern Israel. Its bulbs are covered with fibrous remains of leaves. The leaves are narrow-linear, stem terminating in a single, greenish-yellow flower with lilac-colored veins. The plant belongs to Apogon section, which includes species without beard at the base of outer tepals of the perianth. The population of this species has declined sharply during the recent years and the size of its flowers has also reduced, which is possibly a result of change in moisture conditions.

All the iris species listed above are included in the list of protected plants of Israel.

Elegant iris (*I. elegantissima* Sosn.) is a plant of the Caucasian flora. It is a short, up to 20 cm tall, plant with a short, creeping rhizome and erect stem, bearing a single large flower; the outer tepals are yellow-brown with dark chocolate-brown veins and minute spots; inner tepals are pale, smoky yellow with violet veins. The plant flowers in May.

The distribution area of this species is in Transcaucasia and is confined to a few localities at the junction of boundaries of the USSR, Iran, and Turkey. The plant is found in the areas of stony wormwood population and among arid, sparse forests. It has high ornamental properties and, therefore, is collected for bouquets. Its rhizomes are dug out for cultivation and sale. Natural populations are also declining due to consumption of the rhizomes by rodents. The plant has been included in the Red Data Book of the USSR.

Iberian (Georgian) iris (*I. iberica* Hoffm.) is an endemic of eastern Transcaucasia, known only within the boundaries of the USSR. It is a perennial with slender, creeping rhizome. Stem is 10–20 cm tall with a single flower; inner tepals of perianth are almost white or light bluish with pale lilac veins, and outer tepals are yellowish-brown with dark chocolate-brown veins, a large dark spot and a few smaller spots in the middle. The plant flowers in May and multiplies through seeds. It grows in the lower mountain belt along stony, pebbled, and debris-covered slopes at several places in Georgia and Azerbaijan. The extinction of the plant from natural habitats and reduction in its population are associated with cattle grazing and collection of plants for sale. This species is cultivated in several botanical gardens. It is included in the Red Data Book of the USSR.

Camilla's iris (*I. camillae* Grossh.) is a rare endemic of eastern Transcaucasia. It is perennial with short, creeping rhizomes. Stem is up to 40 cm tall with a single flower; leaves are falcate. Perianth is uniformly yellow, dove-blue, or violet, without stripes or spots. The plant flowers during spring, and fruits mature in summer. The only habitat of the plant known from the Kirovobad-Kazakh plain is in the region of Lake Kazangel' (Kirovobad district of Azerbaijan), where it grows on dry, debris-covered slopes. It multiplies by seed. It is cultivated in the Tbilisi, Baku, and other botanical gardens. The plant is ornamental and, therefore, intensively transferred to gardens by floriculturists, which is the main cause of disappearance of the species from its natural habitats. The plant has been included in the Red Data Book of the USSR.

Winogradov's iris [*Iridodictyum winogradowii* (Fomin) Rodionenko (*Iris winogradowii* Fomin)] is a rare endemic mountainous Kolkhida species. It is a short rhizomatous plant, 10–15 cm tall, growing as isolated specimens. Leaves are green; perianth is pale yellow, outer tepals have dark, minute spots along the middle. The plant flowers during late winter and early spring, and fruits in midsummer. The distribution of Winogradov's iris is confined to the Meskhetskii mountain range near Bakuriani (Georgia), where it grows on grassy slopes and limestone deposits of the alpine belt. The natural populations have declined significantly during recent years because of the digging of rhizomes and transportation to private gardens. The demand for plantation material of this iris is also increasing in the international market (because of the large, yellow flowers and early blooming). The plant has been included in the Red Data Books of the IUCN and the USSR.

Grand juno [*Juno magnifica* (Vved.) Vved.] is a highly endemic species on the verge of extinction. It is a perennial, up to 25–40 cm tall; bulb is about 3 cm in diameter, with thick, fusiform roots. Flowers are large, inner tepals of perianths are pale lilac with 3 weak, violet veins, outer tepals are light lilac, sometimes almost white. The plant flowers during spring, in April, immediately after the snow melts. The distribution area of juno is located in the Samarkand mountains (Uzbekistan), where the plant has been found in the middle course of the Zeravshan river, in the foothills of Urgut, Agalyk, and Aksai regions. It grows in the crevices of rocks and in fine earth areas in lower mountainous belt. Found very rarely, it suffers heavily from trampling by cattle, collection for bouquets, and digging of the underground bulbs for plantation in flower gardens. The species is included in the Red Data Book of the USSR.

FAMILY LARDIZABALACEAE

The Lardizabalaceae family includes 8 genera and about 30 species, mostly woody creepers. Their distribution area is scattered and divided into 3 unequal parts. The majority of species are distributed in Asia where, in the west, along the Himalayas, they reach up to northeastern Pakistan, and in the east up to Japan, Korea, Taiwan, Khainan Islands, and North Vietnam; 2 genera are found in central Chile. The members of Lardizabalaceae do not have much practical value. The fruits of some species are used for food, and a few species are ornamental also.

Chinese sinofranchetia (*Sinofranchetia chinensis* Hemsl.), the rarest species of this family, is a member of a monotypic genus. Its distribution is restricted to a few pockets in central China. It is a climbing shrub with palmately compound leaves. The flowers are unisexual and monoecious, borne in axillary racemes. The plant has maintained several primitive morphological characters (free and flat stamens) which bring it closer to the ancient Magnoliaceae.

The endemic Asian genus *Akebia* comprises 5 species, distributed in continental China, Taiwan, Korea, and Japan.

Chingshui akebia (*A. chingshuiensis* Shimizu) is a creeper. Leaves are tripartite, long-petioled, 2–3 cm long and 1–1.5 cm broad. Male flowers are purple, in groups of 10–20 in the lower part of the raceme; a few female flowers are in the upper part. Corolla is absent. Berries are large, oval. The only known habitat of this endemic of Taiwan is on Chingshui mountain at an altitude of 1000–2400 m.

FAMILY LILIACEAE

One of the largest genera of the family Liliaceae is onion (*Allium*) with 300 species, distributed in the northern hemisphere.* The natural habitats of the bulbs are decreasing due to change in the conditions for their growth owing to the effect of human economic activity as well as use of the bulbs for food.

Turfy onion (*A. caespitosum* Siev. ex Bong. et C.A. Mey.) is endangered. It forms sparse turfs owing to its long branches. Stem is 15–20 cm tall; umbel is hemispherical, with a few flowers. The plant multiplies through seeds as well as vegetatively. This onion is known only from Zaisan basin (riverine sands of Zaisan and valleys of the Irtysh, Kurtu, and Cherniy Irtysh rivers). There is no information on the present status of this species; possibly, it has been conserved in the areas bordering the People's Republic of China along the Black Irtysh valley. A sharp decline in reserves of turfy onion has been noticed as a result of economic exploitation of the territory. This species is of great scientific and applied value for onion breeding. It has been included in the Red Data Book of the USSR.

Dwarf onion (*A. pumilum* Vved.) grows up to 10 cm. The bulbs are covered with reticulate fibrous scales. Leaves are narrow-linear, flat, 1–2 mm broad, shorter than the stem. Umbel is few-flowered, perianth leaves are pinkish, about 4 mm long; filaments are equal to their length, fused at the base. The plant flowers in June. This endemic of southern Altai (USSR) is possibly already extinct. It was collected only once from the

---

*At present, all the onions are grouped under an independent family Alliaceae.

Ukok plateau and has never been found again since then. The causes of its disappearance are unknown.

Turkestan dipcadi (*Dipcadi turkestanicum* Vved.) is a perennial, bulbous herb with leafless stem. Bulb is ovoid with grayish, papery sheaths. Leaves number 2–3, are linear, grooved, glaucous, fleshy. Stem is 15–20 cm tall, cylindrical, light glaucous, terminating in a loose, unilateral raceme of 4–8 flowers. Flowers are dull dirty-green, with a white border along the margin, whitish inside, fleshy; anthers are yellow. The plant flowers in April, fruits mature in May. This endemic plant is the only member of the African genus on the territory of the USSR: it was found in Soviet Central Asia in the Khaudaktau mountains in the southern part of the Shirabad valley on the Barkhan sands in the soil-blowing basins. It was collected for the last time in 1978. Presently dipcadi is included among plants which have, most probably, disappeared from natural habitats. A few specimens are being cultivated in the botanical garden of Dushanbe.

The foxtail lily genus (*Eremurus*) comprises about 50 species. All of them are perennial herbs with fusiform fleshy roots, a basal rosette of liliaceous leaves, and a scape bearing a large multiflowered raceme. Flowers are bisexual, white, pink, dirty red, yellow, or brown, borne singly in the axils of bracts. Perianth is campanulate or cup-shaped, polyphylous. Stamens number 6; ovary is trilocular; fruit is a capsule.

Velvety-bract foxtail lily (*E. lachnostegius* Vved.) is a fast disappearing, highly endemic species. The roots of this plant are fusiform; stem is naked, strong, 30–60 cm tall; leaves are glaucous, grooved; inflorescence is a 10–20 cm long raceme with black, elongated, velvety bract and light yellow flowers opening in March–April.

The distribution area of this species is small, along the southern slopes of the Gissar range (Gazimailik and Aruktau ranges, Kafirnigan river valley). Its habitats are confined to the varied grass-wormwood pistachio outgrowth at an altitude of 950–1000 m. The plant multiplies only through seeds, the young plants developing very slowly: first flowering is observed after 5–8 years of growth. A few of its outgrowths are being conserved in the Ramit reserves. The species is included in the Red Data Book of the USSR.

Pink foxtail lily (*E. roseolus* Vved.) has thick roots, 70–100 cm tall stem, and densely pubescent leaves. Raceme is sparse, almost cylindrical, with numerous pink flowers. The plant flowers during May–June. The distribution area of this species lies in Soviet Central Asia, restricted to the central and southern regions of Tajikistan (Gazimailik, Aruktau, and Tashbulak ranges, southern tip of Pripyanj Karatau). It is found from the upper limits of the meadow grass-sedge semisavannas with tall grasses to the rosaria and sparse pistachio forests, in maple forests (with juniper), on stony-shallow slopes, at altitudes of 900–2000 m. It multiplies through seeds, taking 5–8 years from germination to first flowering. Its natural population is declining as a result of commercial exploitation of the territory and digging out of the plants. The plant has been included in the Red Data Book of the USSR.

Arabian fritillary (*Fritillaria arabica* Gandoger) is a plant of the steppy and semidesert regions of Negev, Edom, and Moav (Israel). It bears an ovoid bulb with thick fusiform and slender roots. The short stem bears leaves only in the upper half. Leaves are broad-elliptical, semiamplexicaul; 7 grayish-purple flowers are borne in terminal raceme. The Arabian fritillary is disappearing due to changing environmental conditions and digging out of the bulbs. The species has been included in the list of plants of Israel which require protection.

The lily genus (*Lilium*) is characterized by maximum variability in East Asia. The **pretty lily** (*L. speciosum* Thunb.) produces bulbs which are 7–10 cm long and across. Stems are 1–1.5 m tall, glabrous, smooth. Leaves are short-petiolate, broad lanceolate, 10–18 cm long, and 2–6 cm broad, with 5–7 veins. Flowers are reddish-pink or white with dark red, minute spots, upto 10 in a raceme. The plant flowers in August. It grows on the islands of Shikoku and Kyushu (Japan); the northern boundary of its distribution area in the south reaches up to the Primorye territory of the USSR. It is found extremely rarely and is widely cultivated (Plate 18).

Brilliant lily (*L. philippinense* Baker) is an unusually attractive plant with hard, erect, leafy stem. Leaves are lanceolate, dark green, sessile, narrowed toward the apex, falcate. Stem terminates in a single, large, pure white flower, tubular below, and with a broad limb above. Stamens are orange; pistils are greenish (Plate 18). This lily grows in pine forests consisting of *Pinus insularis*, subjected to frequent fires. Large areas of these forests have been cut. Digging out of the bulbs and collection of flowering plants are the main causes of its disappearance from natural habitats. At present, the plant is endangered.

The genus *Protolirion* has 3 species. All of them are small, pale yellow, saprophytic herbs with slender rhizomes, simple stems, and scaly leaves. Flowers are small, arranged in racemes.

Sakura's protolirion [*P. sakuraii* (Makino) Dandy] is a very rare plant in Japan. The stem does not exceed 7–12 cm in length; scales compactly cover its lower part, and upper part is relatively free. The racemes bear 10 or more drooping flowers up to 4 mm in diameter. The plant flowers in July. It grows in the forests on Honshu Island (Mino province) as well as in Taiwan.

The tulip genus (*Tulipa*), distributed in the temperate zone of the Old World, especially in Central Asia, is characterized by considerable variability and high ornamental value. It includes about 100 species of bulbous plants with large and medium-sized red, pink, yellow, or white flowers of various shades and their combinations. The population of practically all the tulip species is decreasing every year, due to intensification of agriculture and industrial use of the territory. Mass digging out of the bulbs for private collections and sale is playing a significant role in elimination of the highly ornamental species. The extent of their exploitation at present has reached a very high level. With a view to saving the natural populations, export of tulips must be regulated. Among the most ornamental Soviet Central Asian species are **Greig's tulip** (*T. greigii* Regel), **Albert's tulip** (*T. albertii* Regel), **Kaufmann's tulip** (*T. kaufmanniana* Regel), and **Ostrowskii's tulip** (*T. ostrowskiana* Regel). They are not yet included in the category of disappearing or rare species, as they have a fairly large distribution area and high population levels. However, with the present rate of decline in their population, they may come under these categories fairly soon. One of these species is described here as an example of a highly ornamental tulip.

Greig's tulip (*T. greigii* Regel), an endemic of the USSR, is one of the most ornamental tulip species in the flora of Soviet Central Asia. It has large, bright red, orange-red, sometimes yellowish flowers with red spots in the middle and spotted leaves. It flowers during April–May, first flowering after 6–8 years; it multiplies through seeds and bulbs. The distribution of this species is restricted to western Tien-Shan and includes the Karatau, Talasskii, and Kirgizian ranges right up to the Kurdi paths and western parts of the Chu-Ilian mountains. It grows on debris-covered and loamy steppy slopes of foothills and lower hills at altitudes of 500–1100 m. Information on the population strength of this species is not available, but it is known that every year it is collected for bouquets and sale, and the bulbs are dug out for transfer to private gardens. Its population is also declining because new areas in the foothills are being brought under agriculture. The Greig's tulip is widely cultivated for hybridization and developing new varieties in the USSR as well as in other countries. At present, a small area of its natural population is being conserved in the Aksu-Jabagly reserve. The plant is included in the Red Data Book of the USSR.

Besides such widespread tulips, there is also a whole range of highly endemic species which need protection measures and constant control on their environment.

FAMILY MAGNOLIACEAE

The family Magnoliaceae comprises 12 genera and about 230 species. They are mainly distributed in the subtropical regions of the northern hemisphere, in east and southwest Asia as well as southeast North America, Central America, and the West Indies.

PLATE 21. 1—Sukhakul's paphiopedilum; 2—Tankarvill's limodorum; 3—blue vanda; 4—great scarlet poppy.

PLATE 22. 1—Cedar-of-Lebanon; 2—aretiodes dionysia; 3—Patma-wort; 4—lanceolate Chinese fir.

Red magnolia (*Magnolia campbellii* Hook. et Thoms.) is one of the most beautiful deciduous trees. Its height reaches up to 50 m; leaves are large; flowers are aromatic, from pink to deep pink (very rarely pure white), with a large number of yellow-red stamens (Plate 19). The area of distribution of red magnolia is in the Himalayas at an altitude of 2400-3000 m (Nepal, Bhutan, Manipur, southeastern Tibet, northern Burma, and western China). Till now only a few specimens of this species have been found, therefore, it belongs to the category of disappearing plants. It is being cultivated in the parks of Sukhumi.

Magnolia kachirachirai (Kanehira et Yamamoto) Dandy is an endemic of Taiwan. It is an evergreen tree up to 20 m in height and 1.2 m in diameter. The bark is dark brown, smooth; leaves are leathery, thick, lanceolate, 6.5-12 cm long, dark green on the upper surface and glaucous on the lower; flowers are yellowish-green. The plant grows in broad-leaved evergreen forests at an altitude of 500-1300 m in the southeastern part of Taiwan near its southern tip.

FAMILY MYRISTICACEAE

Myristicaceae is a tropical family comprising 16 genera and about 400 species distributed in Asia, Australia, Africa, and America. The largest genus, nutmeg (*Myristica*), includes more than 100 species.

Common nutmeg (*M. fragrans* Houtt.) is a native of the Malaccas (Malaysia). This is the only place in the world where it has been preserved in the wild. It is being cultivated on the islands of the Malayan Archipelago and the Antilles. It is an evergreen, tropical tree with simple, leathery leaves; flowers are yellowish, aggregated in leaf axils in small inflorescences; fruit is a fleshy, ovoid-globose berry of orange-yellow color with thick rind (Plate 19). On maturation, the rind and flesh of the fruit split into 2 parts exposing the dark brown seed, which is covered with a bright red juicy aril at some places. The seed is the nutmeg, a spice, and the aril is the nutmeg color or mace. A single tree produces 1500-2000 fruits. The present status of its naturally growing populations needs to be determined precisely. These populations have a great significance for conserving the plant germplasm and evolving better varieties for cultivation.

FAMILY NEPENTHACEAE

The family Nepenthaceae includes a single genus of the pitcher plant (*Nepenthes*) with 65 species, distributed all over the humid tropical regions of the Old World. The center of their distribution is Kalimanthan Island with humid tropical forests covered with lianas. The pitcher plants also belong to the category of lianas. They twine around the trunks and branches of trees over a few meters upward with the help of petiolar tendrils formed between the leaf-like bases and their lamina are modified into pitchers. The pitchers are always vertical in position, with a lid over each. Pitchers of different species differ in their size, shape, and color. The throat of the pitchers is covered with a waxy layer (so that the prey can easily slip down) and hairs. Each pitcher contains up to one liter of fluid inside which digests the captured animals.

Nepenthes rajah Hook. is a rare endemic plant. Its area of distribution is in eastern Malaysia (Sabah) and is restricted to Kinabalu Mountain. At present only 4 habitats of this plant are known: the Maria-Paraya plateau, eastern part of the Mesilau Greek, upper reaches of the Kolopis River, and eastern slopes of the Tambuiyucon Mountain. The total population of the species does not exceed a few hundred plants. The pitcher of this species is the largest among this genus with an ellipsoid opening and a broad lid. The plant is found on open, sunny, and strongly marshy areas, at the upper boundary of the mountain rainforests at an altitude of 1650–2560 m. Young plants have been found on steep, eroded slopes along the Mesilau Creek, and isolated specimens found near waterfalls. Regeneration of the plants through seeds is poor.

All the known habitats of this pitcher plant are included in the Kinabalu National Park. The reduction in the population level of this pitcher plant, like that of other insectivorous plants, is related to increasing demand in the world market. The interest in this species has increased because of its very large pitchers. At present, 5 specimens of this species are present in private collections in the United States, 2 in Australia, 3 in France, and 6 in Japan. They were all collected and imported illegally, as they were not meant for scientific collections, without official permission for collection and export. The non-correspondence between the number of plants dug out and the specimens in private collections, as well as in the number of plants in botanical gardens over the world is because of the difficulty in cultivating these plants from seeds and maintaining the pitcher plant in artificial conditions. That is why the amateur collectors of exotic plants prefer to acquire adult plants with roots, replacing the dead plants with them. It should also be taken into account that not all the plants dug out reach the collectors; many of them perish in transit. Considering the critical condition of this pitcher plant, it has been included in Appendix 1 of CITES.

Raffles' pitcher plant (*N. rafflesiana* Jack) is a creeper climbing trees up to a height of 16–20 m. It has large, greenish pitchers with a striped-red limb (the neck) and red spots on the outside (Plate 19). This species is distributed on the Malacca peninsula and Kalimanthan Island. The rhizomes of this pitcher plant or pieces of aerial branches, which readily root, are marketed. Multiplication through seeds is difficult. This pitcher plant is widely cultivated and its hybrids with other pitcher plants have been developed.

Khasi pitcher plant (*N. khasiana* Hook. f.) is known after its habitats in the Khasi and Jaintia mountains (northeastern India), where it has now almost disappeared. It is a tall, up to 6 m long liana with pitchers about 15 cm, green with red streaks. It is cultivated, and is successfully multiplied through cuttings and seeds.

FAMILY ORCHIDACEAE

The members of the air plant genus (*Aerides*) are caulescent epiphytes with leaves in 2 rows. Perianth lobes are almost equal, lip tripartite.

Multiflorate air plant (*A. multiflorum* Roxb.) has a 10–25 cm long stem, deeply grooved 15–20 cm long, keel-shaped leaves. Racemes are longer than leaves; flowers are 2–5 cm long, white with amethyst streaks; spur is green. The plant flowers during June–July; it grows in tropical forests from the Himalayas to Tenasserim (India, Burma). At present, this species is included in India in the category of plants which have, most probably, disappeared.

Curly air plant (*A. crispum* Lindl.) is an epiphyte with bright pink, strongly scented flowers. The pendulous raceme is the longest among all air plants (Plate 19). The numerous flowers open in June–July, blooming continuing for 2–3 weeks. The area of distribution of this species includes the Western and Eastern Ghats of India as well as Sri Lanka and Burma. Collection of living plants for private collections and for sale has significantly reduced the natural populations. At present, the plant is endangered in the entire area of its distribution.

All the orchids in the jewel orchid genus (*Anoectochilus*) are terrestrial plants with almost round, velvety leaves of emerald color with metallic gleam.

Short-lip jewel orchid (*A. brevilabris* Lindl.) has leaves green on upper surface with a white stripe in middle and golden veins, reddish on the lower surface. Spiculate inflorescence is 10–15 cm long; perianth is greenish with pink streaks, lip is white with sacciform spur. The plant grows on highly moist areas under tree cover at an altitude of 1200–1500 m. It occupies the mountainous regions of India (Sikkim, Khasi) and Burma.

Sikkimese jewel orchid (*A. sikkimensis* King et Pantling) differs from the former species by its olive-green with white perianth and green spurs, collum and tooth on the lip. Leaves are dark red, velvety, with golden yellow veins. The area of distribution of this species is in northeast India (Sikkim, Arunachal Pradesh). The reduction in its population and disappearance of some populations are due to clearing of forests, which changes its growth conditions and prevents fruiting.

The genus *Corybas* comprises about 60 species. Its area of distribution stretches from the Himalayas and China through Malaysia, to the Pacific islands, Australia, and New Zealand. The corybases are perennials perennating in the soil in the form of tubers. Their aerial parts consist of a short stem, single leaf and flower. The flower has an unusual structure for the orchids. Almost all the perianth lobes are reduced, as a result of which the upper sepal and lip are highly enlarged. Flowers are purple, pink, or white.

Arched corybas [*C. fornicatus* (Blume) Kuntze] is a small terrestrial plant with red-white flowers. It grows on well-drained and strongly humus-rich areas in wet mountain forests, usually among mosses. The precise requirements for its growth are the main reason of the rare occurrence of this plant in nature. The species is distributed in Indonesia and Malaysia. Its known habitats are the Gunung-Salak mountains (western Java) and Gunung-Dorovati-Kukusan (eastern Java), where it occupies small areas together with the widely distributed colored corybas. This corybas has not been found in the Gunung-Salak mountains in recent years. Most probably, 2 other species of the same genus also disappeared along with it: the **keel-shaped corybas** [*C. carinatus* (J.J. Smith) Schlechter] and **wine-red corybas** [*C. vinosus* (J.J. Smith) Schlechter]. The arched corybas was found on the Malaccas peninsula in 1930 on Fraser's Hill. The search for this orchid in this place as well as in other similar areas did not succeed. In reality, only one population of this species has been conserved, and that is also endangered due to cutting and burning of forests. Possibly, this small plant can still be found in other areas. The known habitats of corybas on the Gunung-Salak and Gunung-Dorovati-Kukusam mountains have been declared protected areas.

The genus *Cymbidium* is distributed in tropical Asia, Africa, and Australia. The plants of this genus are often cultivated for their beautiful inflorescence.

Large-flowered cymbidium (*C. grandiflorum* Griff.) is a giant, epiphytic orchid, with 60–80 cm, sometimes 120 cm long inflorescence and flowers numbering 10–20, 7.5–10 cm in diameter. Flowers are greenish, paler towards margin, lip is ochre-yellow with purple streaks and long ciliate hairs along margin; leaves are 40–60 cm long and 2–2.5 cm broad. In natural conditions, it flowers in February and blooming continues for 2 months. It grows in the mountains of Sikkim (eastern Himalayas) at an altitude of 1500–2500 m. Its area of distribution is restricted to the western part of eastern Nepal, and eastern Bhutan. The population is decreasing because of excessive collection of living plants for sale mainly in Europe and America. This orchid can be saved from extinction only by complete stoppage of collection and export. In India it has been included in the list of protected plants. The duration of flowering of the large-flowered cymbidium increases and shifts to later periods, April to June. The flowers become larger with age.

Giant cymbidium (*C. giganteum* Wall. ex Lindl.) is a protected plant of India. It resembles the large-flowered cymbidium, but its flowers are ochre-yellow, with numerous vertical dark brown streaks, and are smaller: diameter is 6–7.5 cm. The plant flowers in October–November, growing at an altitude of 1200–2000 m. Its area of distribution is limited to the mountainous regions of northeastern India and Bhutan.

Aloe-leaved cymbidium (*C. aloifolium* Sw.) sometimes forms unique outgrowths on trees. Its racemes are long and multiflorate; flowers are yellow, purple in the middle, 4–5 cm in diameter (Plate 20). Once it was a common plant of the northeastern regions of India (its area of distribution extends further towards the southeast through Burma and Java); it has now become rare, almost disappearing as a result of digging of tubers for medicinal purposes.

Dactylostalix ringens Reichenb. is a member of an endemic monotypic genus of the flora of East Asia. It grows in the southern Kurils and in Japan. This is a terrestrial orchid with a short, slender rhizome and solitary, basally roundish, 3–6 cm long and 3–4 cm; broad, petiolate leaves; stem is 10–20 cm long. Flowers are few; bracts are thin, rectangular-elliptical, 2–3 mm long; perianth is pale green; lip is cuneate, trilobate, white with purple streaks. The plant flowers from May to July, growing in mountainous forests of Hokkaido, Honshu, Shikoku, and Kunashir Iturup islands. It is found very rarely.

Dendrobium is one of the most widespread genera in the eastern tropics. It includes 1400 species, many of which are exceptionally beautiful and are highly valued for collection; some plants contain alkaloids, which double their value. All the plants of the genus are epiphytes.

Pauciflorate dendrobium (*D. pauciflorum* King et Pantl.) is a rare species with limited distribution; it has probably now disappeared. It is an epiphytic orchid with very small pseudobulbs, linear leaves, and small golden-yellow flowers with purple margins, numbering 1–5. It flowers in June. The plant was discovered at the end of the last century in India from only 2 places in the border states of West Bengal and Sikkim. Its habitat in West Bengal is in the valley of the Tista River, very close to the border with Sikkim where dendrobium grew in coastal jungles at an altitude of 600–1000 m. During recent excursions for this plant in the former habitats to determine the status of the species, a single specimen was found and an attempt was made to bring it under cultivation. However, the plant died. There is little hope that this dendrobium could have survived in nature in still unknown places. Its search is essential so as to collect as many seeds as possible, establish its culture, and reintroduce it in nature.

Dense-flowered dendrobium (*D. densiflorum* Wall.) has pale yellow flowers; lip is deep orange in the center and paler along margin (Plate 20). This orchid is found in the tropical belt of the Himalayas (Sikkim, Assam, the Khasi mountains) in forests at altitudes up to 1200 m.

Golden dendrobium (*D. chrysanthum* Lindl.) is endemic to India. It is covered with bright-colored inflorescence at the time of blooming in August. It grows in Arunachal Pradesh. Both these species of dendrobium are being cultivated in the botanical garden of Tipi (India).

Mabele dendrobium (*D. mabelae* Gammil) is endemic to India. It is distinguished by tender, pure white inflorescence. This attracts orchid lovers and increases the demand for living specimens. Its distribution is restricted to the Western Ghats.

The genus *Didiciea* is endemic to East Asia. It comprises only 2 species, of which one grows in Japan and the other in the eastern Himalayas (Sikkim). These are small, terrestrial orchids with tuberous structures and few leaves; inflorescence is a raceme; flowers are small, yellowish-green or dark purple.

Japanese didiciea (*D. japonica* Hara) is a rare plant, whose only habitat is known at Tara mountain in the Khidsen prefecture (Kyushu Island). This plant, almost unnoticeable during the rest of the year, attracts attention in May when it starts blooming with 4 cm long raceme, consisting of 8 chocolate-brown to purple flowers. Leaves are cordate with pointed tip, 5-veined, chocolate-brown to purple on the lower side, and 4.5 cm long. Stem is slender and long.

Cunningham's didiciea (*D. cunninghamii* King et Pantl.) has broad ovate, 3-veined, 4 cm long leaves; raceme is more than 20 cm long; bract is small. It flowers in July. This rare orchid was collected at a height of almost 3500 m near the village Lachen (Sikkim), which was for a long time considered to be its only habitat. Subsequently, it was also found in Garhwal. Nevertheless, Cunningham's didiciea was included in the list of most probably endangered plants of India, and was also included in Appendix 1 of CITES.

The genus *Diplomeris* includes only 2 species, one of which, the **pretty diplomeris** (*D. pulchella* D. Don.), is still fairly widely distributed. The **hirsute diplomeris** [*D. hirsuta* (Lindl.) Lindl.] is endangered. This beautiful orchid with pure white flowers is being cultivated. The cultivated plants are raised from seeds; they flower for 3–4 years. The plant flowers during July–August. Under natural conditions, diplomeris was fairly common over a large area in the Himalayas during the 19th century. It was found in India (Sikkim, West Bengal), Nepal, and Bhutan. At present, it has been preserved on a small area of 1 km length in West Bengal (in the

Mahanadi reserve) and in Darjeeling district. It grows on sandstone rocks at an altitude of 300–500 m together with different species of legonia, colocasia, globba, mosses, and liverworts. The rocks are directly facing the monsoon winds and rains. This ecological peculiarity of diplomeris may not always favor its multiplication and spread: for example, an area of 100 m^2 under this plant was washed by rains in 1976. Road construction has also caused a significant damage to the population of this species. At present, collection and sale of living plants are prohibited. The hirsute diplomeris is included in the Red Data Book of IUCN.

The area of distribution of the endemic monotypic genus *Eleorchis* covers Hokkaido and Honshu islands (Japan) as well as southern regions of the Kurils (USSR).

Japanese eleorchis [*E. japonica* (A. Gray) F. Maekawa] is a terrestrial orchid with tubers up to 6 mm in diameter. Stem is 20–30 cm long, slender, glabrous, with 1–2 scaly leaves at the base. Leaves are linear-lanceolate, 15 mm long and 8 mm broad, narrowed towards both ends. Flower is solitary (rarely 2), pink-purple. This eleorchis flowers in July, growing in marshes.

The genus *Ephippianthus* with 2 species is endemic to East Asia; its area of distribution includes the Far East (regions of the RSFSR), Korea, and Japan. These are very delicate, terrestrial plants with slender, long rhizomes, a few leaves, erect stem covered with scales in the lower part, a few-flowered raceme, and small, greenish-yellow flowers.

Schmidt's ephippianthus [*E. schmidtii* Reichenb. f. (*E. sachalinensis* Reichenb. f.)], like all other orchids, is found very rarely in natural conditions; its population is very limited. Its main habitats are confined to the mountainous coniferous forests with moss cover and Erman's birch forests with bamboo. Its area of distribution includes Sakhalin, Kuril, Lower Coarse of Amur, Primorye territory (USSR), Hokkaido Island and central and northern regions of Honshu Island (Japan), and the northern part of Korea. This delicate and slender plant can be found during the flowering period in July–August, when a loose raceme forms of 3–7 yellowish-green flowers, which are somewhat horizontal.

Sawadan's ephippianthus [*E. sawadanus* (F. Maekawa) Ohwi] has a flat column with wing-shaped processes in the upper part and dentate lip. This is a very rare plant. It flowers in June, growing in the hilly forests only on Honshu Island (Hakone, Fiji, Chichibu, and Yamato prefectures). The present status of this species is unknown.

The large genus *Paphiopedilum* includes 9 species, which have long been cultivated for their ornamental value. But there are some species which have disappeared from the natural habitats and there is practically no hope of establishing their culture.

Drury's paphiopedilum [*P. druryi* (Bedd.) Stein] belongs to the group of endangered plants and is possibly already extinct. It was first found in 1865 in the state of Kerala in the Travancore mountains and in 1875 in the Calicut mountains. It was seen for the last time in 1972. In 1976, a special expedition was organized to determine the condition of the former habitats and occurrence of the orchid there. However, all attempts to find it in the Travancore mountains failed. The only specimen was found on the Agastia-Malai mountains. So far this is the only place where paphiopedilum has been conserved. It is quite possible that it exists in the form of rhizomes or young plants, therefore, it is very important to prevent the destruction of the areas or change in their conditions.

Drury's paphiopedilum produces a rhizome up to 1 m long, bearing 5–6 aerial, leafy branches and large, up to 7.5 cm in diameter, greenish-yellow or yellow flowers with purple streaks. Their lip is bright yellow, resembling a lady's slippers in shape. The orchid grows in well-lit areas, sometimes as an epiphyte on different spurge species, sedges, and cereals. During its vegetative growth, from April to November, the grassy stand covers it from direct sun rays, and after it dries out, when the sun rays are not so hot, the branches of paphiopedilum are completely exposed to light. Another orchid, **spotted air plant** (*Aërides maculosus* Lindl.), is often found together with this plant (Plate 20).

Drury's paphiopedilum is the only species of this genus found in complete isolation near the southern tip of India, which is possibly related with the change in the conditions over a large territory of the Indian peninsula during the Pleistocene. Taxonomically this species is closest to *P. villosum* (Lindl.) Pfitzer. A long-term program has been initiated in India to restore the natural population of Drury's paphiopedilum. It envisages artificial cultivation of the plants from all the seeds that can be obtained from private and institutional collections, and their plantation in the earlier habitats. If their return to nature becomes successful, more rigid control can be imposed on collections of the plants. Its export is already prohibited by government order, including that of the cultivated plants, as there are only 12 specimens of this species in India. The plant has been included in the Red Data Book of IUCN.

Fairey's paphiopedilum [*P. fairieanum* (Lindl.) Pfitzer.] belongs to the category of disappearing plants of India. It is often called the 'last orchid.' Its flowers are very similar to those of lady's slipper: equally broad lip and remote lateral bracts (Plate 20). This species was first described in 1857. In 1905 it was rediscovered on the Torso River (Chumbi, western Bhutan). On the basis of herbarium specimens it is known that this orchid grew in the eastern part of the Kameng region and in the Salera valley (Sikkim, India). At present, 2 reserves have been created for protecting the habitats of this orchid in Sikkim.

PLATE 23. 1—Glyptostroboid metasequoia; 2—parasol fir; 3—granthamian camellia; 4—aralioid trochodendron.

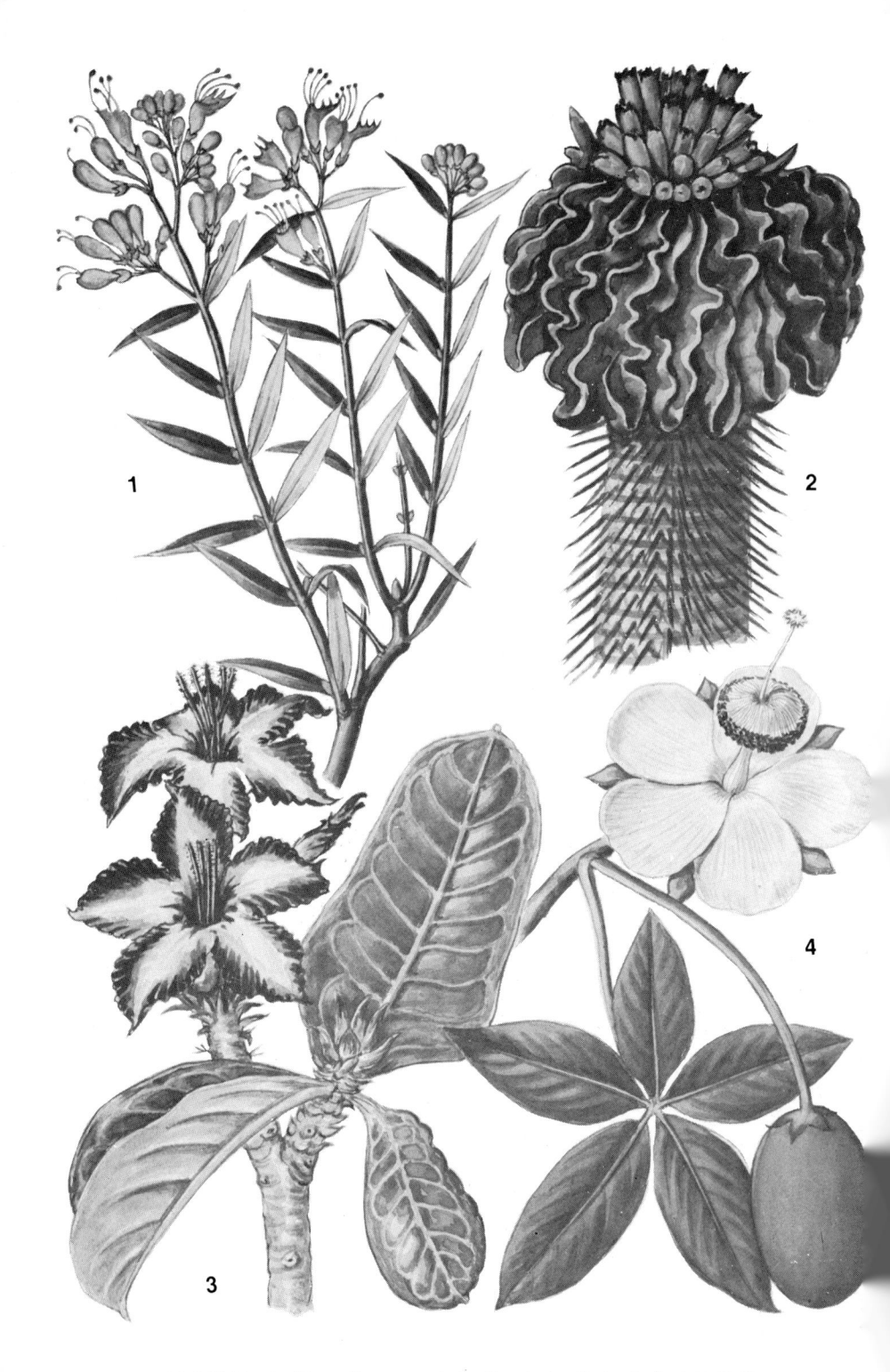

PLATE 24. 1—Mongolian caryopteris; 2—elephant's trunk; 3—impala lily; 4—baobab (monkey bread).

Sukhakul's paphiopedilum (*P. sukhakulii* Schoser et Senghas) is endangered in Thailand (Plate 21). It was discovered very recently in evergreen hilly forests; nevertheless, its demand in the international market is very high, which is depleting even otherwise limited natural reserves. The only way to save this orchid is to prohibit its export. Formally, this plant and several other orchids are considered protected by the Law of Forests Conservation in Thailand.

The genus *Renanthora* includes epiphytic orchids with leafy stems. The flowers are large and sparse.

The species *R. imschootiana* Rolfe has a short, twining stem. Its inflorescence consists of many flowers, in which the petals are yellow with red spots and sepals are red. It is distributed in the northeastern regions of India (Himalayas) and Indochina and is cultivated in the Thimpu botanical garden (India). As a special item, it is included in Appendix 1 of CITES.

Blue vanda (*Vanda coerulea* Griff. ex Lindl.) is disappearing from its natural habitats. It is an epiphytic orchid with large, delicate, dove-blue flowers (Plate 21), up to 10 cm in diameter. The plant blooms in October and flowers almost up to the new year. Its beauty was the cause of a reduction in its population in nature, and subsequent disappearance from all its earlier habitats. Plant collectors transported this orchid for sale and cultivation in glasshouses.

The area of distribution of this species covers northeastern India: the Khasi and Gentea mountains (Assam state), Naga hills, and Thirap regions (Arunachal Pradesh). These areas need a very strict protection. At present the plant is being conserved only under cultivation.

FAMILY PALMAE

Red areca palm (*Areca concinna* Thwaites) is an endangered plant. It is a short palm, 2–4 m tall and 3–4 cm in diameter. The leaves are pinnate, 1–1.5 m long. The fruits are bright red, from where the palm derives its name. The only population of this plant comprises about 1000 specimens occupying an area of 2–4 ha in a marshy forest in the Kalutara region (Sri Lanka). The surrounding vegetation is completely cleared and this territory is now occupied by agricultural crops. It is proposed to develop a reserve including the habitat of this areca palm so as to protect the plant and its habitat. Reclamation work is the main threat to the species. This palm has been included in the Red Data Book of IUCN.

The members of the fish-tail palm (*Caryota*) are distributed in tropical Asia and Australia. The English name of the genus was given on the basis of its paripinnate leaves having an appearance of the caudal fin of fishes.

No fish-tail palm (*Caryota no* Becc.) grows in Indonesia and Malaysia.

It has a fairly wide area of distribution: the Borneo and Kalimanthan islands, Sabah and Sarawak, and western Malaysia, but the number of its habitats is limited and each habitat does not have more than one plant. This giant palm, with a height up to 20 m, basally thickened trunk, and large leaves, up to 8 m long, is found in the lowland tropical forests at an altitude up to 400 m. It prefers areas at the foot of limestone deposits. The critical position of the species is mainly due to destruction of its habitats, cutting of the tropical forests; the local population is causing great damage to this plant, using its apical buds, or 'cabbage', as vegetable (the palms die in the absence of apical buds). This tree gives sago. It must be specifically emphasized that this palm is a monocarpic plant, i.e. flowers once, after which the tree dies. Each fallen tree is a nonfruiting plant; therefore, felling has a definitely harmful effect on regeneration of the species. Its habitat on Sarawak is a part of the Mulu National Park; it is proposed to organize a national park on Borneo also including the area where fish-tail palm grows. Harvesting of 'cabbage' is prohibited at the prohibited territories. Its culture has been developed in Bogor on Java. The plant is included in the Red Data Book of IUCN.

Glaucous ceratolobus (*Ceratolobus glaucescens* Blume) is a climber-like palm. Stem is 6 m tall; leaves are pinnate, up to 1.5 m long, terminating in a tendril and resembling fish tail in shape, dark green on upper surface, lower surface densely white-pubescent. The plant is dioecious. Often ants inhabit the inflorescence after development of ovary, which protects the fruit from other animals and facilitates its maturation. This species has great scientific importance for studying the evolution of all palms. As was mentioned above, it is a climber, a rare type of living form for palms, and it has a very unusual type of inflorescence, typical only for the genus *Ceratolobus* (the genus comprises 7 species). Each inflorescence is enclosed by a single bract which opens from the top. The pollinators reach up to the flowers through this opening. The biology of pollination of ceratolobus is still incompletely studied.

The condition of this palm is such that in the absence of strict protection measures, it would soon disappear. Only one population has survived near the city of Palabuhanratu at the southern tip of the coastal region in western Java; the total number of male and female plants here does not exceed 30 (both types of plants are approximately equal in the population). The earlier known habitat, 70 km north of this, near Bogor on the fossil coral limestone deposits, has already disappeared due to fire, tree felling, and destruction of the habitat. The area near Palabuhanratu forms a part of the natural Sukawaian reserves. Here ceratolobus grows in a coastal rain forest. Unfortunately, the protection regime in the reserves is often disturbed, which cannot help in the survival of the palm. At present, it is cultivated in Bogor and several other botanical gardens. Its seeds germinate easily, and the palm produces flowers in the third year. The species has been included in the Red Data Book of IUCN.

The genus *Johannesteijsmannia* comprises 4 species, 3 of which are distributed in western Malaysia and belong to the category of disappearing plants due to limited distribution and small populations: *J. perakensis* Dransfield is found on 2 parallel mountain ranges in Perak; *J. magnifica* Dransfield grows in Ulu-Semeni (Selangor) near Ipoh (Perak) and on several elevations in the northeastern part of Negri-Sembilan; *J. lanceolata* Dransfield has also been found in Ulu-Semeni together with *J. magnifica* at a distance of 1.5 km from the Taman Negara National Park. All these species are confined to rainforests. Several habitats are being conserved in forest reserves.

The long-leaf johannesteijsmannia [*J. altifrons* (Reishenb. f. et Zoll.) H.E. Moore] belongs to the category of species with uncertain status, as its present condition is not fully understood. The area of distribution of this palm covers Malaysia (northern regions, eastern Johor, western Sarawak) and Indonesia (Sumatra). In spite of such a wide distribution, its habitats are confined exclusively to the primary rainforests, which are located up to an altitude of 300 m. These forests are presently being intensively exploited for timber. This species can tolerate only selective felling, but is strongly damaged by the falling trees and wilts under bright sunlight. This palm was never found in the secondary forests.

The species name was given to this plant due to its unique external appearance; its stem is underground and a bundle of 20-30 very large, erect, up to 6 m long leaves on long (2.5 m) petioles emerge from the soil. At present, this palm is being conserved in the national parks of Taman Negara (Malaya) and Bake (Sarawak) and in the Langkat natural reserves (Sumatra). Repeated attempts were made to cultivate species of this genus, but almost without any success, as the seeds were undeveloped. All species of the genus are included in the Red Data Book of IUCN.

Maxburretia rupicola (Ridley) Furtado belongs to the category of relict plants of western Malaysia. Only 3 habitats of this palm are known. One of them is on the rocks in the Batu caves, where stone mining is common; as a result, the entire vegetation is endangered. The second habitat in Bukit-Takun suffered from incidental fire caused by the inhabitants of Templer Park, on the territory of which it is located. And the third habitat, Bukit-Anak-Takun, is also within the boundaries of Templer Park. The total number of palms in these 3 areas does not exceed 1000. It is assumed that it also grows on the limestone deposits in southern Thailand. The herbarium specimen in Bogor (collections near Surat Thani in 1963) is identical to this maxburretia.

Maxburretia rupicola is a squat, acaulous palm or has a short, up to 1 m tall, tomentose stem bearing a cluster of up to 1 m long, petiolate leaves. The straight paniculate inflorescences are of two types: bifurcated with bisexual flowers and thrice zig-zag shaped branch with male flowers. The flowers are yellow and emit a strong smell. This palm grows on humus-rich soils on limestone deposits in well-lit forests.

The Malayan Society of Naturalists is proposing to declare the Batu caves a natural national memorial so as to control the habitat of maxburretia as well as other rare and endemic species.

Nenga gajah Dransfield is a very rare palm. Its stem is up to 2 m tall and 15 cm in diameter; the crown consists of 8–10 divergent, up to 3 m long leaves. The erect, branched inflorescences are borne between leaves. Nenga grows in the litter of tropical rainforests on the slopes or in the bottom of valleys at an altitude of 800 m. This is mainly a Dipterocarpaceae forest, distinguished by rich and variable species composition: rafflesia and giant arum (*Amorphophallus titanum* Becc.) from the family Araceae also grow here.

Only 2 habitats of Nenga are known, which are in Sumatra at a distance of 500 km from each other: Kepahiang, 50 km northeast of Bengkulu on the main watershed between the eastern and western coasts, and Gunung-Talakmau Belkan, 70 km north-northwest of Bukittingi. Most probably, this palm had a wider distribution in the past, but it is reduced as a result of destruction of the tropical forests and development of coffee plantations and other crops. The Kepahiang area is located on the territory of a protected forest. This species has been included in the Red Data Book of IUCN.

FAMILY PAPAVERACEAE

111 The largest and most widespread in the family Papaveraceae is the poppy genus (*Papaver*), which includes about 100 species.

Great scarlet poppy (*P. blacteatum* Lindl.) is a disappearing species. It is a perennial with a height of 60–120 cm and erect, thick stem, large (up to 45 cm long) pinnatisect leaves, and huge (diameter up to 20–25 cm) blood-red flowers with large black spots at the base (Plate 21). This is an endemic of Ciscaucasia (Pyatigore', Tersk, and Sunzhensk ranges). It grows on the foothills on clayey slopes at an altitude of 200–700 m. The area of its distribution covers a few hundred hectares; within the limits of its natural area of distribution, this poppy is found extremely rarely as isolated specimens. Natural regeneration is practically absent. The main causes of reduction in the area and population of this ornamental species is the collection of living plants for bouquets and use of the habitats for agriculture. At present, several natural habitats of this poppy have been included in the protected territory of the Beshtaugorskii forest park. This species has been included in the Red Data Book of the USSR.

FAMILY PINACEAE

In Asia, the fir genus (*Abies*) is represented by a few species.

Kamchatka fir (*A. gracilis* Kom.) is the most rare species. Some botanists consider it a subspecies of khingan fir, but in this case also the relict habitat of the fir on Kamchatka Island has maintained its unique properties. The Kamchatka fir is a tree up to 15 m tall with dense, oval-pyramidal crown and gray bark. Its needles are up to 3 cm and cones 2.5–5 cm long. It grows in the southwestern parts of the marshy valley at the Semyachik (Semlyachik) river, where it occupies the hillocky border of the ancient lava plateau covered with loose volcanic deposits. The area of fir plantations is about 22 ha, surrounded by forests of Erman's birch. The fir is easily rejuvenated, but has slow growth. The grove of Kamchatka fir is included in the Kronotski Reserve. This species is included in the Red Data Book of the USSR.

Kawakami fir [*A. kawakamii* (Hayata) Itô] is an endemic of Taiwan. It is a tree up to 16 m tall and 1 m in diameter; its bark is grayish chocolate-brown, scaly, and branches are horizontal; leaves are broad-linear, flat, 10–15 mm long, obtuse. The mature cones are erect, cylindrical or oval, with a diameter up to 4 cm. The plant grows at an altitude of 2800–3000 m along the central range, forming pure plantations on a small territory.

The tanne genus (*Keteleeria*) includes 3 or 4 species. Their area of distribution covers the southern and central provinces of China, Taiwan, Vietnam, and Laos. In natural conditions, tanne very rarely forms pure plantations, often it is found mixed with other conifers.

David's tanne [*K. davidiana* (Franchet) Beissn.] is a tree with 25–35 m height and 2.5 m diameter. The crown of the young plant is conical, with a flat top. Leaves are flat, dark green, lustrous, 2–4 cm long and 3–4 mm broad. Cones are 6–15 cm long, erect, cylindrical, young cones are green, adult cones pale chestnut-colored. The area of distribution of this species covers the central and western regions of China. In Taiwan, it grows in the northern part at an altitude of 300–600 m and in other regions at 500–900 m. Usually it is found in the plantations of broad-leaved trees on open areas. At present, it is disappearing in the mainland as well as on the island.

China tanne or **Fortune's tanne** (*K. fortunei* Carr.) is a tree up to 30 m tall and 2 m in diameter with horizontal branches. Young branches are orange-red; cones are 8–12 cm long.

The present status of tanne is little known. In nature, it is found extremely rarely and does not form any noticeable plantations. It is already being cultivated for more than a century. It does not have much ornamental value and is cultivated in the botanical gardens and dendraria as a rare plant. Both these species are under cultivation at the Black Sea coast of Caucasus and at the southern coasts of Crimea.

The hemlock genus (*Tsuga*) includes 4–18 species, which are distributed in North America and Asia.

The **Japanese hemlock** (*T. blaringhemii* Flous) is a 20–30 m tall tree. Its bark cracks longitudinally; branches are naked, yellowish-gray, dark-colored at the point of the formation of the leafy branches; needle is 2–2.5 cm long and 1.5–1.7 mm wide, with a fairly deep groove on the upper side and 2 white stripes on the lower side; cones are 2.5 cm long and 2.5 cm in diameter. The tree grows in the mountains of Honshu (the Kiu and Yamato prefectures) and on Shikoku Island. It is found extremely rarely. The other rare species of this genus are the **calcareous hemlock** (*T. calcarea* Downie) and **Forest's hemlock** (*T. forrestii* Downie), alpine species from the Yunnan province (China), almost unknown under cultivation in Europe and rare even in their homeland.

The spruce genus (*Picea*) is distinguished by the diversity of species and includes about 45 species distributed all over the northern hemisphere.

The **Koyama** or **Japanese spruce** (*P. koyamai* Shiras) is among the rare species. This is a tree with reddish chocolate-brown branches; leaves are somewhat sickle-shaped, 8–12 mm long; cones are short cylindrical, 4–10 cm long. The tree grows in the hilly regions of Honshu Island (Yatsugatake). Half a century ago, only a small grove of 100 trees remained, which was even then endangered. The present status of Koyama spruce is unknown; most probably, it should be included in the category of endangered plants.

The genus *Cathaya* is the least studied. It derives its name from the ancient Latin name for China, 'Kathai'. This genus was described in 1958 by Chinese botanists based on the material collected in southern and western China. It includes 2 species with limited distribution.

Silvery cathaya (*C. argyrophylla* Chun et Kuang) is a tree with a height up to 20 m and diameter 40 cm. The bark of old trees is ash gray and splits in thin plates; cones are ovoid, almost sessile, remaining on the tree for several years; leaves are green, broad, sometimes sickle-shaped, bent. This species was found in the Guang-zi (Huansi) province (southern China).

There is very little published information on the **Nanchuang cathaya** (*C. nanchuanensis* Chun et Kuang). In particular, it has been indicated that it grows under natural condition on the Kingfu mountains in the Szechwan province (western China).

The members of this genus are not found in botanical gardens or dendraria, not even in the herbaria of the world. All these plants deserve special protection and study.

Golden larch (*Pseudolarix*) is a genus with a single species.

Chinese golden larch (*P. kaempferi* Gord.) was discovered by Scottish traveler R. Fortune. The Chinese called it 'golden pine with deciduous leaves'. This is a tree up to 40–50 m in height and 1.5 m in diameter with a broad, low, conical crown, slender branches, and golden-yellow leaves. The leaves acquire golden-yellow color only in autumn, before falling, having their usual light green color in winter. Mature cones open in the first year after maturation. At one time, golden larch grew in the Zhejiang mountains (eastern China), towards southwest Ningbo at an altitude of 300–1200 m. Its present condition is unknown. It is clear that this rare plant with limited distribution requires constant control and measures for protection of its natural plantations. The Chinese golden larch is an ornamental tree. In the Soviet Union it is cultivated south of Sochi.

The cedar genus (*Cedrus*) comprises 4 species, 3 of which are adapted to the ancient Mediterranean; the fourth species grows in the Himalayas.

The **cedar-of-Lebanon** (*C. libani* A. Rich.) is a 25–40 m tall tree with widespread branches and umbellate or pyramidal crown. Its leaves are hard, needle-shaped, dark green to silvery gray, sometimes with glaucous tinge (Plate 22). The cones are ovoid, maturing in the second to third year and disintegrating soon after. The tree grows in Asia Minor, in the Tavr range, forming mixed forests with Cilician fir and juniper at an altitude of 1300–2000 m. Although, at these heights, the snow sometimes persists up to 5 months, and there are severe frosts during winter and strong drought in summer, the cedar-of-Lebanon grows fairly well.

This tree has been valued since ancient times for its timber, which possesses not only excellent building qualities and beautiful color but also pleasant smell, resembling the juniper wood. It was used in construction, for fencing, and in making the Egyptian artefacts and other decorative pieces. Till today, the cedar-of-Lebanon attracts the attention of local people as a traditional holy tree. It is a symbol of the country and is included in the state emblem, postal stamps, and coins of Lebanon. Prolonged exploitation of this cedar has reduced its natural reserves so much that it is strictly protected in Syria and Lebanon (only a few plantations have survived). The Bsherr grove is the most famous, where trees up to 7 m circumference are found.

The cedar-of-Lebanon is one of the first trees introduced in western Europe, and plantations are known from the 17th century. Individual specimens have been preserved since then in Italy and in southern France. It was first introduced in cultivation in Russia in 1826 in the Nikitskii botanical garden. Now it is being cultivated on the southern coasts of Crimea, Black Sea coasts of Caucasus, Transcaucasia, and Central Asia.

Cyprian cedar (*C. brevifolia* Henry) is a tree with a height up to 12 m, much shorter than other cedar species. Its crown is umbellate; needles are very short, glaucous green. It grows only in the mountains in Cyprus. Its wood was also in great demand till natural resources were depleted. Its present status requires precise determination. An area of natural plantation is being protected in the Pavos forest reserves. It is rarely found in cultivation, mainly in the botanical gardens.

Ducampopinus is a monotypic genus. The **Krempf's ducampopinus** [*D. krempfii* (Lecomte) A. Cheval] is a rare endemic of Indochina. It was discovered in the beginning of the 20th century, when the German botanist M. Krempf, traveling in Vietnam, found a unique plant resembling pine but with very broad, up to 7 mm, needles. Since then the generic status of this plant is being discussed: some workers separate it as an independent genus *Ducampopinus*, whereas others include it under *Pinus*. It is a medium-sized tree, from 12 to 30 m tall, with reddish bark and flat, serrulate leaves, arranged in pairs. The cones have deciduous scales. Ducampopinus grows in the mountains of Trunbo (formerly Annam) province of Vietnam. It is found rarely as an individual plant and does not form significant plantations. The present status of the species is unknown.

The pine genus (*Pinus*) is the second largest genus in the family Pinaceae after *Podocarpus*: it comprises about 100 species.

The **erect-cone Aleppo pine** (*P. brutia* Ten.) has a fairly wide area of distribution. It covers Asia Minor and the adjoining islands of Cyprus and Crete, enters Syria and Lebanon; isolated groves are found in Iraq and the Caucasus (the Pitsunda and El-Dar subspecies). From the standpoint of conservation, only its subspecies are interesting. They represent relict formations that were separated from the main area of distribution even in the Tertiary period.

Eldar pine [*P. brutia* Ten. ssp. *eldarica* (Medw.) Nahal] is a rare endemic subspecies. The tree, 15 m tall, has long, spreading branches and a broad crown. The needles are hard, green, 6–12 cm long; cones are 5–8 cm long and 3–5 cm across. Fruiting begins in the eighth to tenth year, the seeds maturing after 2 years. The only known habitat of eldar pine (area 400 ha) is the Eller-Oukhi range at an altitude of 400–600 m along the right bank of the Tori river (a right tributary of Alazan). Here the pine forms sparse plantations with arborescent junipers; several endemic species also grow with it. There is practically no regeneration. Seedlings suffer due to erosion, wind, and cattle grazing, as well as collection and marketing of seeds. Earlier, this pine was distributed more widely, for example, in the 20th century it was found in the Gryanjha (Kirovabad) region, i.e. 60–80 km south of the present area of distribution, and in the recent past it moved down along the Ellor-Oukhi range. At present, it has been destroyed by humans in accessible places. Only about 700 adult trees have survived.

This tree is salt-tolerant and least demanding of soil fertility; as a result, it is widely used for afforestation of the mountain slopes and green belt formation in the arid regions of Transcaucasia, Crimea, and Central Asia.

At present, an area of eldar pine is included in the Gek-Gel reserve forests (Azerbaijan), and the subspecies is included in the Red Data Book of the USSR.

Pityusa pine [*P. brutia* Ten. ssp. *pityusa* (Stev.) Nahal] is a rare endemic subspecies. The tree is up to 25 m tall with brown-gray bark and broadly conical or broadly spreading crown. Needles are light green, 10–15 cm long cones are roundish, sessile, 6–10 cm long. Fruiting begins at the age of 20–30 years. The area of distribution of this species is along the Black Sea coast of the Caucasus from Anape to Abkhazia, but in this belt it grows in groups or in isolation, forming a few large plantations. It usually grows on rocks and steep slopes together with Iberian oak and Hartwis oak. In the mountains, it rises up to 300–400 m. Cape Pitsunda is the only place where it grows in the plains. The total area of its plantation is 1540 ha.

The pityusa pine is ornamental, less demanding of particular soil conditions (it grows on limestone rocks without any soil) and humidity, and salt-tolerant. It is widely cultivated. It is being conserved in the Pitsunda-Myusser reserve (Georgia). It is included in the Red Data Book of the USSR.

Lace-bark pine (*P. bungeana* Zucc.) is a rare endemic of China. Trees are 25–30 m tall with short, often branched trunk up to the base and crown peeling off thin layers of newbark on white surface (in pine white bark is unusual). Needles are soft, light green, finely serrate and arranged in threes. Under natural conditions, this pine grows in the mountains of central China, but it is cultivated throughout the country near temples and palaces. The present status of this rare species needs to be determined precisely.

The Douglas fir genus (*Pseudotsuga*) includes about 20 species, some of which grow in North America and others in East Asia. These are 20–50 m tall, evergreen trees resembling spruce, with flat needles protruding in all directions and pendulous cones. They live for a very long time.

Wilson's Douglas fir (*P. wilsoniana* Hayata) is an endemic of Taiwan. Height of trees is up to 30 m and diameter 2 m (at chest height) with thick bark having deep fissures. Needles are linear, flat, with crispate margins, 15–25 mm long and 1–1.5 mm broad. Mature cones are ovoid, up to 6 cm long and 3–3.5 cm in diameter. This fairly rare plant is found in mountain forests at an altitude of 800–2500 m.

FAMILY POACEAE

The family Poaceae includes heptanervate hubbardia (*Hubbardia heptaneuron* Bor), a tender annual. Its only habitat was known in the region of Gersoppa Falls on the Sharavati River (India), in the zone of scattering splashes. It grew on rocks and could be found only during the monsoon rains (June to October). For the remaining part of the year, this hubbardia perennates through seeds. Change in its growth conditions due to construction of a hydroelectric power plant led to its disappearance. Attempts to find this hubbardia in identical conditions of the same region have so far not succeeded.

Heptanervate hubbardia is of great taxonomical interest as it is the only member of the independent tribe of cereals Hubbardieae, distinguished by an unusual structure of the spikelets. It is interesting for morphologists and ecologists for the specific conditions of growth and structure of tissues, especially leaves, indicating that such plants can exist only in a moisture-saturated atmosphere. The species is included in the Red Data Book of IUCN.

FAMILY PODOCARPACEAE

The family Podocarpaceae includes about 140 species belonging to 8 genera. The members of the family grow in humid, sometimes marshy territories of the southern hemisphere. All species of this family are trees or shrubs with evergreen, simple leaves, often arranged spirally. Male and female cones usually form on different trees.

The genus *Falcatifolium* includes 4 species. The name means 'sickle-shaped leaf'. *Falcatifolium angustum* de Laubenf. is the rarest species of this genus and was discovered in 1969 on Kalimanthan Island. It is a shrub or a short tree. It was found at some height along the Sarawak coast (Malaysia). A peculiarity of this plant and other members of the genus is the presence of 2 types of leaves: one is vegetative (photosynthesizing), covering the branches of adult plants in 2 rows, and the other, very small and scaly, is found at the base of branches as well as on short, fruiting shoots. The present condition of the species is unknown. Possibly, the plant is being conserved in one of the national parks of Malaysia.

FAMILY POLYGONACEAE

Goat wheat (*Atraphaxis muschketowii* Krasn.) is a 50–100 cm tall shrub with thick stem and woody branches covered with reddish-brown bark. Leaves are green, 3–6 cm long. Flowers are terminal on annual branches in short, lateral, multiflorate racemes; perianth is pale pink or white with pinkish margin. The plant flowers in May–June, fruits maturing in July. It was described at the end of the 19th century by the Russian florist A.N. Krasnov and named in honor of the Central Asian researcher I.V. Mushketov. Even at that time, the hypothesis was proposed of the relict origin of this species, which maintains mesophilous characters. All the desert species of *Atraphaxis* originated from it as a result of further selection and evolution. The goat wheat grows in the mid-mountain belt in apple, hawthorn, and apricot plantations and among shrubby thickets. A few of its habitats are concentrated in the environs of Alma-Ata, in the Trans-Ilian Alatau (from the Kaskelen ravine to the Talgar ravine). A small part of its area of distribution is included in the territory of the Alma-Ata reserves. The species is included in the Red Data Book of the USSR. A few specimens are also present in the Central Botanical Garden, Academy of Sciences of Kazakhstan.

Rothschild's sorrel (*Rumex rothschildianus* Aarons ex Evenari) is endangered. A few habitats of this plant are concentrated at 8 km north of Tel-Aviv (Israel); isolated plants are found along the entire Mediterranean coast between the Gaza and Haifa mountains. This dioecious annual with a height of 10–45 cm was described in 1941 from a single specimen found at the beginning of the 20th century. It was mentioned on the label of this herbarium specimen that the plant was found in the coastal belt.

Rothschild's sorrel was always included among rare endemic plants. At present, only 2 populations of this plant are known, one of which comprises about 30 specimens on the territory of Udim natural reserve, but even here many annual plants cannot stand competition from perennial fast-growing species of *Retama* and surrender their position to them. Effective control measures against the dominant species have not yet been taken. The condition of the other population and individual plants as well as the entire coastal flora and vegetation of Israel causes apprehension because of construction activity along the coast or the use of the coast for recreation. This species has been included in the Red Data Book of IUCN.

FAMILY PRIMULACEAE

The genus *Dionysia*, distribution of which is mainly restricted to the countries of the Middle East, belongs to the family Primulaceae. These are mostly mountainous, less attractive, pulvinate plants. When they flower, they are covered with numerous flowers of variable color and form checkered carpets over the slopes. The area of distribution of most species is small and confined to mountain, slope, or range.

Aretiodes dionysia (*D. aretiodes* Boiss.) grows only in Iran on the mountain slopes in rock crevices. It is a small semishrub with bright yellow solitary flowers (Plate 22). It is being intensively destroyed by collectors.

Gisser dionysia (*D. hissarica* Lipsky) is distributed only along the southern slopes of the Gissar range within the Sangardak River basin (USSR). This semishrub has fragile stems covered with sparse leaflets with glandular pubescence and mild aroma. Flowers are yellow, in umbels of 2–3; flowering takes place in June and fruiting in July. The specific habitats are moist crevices of suspended rocks, niches, and ledges in the belt of tree-shrub vegetation, which restricts spread of this species.

Involucrate dionysia (*D. involucrata* Zapr.) also grows on the southern slopes of the Gissar range, but in the Varzob River valley (USSR), where it is found on the rocks facing north and northwest at altitudes higher than 1000 m. It differs from *D. hissarica* by imbricate leaves along the stem. Light violet flowers are borne in umbels of 3–5. The population of this species is decreasing as a result of cattle grazing.

Kosinskii's dionysia (*D. kossinskyi* Czerniak.) is distributed in Iran; isolated plants have also been found in east of Kopetdag in Gaudan (USSR), but it has possibly disappeared from this habitat. This plant forms very dense pads; leaves are 2.5 cm long, flowers are solitary, mildly pubescent; corolla is violet; the plant flowers during April–May. Its usual habitats are crevices in suspended rocks in high mountains.

The Gissar, involucrate, and Kosinskii's dionysias are included in the Red Data Book of the USSR.

Marvelous dionysia (*D. mira* Wandelbo) has moved further south than other species. Its area of distribution is in the Arabian peninsula, Oman (mostly in the mountains, on the northern slopes of a range), in the administrative regions Jabal-Akhdar and Jabal-Ashbad (between Muscat and Sur). It grows at an altitude of 1200 m on dry vertical surfaces of limestone rocks. Very often the plant can be found on shady areas with high humidity. This is a delicate semishrub with yellow flowers, in groups of 5–7 in umbellate inflorescences (the only plant with beautiful flowers in this region). Its areas of distribution often coincide with the habitats of a recently discovered Ceratonia species in Oman. Both plants were found in the natural reserve Vadi-Serin. This dionysia is collected intensively for cultivation in parks and in the Alps. In the local dialect, this plant is called 'tar shrub', which indirectly indicates its consumption by the Arabian tar, a unique hilly wild goat. It is being multiplied in the Kew Botanical Garden

(England), where it flowers almost round the year. The species is included in the Red Data Book of IUCN.

FAMILY RAFFLESIACEAE

The members of the endemic genus of Malaysia *Rafflesia*, grow in the tropical rainforests and parasitize on the roots of plants of the family Vitaceae. The plants of this genus do not have chlorophyll, and their vegetative part is reduced, externally resembling fungal hyphae. Similar to the parasitic fungi, the vegetative organs penetrate the tissues of the host plant and feed upon its nutrients. The reproductive organs (flowers) are also initiated inside the host tissue. A tissue outgrowth develops on the roots or basal part of the trunk of the host plant before flowering, from which subsequently large flowers develop. They are very large (up to 1 m or more in diameter), with 5 fleshy bright red or chocolate-brown to red perianth leaves arising from the middle cup-shaped part, surrounded by a thick ring. The flowers emit strong odor of a dead body, which attracts flies and beetles feeding on rotten matter. The insects surround *Rafflesia* in large swarms and help in pollination. Fruits containing minute seeds develop after fertilization. The seeds falling on the host tissue germinate and give rise to new plants. However, the conditions for development of *Rafflesia* are so specific and observations on them so difficult that information on its biology, life cycle, and relationship with the host plant is insufficient even now.

Patma-wort (*R. arnoldii* R. Br.) was discovered in 1818 by European botanists Josef Arnoldi and Thomas Stamford Raffles. The entire family was named after Raffles, and the flower they discovered was given the species name in honor of Arnoldi (Plate 22).

The distribution of this species is restricted to Sumatra island. The most famous habitat is the Batang-Polupukh reserve, where other species of this family are also found, including *R. hasseltii* Suring. It mainly grows at the bottom of humid valleys or on steep but humid slopes at altitudes ranging from 500 to 1000 m and parasitizes on creeping and climbing plants of genus *Tetrasperma*. The latter prefer old, well-lit places and landslides and can also grow on long-fallow cultivated fields. Together with Patma-wort, the rare species *Rhizanthes lowii* (Becc.) Harms and *Amorphophallus titanus* Becc. (giant arum) are found frequently.

The flower of Patma-wort is the largest not only among the members of this family but among all the higher plants of the world. This is a dioecious plant. The method of seed dispersal and its entry into the host body are unknown. It has been suggested that the seeds enter through cracks in the bark. Cracks in bark are often found in plants on the paths where large animals like elephant, tapir, and rhinoceros move.

At present, Patma-wort is endangered. This is due to the loss of stability of the tropical forests as a result of tree felling and to collection of flowers for medicinal purposes. At some places this plant has completely disappeared or its flowering has not been recorded for several years. It is known that this or other closely related species have been conserved in a few reserves developed for conservation of the entire natural complex (for example, Sugai Jernikh, Batang-Polupukh, Gunung Leser) as well as in special reserves for protecting Patma-wort and its habitat (Bengkulu, Despetakh, and Kawang).

Extensive study of the distribution of Patma-wort, its habitat, morphology, phenology, and interactions with the plants on which it parasitizes will help in developing more effective methods for its conservation not only in the specially developed reserves but also outside. This plant has been included in the Red Data Book of IUCN.

FAMILY ROSACEAE

The burnet genus (*Sanguisorba*) of the family Rosaceae comprises only 15 species, which are distributed in the temperate and cold regions of the northern hemisphere. All of them are herbaceous perennials with small capitate inflorescence.

Salad burnet (*S. magnifica* Schischk. et Kom.) is a plant up to 60 cm tall with thick rhizome, simple, sparsely branched stems, and imparipinnate leaves. Peduncles are axillary, inflorescence is long, dense, bright pink. The area of distribution of this species is restricted to the Partizanskaya valley within the limits of Lozovogo range (Partizansk regions of the Primorye territory), where salad burnet grows in the crevices of limestone rocks and on exposed rocks.

The present situation is threatening for the species: limestone mining and explosions destroy the plant and its habitat. The species is included in the Red Data Book of the USSR.

The false spiraea genus (*Sorbaria*) includes about 10 species of deciduous shrubs with large paniculate inflorescences.

Olga's false spiraea (*S. olgae* Zinserl.) is a small shrub with imparipinnate leaves (having 7–11 leaflets) and a large, multiflorate, dense panicle. It is found along the ravines in the environs of Shahimardan on the northern slope of the Alai range (Kirgizia). It has not been tested under cultivation, but owing to its ornamental properties (particularly at the time of blooming) it deserves attention for ornamental gardening. The species has been included in the Red Data Book of the USSR.

FAMILY SIMARUBACEAE

The tree of heaven genus (*Ailanthus*) includes 15 tree species distributed in southeast Asia and Australia.

Ford's tree of heaven (*A. fordii* Nooteboom) was till recently known only in the southwestern part of Siangang island (Hong Kong), where only 10 plants were found. Later it was also found on the New Territory of Siangang (Hong Kong). The total population of all the 9 habitats consists of 140 trees. This is a tree up to 10 m height with naked straight trunk, terminating in a crown, which externally resembles a palm. Leaves are pinnate, up to 40 cm long, with 6–13 paired leaflets, the plants are dioecious. Inflorescence is a large raceme, up to 40 cm long and 20 cm across at the base. It grows in shady ravines, gorges, or gullies up to an altitude of 400 m. At one of the places, near a road, a few young plants were found. Most probably, seeds of this tree were washed away by rain currents from the upper reaches of the slopes and, finding favorable conditions, they germinated. At present, some of the habitats of Ford's tree of heaven are included in the territory of the Country Park. One other protected area has been organized for conservation of this species. It has also been decided to include this species in the list of endangered plants of Hong Kong. Tree of heaven is under cultivation in the Hong Kong Botanical Garden, but can be raised in plantations and for timber. It is included in the Red Data Book of IUCN as a rare plant.

FAMILY SOLANACEAE

The family Solanaceae includes about 80 genera and 2300 species distributed predominantly in the tropical and subtropical regions. These are herbs, semishrubs, or shrubs with erect or climbing stems. The area of distribution of nightshade (*Atropa*) covers Eurasia. **Komarov's nightshade** (*A. komarovii* Blin. et Schalyt) is a perennial herb with short, lignified rhizome. Stem is up to 1.5 m tall; leaves are large, entire; flowers are solitary, yellow; berry is black, shining, with bluish tinge. It differs from other species in the absence of anthocyanin pigmentation of stem. It flowers in July, seeds maturing in August. Repeated estimations revealed high seed productivity of this plant in nature and poor regeneration, i.e., almost complete absence of young plants. This is due to soil erosion as a result of cattle grazing, felling of trees and shrubs, mass washing away of seeds at the time of spring rains and reduction in the volume of water from the mountain springs, and decline in soil and air humidity.

Komarov's nightshade is an endemic plant of western Kopetdag. Places of its growth are known in several ravines: Khozly, Teze-Taplan, and Kardan Dere on the southern slope of the Syunt-Khasardag ridge. It grows on the bases of ravines under the shade of trees and shrubs on moist soil. It never forms large plantations but is found as isolated plants or in small groups. Its total population strength was small and has further decreased over the last 30 years. The plant is endangered. It is included in the Red Data Book of the USSR and is being conserved on the territory of Syunt-Khasardag reserve.

Komarov's nightshade contains more alkaloids than other species; therefore, it is a promising medicinal plant. The first experiments on its cultivation gave positive results. It is being cultivated at the Kara-Kala Experimental Station of the All-Union Institute of Plant Industry.

The mandrake genus (*Mandragora*) includes 5 or 6 species, distributed in the Mediterranean region, northwest and Central Asia, and in the Himalayas.

Turkmenian mandrake (*M. turcomanice* Mizgir.) is a large perennial, acaulescent plant. Leaves are in rosettes, up to 160 cm long. Flowers are axillary, in groups of 3 in a spiral; corolla is campanulate, violet; berry is orange-yellow. Mandrake has a complex structure of underground organs; the upper part consists of 1–8 caudices, with dormant buds, the thick middle part (up to 25 cm in diameter) is the starch-storing region, its weight reaching up to 5 kg, and a cavity forming inside after 8–10 years. Up to 2 m long branched roots originate from the upper part. The developmental rhythm of Turkmenian mandrake is similar to the Mediterranean species and differs sharply from that of most plants of Turkmenia. It remains dormant during summer, regaining its vegetative growth during autumn, winter, and spring. It flowers in October–November and is pollinated by insects, it is self-pollinated with the beginning of cold weather. The seeds have a long period of dormancy, up to 4.5 months.

The Turkmenian mandrake was first found in 1938 by O.F. Mizgireva at the foot of Syunt mountain (western Kopetdag). This region is one of the warmest places in central Asia; it is characterized by a dry hot summer and a warm winter with fluctuating cold weather. The Turkmenian mandrake grows in the upper part of the semidesert at its border with hilly semidesert. All its habitats are on the southern slopes at an altitude of 600 m. The soils are loam, well drained, with sufficient humus content, due to a large quantity of falling leaves and high moisture content. The plants form clusters (of 10–15 plants) at the borders of dewberry thickets and under sparse shrubs of pomegranate and Christ's thorn. These thickets have a dense cover of fallen leaves and dried grasses which protects the soil heat and moisture very well; the prickly shrubs protect mandrake from being grazed by animals.

The distribution of this endemic plant of western Kopetdag is confined to a few ravines near the southern foothills of the Syunt and Chokhagach mountains: Shevlan, Shepli, Altybai, Eke-Chinar, Daganli, Sarym-

Sakli, and Keriz. All the populations taken together comprise a few hundred plants. Predominant among them are old and reproducing plants; young plants and seedlings form an insignificant proportion which indicates poor reproduction through seeds under natural conditions. Laboratory experiments on seed germination gave 100 percent success. It has been established that under natural conditions, the seedlings appear in October–November, and many die at night temperature of $-10°C$ and up to $15-20°C$ during the day. Rejuvenation through vegetative means has also been recorded in mandrake. Since ancient times, mandrake was a souvenir, because of its funny-looking underground organs, often resembling a human figure. Mature fruits contain vitamin C and alkaloids. The fruits are eaten by porcupines and birds and are also damaged by insects and collected by humans.

The limited distribution of mandrake, noncorrespondence of its developmental rhythm to the climatic conditions, digging of its rhizomes, collection of fruits, and other situations are adversely affecting its already small population. The plant is endangered and is included in the Red Data Book of the USSR.

Work is in progress at the Kara-Kala Experimental Station of the All-Union Institute of Plant Industry on cultivation of the Turkmenian mandrake as a fruit and medicinal plant.

FAMILY TAXODIACEAE

The members of the family Taxodiaceae can be termed 'living fossils'. They were predominant on the earth during the Tertiary period. Now only small patches have remained in North America and East Asia. The family comprises 10 genera and 14 species. Many species are cultivated in different countries of the world, including the USSR.

The Chinese fir genus (*Cunninghamia*) has only 2 species. **Lanceolate Chinese fir** (*C. lanceolata* Lamb.) grows in central and southern China, rarely in North Vietnam (Plate 22).

Konisha Chinese fir (*C. konishii* Hayata) is an endemic of Taiwan. It is a huge tree, up to 50 m tall and 1–2.5 m in diameter; branches are spreading; bark is chocolate-brown, forming cracks on the surface; leaves are linear-lanceolate, hard, leathery, 2 cm long and 2.5 mm wide; cones are ovoid. It grows in the mountain forests at an altitude of 1300–2000 m in northern and central Taiwan together with false cypress. It sometimes forms poor plantations.

Taiwania cryptomerioides Hayata is an endemic of Taiwan. It is a well-built tree up to 60 m tall and 2–3 m in diameter with dark green pyramidal crown. Bark is fibrous; leaves on young trees are linear, 2 cm long, and on adult trees setaceous, triangular, 4–5 mm long; cones are terminal on branches, up to 2.5 cm long. The plant grows in mountainous forests with Formosan false cypress at an altitude of 1800–2600 m.

Parasol fir [*Sciadopitys verticillata* (Thunb.) Sieb. et Zucc.] is an evergreen tree up to 40 m tall with beautiful pyramidal crown. Because of the unique arrangement of its leaves, it is called 'umbrella pine'. A false verticil of 20–30 minute, scaly leaves is located at the end of long shoots, and one branch originates from each of their axils. Each branch consists of a short reduced stem, from which one long (up to 12 cm) 'duplicate' leaf originates, formed by the fusion of 2 leaves. The dark green, lustrous, and leathery leaves of parasol fir encircle the branches like an 'umbel' because of the dense arrangement of dwarf shoots. Cones are 8–12 cm long and about 2.5 cm across (Plate 23).

Parasol fir grows in mountain forests, remote and dense ravines, and on the slopes protected from wind, on the Japanese islands of Honshu, Shikoku, and Kyushu. Because of its ornamental nature, this plant is widely cultivated in its homeland and in many other countries. It was introduced into England from Java in the middle of the 19th century. In the USSR, it is known in cultivation at the Black Sea coast of the Caucasus (Sochi, Sukhumi, Batumi) as well as in Crimea (since 1852). Parasol fir is known as a potted plant, used for creating the famous Japanese bonsai gardens. Large-scale cultivation in Japan saved the plant from extinction.

The genus *Metasequoia* was first described by a Japanese paleobotanist in 1941 on the basis of fossil remains of cones and imprints of leafy branches. Three leafless trees with reddish bark were found in the winter of the same year in China at the border of Hupei and Szechwan provinces, but their taxonomic position could be determined only after 5 years and they were included under the fossil genus *Metasequoia*. The scientific expeditions organized specially in these years established its present area of distribution, conditions of growth, and status.

Glyptostroboid metasequoia (*M. glyptostroboides* Hu et Cheng) is a rare relict species. It is a tree with pyramidal crown, up to 30–35 m tall with reddish chocolate-brown trunk and a diameter of more than 2 m (Plate 23). At present, it has been preserved over an area of 0.8 ha in the mountains of the northeastern Szechwan province and in the neighboring province Hupei at an altitude of 700–1350 m. The total number of trees of this species is unknown, but about 1000 adult trees have been estimated in the valley of Water Fir (so named by the local people), its main habitat (in Hupei).

A widespread cultivation of metasequoia has started since the time of its discovery. At present, this plant can be found in the countries with hot continental and marine climate, but it develops best in humid subtropics.

It can be easily multiplied by seeds and cuttings, grows fast, and is less demanding of particular soil and temperature conditions. In the USSR, it is cultivated in the Nikitskii Botanical Garden.

FAMILY THEACEAE

The family Theaceae includes trees, mostly evergreen plants. About 400 species are distributed in the tropics and subtropics. Two of its genera are most famous: tea (*Thea*) and camellia (*Camellia*). Tea cultivation is known since ancient times. The ornamental camellias (about 60 species) also originated from southeast Asia and started spreading in cultivation slightly later.

Crapnellian camellia (*C. crapnelliana* Tutcher) is an evergreen tree 5–7 m tall with reddish bark and dark green leaves. Flowers are large, white, up to 10 cm in diameter, with numerous yellow stamens; the plant flowers in November–December. It was first described in 1903 from a single specimen found on the southern slope of Mount Parker on Siangan Island (Hong Kong). Later, this tree was cut and for a long time the plant was considered extinct. In 1965, a few plants were found at the same place, and 58 additional adult plants with numerous seedlings were found in 1967 on the mainland near Mau Ping on the New Territory. Camellia grows along the mountain streams in dense forests at an altitude of 200–300 m. Beacuse of restricted distribution and low population as well as high interest of amateur floriculturists in this species, sale and purchase of any plant of this species, except the specimens cultivated in gardens or introduced into Hong Kong, is prohibited by the existing law in Hong Kong. At present, camellia is being intensively cultivated in many countries; it is reproduced by seeds or cuttings. The species is included in the Red Data Book of IUCN.

Granthamian camellia (*C. granthamiana* Sealth) is a tree up to 3 m tall with dark green bark. Leaves are large, 8–11 cm long, flowers are pure white, 14 cm across, with numerous stamens with golden anthers (Plate 23). The plant flowers in October; seeds mature in November. It hybridizes easily with other species of this genus (those hybrids are widely used in floriculture) and also forms intergeneric hybrids with *Franklinia alatamaha* Marshale and *Tutcheria virgata* (Koidz.) Nakai. in which the characters of camellia are predominant.

This plant is cultivated in the Hong Kong Botanical Garden. Experiments on hybridization are also being carried out there, and plantation material is distributed widely from there. Granthamian camellia grafted on Japanese camellia is cultivated in the temperate zone.

The discovery of camellia species, its introduction into cultivation and development, cultivation and distribution of hybrids took place over a relatively short period. The only specimen of this species was discovered in 1955. In 1972, a group of 18 adult trees and a few seedlings were found at a distance of 1 km from the first location, and one more tree with seedlings was discovered at a distance of 2–3 km from the previous place. All the known plants are confined to the catchment area of the Shung Ming reservoir and grow along the riverbanks in the valleys under the cover of forest plantations at an altitude of 600 m. The trees are located on a plot for special scientific study. In accordance with the existing law on rural areas and forests, trade in any part of this plant, except cultivated specimens, is prohibited. The species is included in the Red Data Book of IUCN.

FAMILY THYMELAEACEAE

Baksan spurge laurel (*Daphne baksanica* Pobed.) is a shrub with light brown bark and leafless branches. Leaves are oblanceolate, bunched at the pubescent tips of branches, white-pubescent on both sides; flowers are in heads of 3–5. This spurge laurel is a narrow endemic of the Baksan valley (central Caucasus), known from one location in the environs of Bylym where it has been found in stony ravines at an altitude of 1100–1235 m. It was collected twice in 1896, after which it could not be found and collected. The causes of disappearance of this species from natural habitats are unknown. It is being conserved only under cultivation in the botanical garden in Nal'chik. The species is included in the Red Data Book of the USSR.

Caucasian stelleropsis (*Stelleropsis caucasica* Pobed.) is a semishrub with a height of 8–15 cm. Stems (numbering 7–8) are weakly branched, woody at base, and densely leafy; leaves are elliptical, broad. Inflorescence is almost capitate, with 4–5 flowers, terminal on stems; flowers are yellow. The plant blooms in July. This endemic species of the central Caucasus (northern Cis-Al'brus and western Balkaria) grows in the subalpine belt on limestone rocks. Four habitats are known, at which about 500 plants of this species have been found. The plants have poor surviving ability. Natural regeneration is lacking. The species has been included in the Red Data Book of the USSR as an endangered plant.

Pin-headed daphne (*Wikstroemia capitellata* Hara) is a deciduous shrub, about 2 m tall, with thick branches and numerous large scars of old leaves. Leaves are thin, alternate, densely arranged at the spices, glabrous, whitish on lower side; inflorescence is capitate, 5–10 mm long. This endemic of Japan is known from several places in Hyoga province

(Kyushu Island), where it grows in marshy hill forests. It is found extremely rarely.

FAMILY TROCHODENDRACEAE

Aralioid trochodendron (*Trochodendron aralioides* Sieb. et Zucc.) is the only member of the endemic East Asian family Trochodendraceae. It is an evergreen tree, up to 20 m tall, with gray to chocolate-brown bark, woody branches, and greenish branchlets, at the tips of which leaves are clustered. Leaves are 8–15 cm long, leathery, glossy on the upper surface, aromatic. Flowers are bright green, in terminal racemes, pedicellate, bisexual, without perianth; stamers are numerous, with thin filaments around the carpels fused to half their length. Fruits are oval, 7–10 mm in diameter (Plate 23). Trochodendron flowers during May–June. It grows in the mountains up to an altitude of 1600 m, sometimes in marshy places. The area of distribution of the species covers Japan (Honshu, Shikoku and Kyushu islands), southern Korea, Taiwan and Ryuku Island. This relict plant has retained several primitive characters, including the presence of tracheid alone in the wood, which brings it closer to the ancient magnolias. In spite of a relatively large area of distribution, the plant is found rarely and needs universal protection as an ancient member of the tropical flora. In the USSR, it is being cultivated in Adler.

FAMILY ULMACEAE

Wallich's elm (*Ulmus wallichiana* Planchon) had a fairly large area of distribution initially, including the hilly regions of Afghanistan, India, Nepal, and Pakistan. This species did not form pure plantations, but was one of the main components of the mixed broad-leaved forests together with oak or cedar along mountain rivers and streams at altitudes of 800–2000 m, sometimes rising up to 2300–2800 m. This elm is a tree up to 30 m tall, with spreading crown, large elliptical leaves, and small reddish racemes. At one time it could be seen quite commonly in the settlements in the mountains. At present, it is included in the category of endangered plants. Because of its excellent properties and the beautiful pattern of the wood, this tree was much used for furniture making. The fibrous bark was used to make ropes, and the foliage was used for cattle feeding. However, the main cause of catastrophic reduction in the population of this elm is the mass cutting of branches of the fruiting young and adult trees as fodder for domestic cattle. This almost completely prevented fruiting and resulted in the mortality of the plantation. The studies on the status of Wallich's elm in natural habitats and under cultivation established that a few specimens of this plant are still conserved in the game reserve in Kashmir (India), 3 more trees grow near it on inaccessible rocks, and a small plantation exists at the Forest Research Station in Wageningen (the Netherlands). At present, it is proposed to develop a game and forest reserve in the Himalayas, which should include the typical habitats of the Wallich's elm. It is also proposed to multiply elm by sowing seeds near the houses of the forestry officials and observers of the reserves so as to obtain seeds for the forest plantation being developed at different places in the area of its distribution. Besides seed multiplication, cuttings, layers, and root suckers can be used as plantation material. Wallich's elm is included in the Red Data Book of IUCN.

FAMILY VERBENACEAE

The members of the family Verbenaceae are distributed primarily in the tropics and subtropics. Only a few plants enter the temperate zone. The family includes herbs, shrubs, creepers, and a few trees, comprising about 80 genera and 800 species.

The genus *Caryopteris* includes 10 species, distributed in Central Asia (Himalayas and Mongolia).

Mongolian caryopteris (*C. mongolica* Bge.) is a relict of Paleogenic ancient Mediterranean flora. It is a 20–40 cm tall shrub with slender, virgate branches, grayish due to short white hairs. Leaves on the upper surface are green, on the lower surface almost white. Flowers are dove-blue, arranged in corymbose inflorescences (Plate 24). The plant grows on steppe and desert stony and pebbled slopes, peaks of hills and mountains, on rocks, and on dry gravel. It flowers at the end of July and beginning of August. The area of distribution of this species is in Central Asia. Two isolated habitats of caryopteris have been found in the Buryat ASSR: on the Kharaty mountain near the fall of the Jida river into Selenga and on Chernaya Mountain near the village Ust'-Kyakhta. In Mongolia, the distribution area of this species includes central Khalkh, eastern Mongolia, the Valley of Lakes, Gobi Altai, eastern and Alashan Gobi, Ordosa, and Hesi. In China the plant is found in the Pohuashang mountains west of Beijing (the easternmost locations) and in the northwestern parts of Shanxi province. The population of this species is declining as a result of cattle grazing and land reclamation. Mongolian caryopteris is included in CITES.

3

Africa

Africa is located in 3 floristic kingdoms. Its northernmost part belongs to the ancient Mediterranean subkingdom of the Holarctic kingdom. The African subkingdom of the Paleotropical kingdom includes a large part of the African continent. The Cape kingdom includes the southern tip of Africa from Clanwilliam in the west to the outskirts of Port Elizabeth in the east. The Cape kingdom is the smallest among the floristic kingdoms of the earth, but it is distinguished by exceptionally unique flora, unusual floristic wealth, and a high degree of endemism (7 families and 90 percent of species are endemic).

The flora of some regions in Africa has been studied very poorly; therefore, it is possible that some rare plants are more widely distributed than is known at present. The vegetation here is already significantly changed by human activity. Burning, felling, reclamation of land for cultivation and pastures, amelioration, collection and sale of ornamental plants, and perpetual exploitation of some valuable trees all led to a situation in which many species are already rare or endangered. Unfortunately, we do not have a list of rare plants for the entire continent, although such a list has already been prepared. But almost all the African countries have adopted special laws about conservation of individual species and several reserves have been created for their protection. For example, Ivory Coast has already prohibited collection of fruits and seeds as well as digging and damage to 26 plant species of commercial value; in Central Africa, all species of Encephalartos are protected since 1953; in Swaziland, ornamental plants, succulents, sago palms, and some ferns have been placed under protection since 1952, as have the entire *Gladiolus* genus in Lesotho and all species of Encephalartos in Uganda. Information is presented here on only some plants protected in Africa. For want of material, in several cases it was not possible to determine whether a plant is rare for the entire continent or only for the individual region, and the characteristics for some confirmed rare species are not available.

FAMILY ACANTHACEAE

The family Acanthaceae includes 250 genera and about 2600 species in Asia, Africa, and Central America.

African asystasia [*Asystasia africana* (S. Moore) C.B. Clarke (*Filitia africana*)] is a small tree or green prickly shrub, reaching up to a height of 2 m. Its inflorescence forms at the apex (terminal or axillary), is 8 cm long and finely pubescent. Flowers are alternate; corolla is white, up to 20 mm long, bilabiate. The plant is found in Angola and Gabon at an altitude of 1000 m, very rarely in Cameroon. The species needs to be protected in Cameroon.

FAMILY ANACARDIACEAE

The family Anacardiaceae mainly consists of trees and shrubs, rarely semishrubs, sometimes woody creepers. The family comprises about 80 genera and 600 species, distributed in the tropical countries, but partly entering the temperate regions of southern Europe, Asia, and America. The genus *Lannea* is characterized by trees or shrubs with resinous barks and simple leaves. The spiny lannea (*L. spinosa*) needs to be protected in Nigeria. Its population has significantly declined in recent years.

Small trees or shrubs with compound leaves and less conspicuous flowers in the inflorescences are characteristic features of the sumach genus (*Rhus*). Some sumaches are valuable as industrial and medicinal plants; almost all of them are ornamental.

Lancea sumach (*R. lancea* L. f.) is an evergreen shrub or tree with a height up to 9 m and dark chocolate-brown bark and reddish branches. Leaflets of compound leaves are narrow, with dark upper surface and pale green lower surface. Flowers are small, yellowish-green, borne in an elegant inflorescence. Fruit is roundish, with large seed and a thin layer of flesh (it is used to make beer; the berries are eaten by birds). Its timber is heavy, reddish chocolate-brown, and gives a fine finish and is used for various purposes. The plant is found in arid regions of South Africa along riverbanks and in low-lying areas.

Caffr's sclerocarya (*Sclerocarya caffra* Sond.) is a tree 9–18 m tall and up to 1 m in diameter with a light-colored bark. Leaves are 15–30 cm long, consisting of 3–6 or more pairs of opposite leaflets. Male and female flowers form on different trees; male flowers are arranged in terminal spikes or in leaf axils, female flowers are solitary or in small groups at the tips of branches. Wood is pinkish-white or light red. Fruits are edible. The plant is cultivated from seeds. This sclerocarya grows in outgrowths of shrubs and in forests of Natal, Zululand, Swaziland, partly in Transvaal, Botswana, Mozambique, and tropical Africa. In spite of fairly large distribution, the population has significantly declined because of intensive felling. At present, it is protected in southern Africa.

FAMILY APOCYNACEAE

Impala lily [*Adenium obesum* (Forsk) Roem. et Schutt. var. *multiflorum* (Klotrsch) Codd (*A. multiflorum* Klotrsch)], grows in tropical Africa in the eastern and northern Transvaal, northern Zululand, Kenya, and Swaziland. It is a shrub about 1.2 m tall. Leaves are succulent, light green, fleshy, arranged in rosettes at the tips of branches. Flowers are large, white with pinkish or red fimbria, appearing during winter, often on leafless branches (Plate 24). Fruits are paired, each part pisiform. The juice from the fruit is used as arrow poison. The plant is eaten by wild animals. The impala lily has tree-like shape only in the protected territories of Zambia.

The African and Madagascar genus *Pachypodium* is interesting due to its unusual forms and adaptation to habitats. The plants of this genus grow in the arid regions of South Africa (4 species) and Namibia (2 species). All of them are succulent. Members of the genus have been included in Appendix 2 of CITES and are endangered.

Elephant's trunks [*P. namaquanum* (Wyley ex Harv.) Welw.] is a succulent tree 1.5–2 m tall with cylindrical, spiny, usually unbranched trunk (Plate 24). Grayish-green velvety leaves are crowded at the apices, which drop early. Crown is always inclined towards the north (i.e., towards the sun, as this is a plant of the southern hemisphere). It has a very unusual shape. In early September, its crown is full with tubular reddish chocolate-brown flowers with jasmine fragrance. The plant is found on dry stony hills near the Orange River Namakva land and Namibia. It is a protected plant and a fine is charged on damaging it.

Bottle tree [*P. lealii* Welw. (*P. giganteum* Engl.)] is a shrub or tree up to 7.6 m tall with bottle-shaped trunk, broader at the base and narrowed toward apex. A few lateral branches are present around the main trunk, which in turn are branched near the tip. Bark is grayish-green or light chocolate-brown, often with purple ribs. Leaves are sessile, narrow oblong, velvety, borne at the apices of branches. Spines are lilac-colored, 1.5–3 cm long, usually in pairs. Flowers are petunia-like, in racemes at the tips of branches. The plant grows in dry stony hills in the northern parts of Namibia.

The **two-spine pachypodium** [*P. bispinosum* (L. f.) A. DC.] and **succulent pachypodium** [*P. succulentum* (L. f.) A. DC.] are short fleshy shrubs with a height up to 30 cm and funny-looking, swollen trunks bearing narrow, pubescent leaves, arranged spirally. These species grow in the Cape province and Orange Free State province of the South African Republic.

Saunder's pachypodium (*P. saundersii* N.E. Br.) is found in central Africa on rocks and in forests. It is a shrub or a very small tree with quite a thick base, often almost underground (only tiny branches are visible).

Rutenberg's pachypodium (*P. rutenbergianum* Vatke) grows in Namibia. It is a short, unarmed tree.

Caffr's serpent-wood (*Rauwolfia caffra* Sond.) is a tree with a height up to 21 m and trunk diameter up to 1.5 m. Bark is cream-colored to gray or dark chocolate-brown, usually cracking into small segments. Leaves are exquisite, similar to those of oleander, in whorls crowded near the tips of branches. Flowers are small, white, in dense racemes. Fruits are small, roundish, winged, edible for monkeys and birds, poisonous to human beings and dogs. Fruit juice is used for medical purposes. It can be easily raised from seeds. It is an ornamental plant, found in the forests and shoals in tropical Africa and South Africa.

FAMILY ASCLEPIADACEAE

Asclepiadaceae is a typical family of Africa, the center of its distribution, predominantly comprising woody climbers. It is distinguished by the diversity of forms.

Three of its species are endangered in Africa, all of them included in the Red Data Book of IUCN.

Thick whitesloanea [*Whitesloanea crassa* (N.E. Br.)] is possibly, already extinct. It is a succulent with 4–5 cm long and 5–5.5 cm thick stem. Stem is 4-angled with sharp incised-dentate ridges, convergent at the apex. Leaves are absent. Inflorescence develops at the stem base at soil level on fleshy, 17 mm long peduncles, successively producing a few flowers. Corolla is 33 mm in diameter; petals are lanceolate, whitish-green with purple spots on the upper side and light yellowish with dark red spots on the lower side. The plant grows in Somalia; it was first discovered in 1914 near Odweina; its search from 1942 to 1957, however, did not yield any results. In 1957, the plant was found at another place, towards the north, in the Golis mountain (between Burao and Berbera). The plant is found in stony, very dry semideserts, growing among stones and, in spite of similar conditions being available all around, only in a small area. According to several authors, such a limitation is due to increase in cattle grazing (the plant is very sensitive to grazing) and increase in aridity during the last 10 years.

The whitesloanea has a unique type of corolla which determines its independent generic status. Found only in Somalia, it includes 4 species, all of them rare and endangered. Conservation measures have not yet been taken; it is necessary to search for new habitats in nature and cultivate these plants.

Two species of the genus *Caralluma* are endangered.

Distinct caralluma (*C. distincta* E.A. Bruce) is a herbaceous succulent with noticeably segmented erect stems, pale green or mottled with dark green chestnut spots. Each stem segment has long, opposite pairs of scales which are reduced leaves. One or two flowers arise from the axil of scaly leaves near the stem apex; corolla is campanulate. The area of distribution is Kenya and Tanzania. Three habitats are known, with very small populations, often consisting of not more than 10 specimens. It suffers particularly heavily from grazing and can even die as a result of cultivation for agricultural purposes. The plant is cultivated in private collections and can be raised relatively easily, but is prone to diseases. The species is included in the Red Data Book of IUCN.

Tubular caralluma (*C. tubiformis* E.A. Bruce et Bally) is a herbaceous succulent with erect 4-angled stems, up to 15 cm tall and 2 cm thick, with alternate pairs of long, acute scales along the ridges, which are reduced leaves. One or two flowers arise from the axils of scaly leaves near the stem apex. Corolla is campanulate, pale green, with purple spots on the outside and dark purple spots on the inside. The plant grows in Kenya, on the shallow banks of the Evusao Ngoro River (northern mountains of Kenya), the north of Rumuruti, in the desert between Lodwar and Lokitaung, west of Lake Rudolph. It is known from a few distant places (one or two of them were found recently). All the habitats (except one) are located in the dry savanna where the vegetation is highly disturbed by overgrazing. One habitat is in the arid zone, farther north. The populations are very small (often not more than 10 specimens) and very sensitive to grazing, like other succulents. Unlike other succulents, this plant does not have spines like cacti or acrid milky juice like spurge. The species is included in the Red Data Book of IUCN.

FAMILY ASTERACEAE

In Africa, **Swaziland eumorphia** (*Eumorphia swaziensis* Compton) is a protected plant. It is a highly branched shrub up to 60 cm tall. Leaves are opposite or alternate, numerous, up to 2 cm long, semispreading, usually trilobed above the midpoint, but some leaves are smaller and undivided. Capitula are numerous, arranged in irregular, scattered inflorescences. Ligulate florets, numbering 5, are pistillate, up to 3 mm long, white, tridentate, naked; tubular flowers are bisexual, with tubular corolla, and 5-lobed.

The distribution of eumorphia is restricted to Swaziland, where 2 habitats are known. The population in the northeastern Mbaban mountains had about 100 plants in 1977; the other habitat is 9 km away and has a few hundred plants. This plant is found on stony grassy meadows at an altitude of 1400 m. At present there is no significant threat to its existence, but the populations in lowlands have disappeared; therefore, the control and development of methods to introduce it into cultivation are essential. This is a rare ornamental shrub, flowering in March–April; very suitable for ornamental gardening. It has been included in the Red Data Book of IUCN.

Multicorymbose groundsel (*Senecio multicorymbosus* Klatt) is a shrub or tree up to 8 m tall. Leaves are oblanceolate, acute with serrated margins, lower surface of young leaves is densely pubescent, and lower surface of adult plants is glabrous. Capitula are clustered into large corymbs. Flowers are mustard-colored. This groundsel is found in evergreen forests along the ravines and is protected in Cameroon and Zambia.

FAMILY BIGNONIACEAE

A few members of the family Bignoniaceae are found in tropical Africa. **Kunthian stereospermum** (*Stereospermum kunthianum* Cham.), a tree up to 13 m tall with stem diameter of 20 cm and smooth, pale gray bark, is protected in Chad, Chadian Somalia, and Zambia. Leaves are opposite, paired, up to 25 cm long; leaflets are lanceolate or elliptical-oblong. Flowers are pale pinkish with purple spots, appearing before opening of the leaves or simultaneously with the appearance of young leaves.

FAMILY BOMBACACEAE

Typical of tropical countries, particularly tropical America, the family Bombacaceae includes 20 genera and nearly 190 species. Often these are large trees with thick, barrel-shaped trunks. Parenchyms are strongly developed in the thickened trunks and storewater, enabling the plant to overcome severe droughts.

Baobab or **monkey bread** (*Adansonia digitata* L.) is a tree 10–25 m tall with a strong stem, up to 12 m in diameter and a huge crown. Leaves are large, palmately compound, dropping during dry period, new leaves appearing during rainy season. Bark is very hard, smooth. Flowers have a diameter up to 20 cm and appear during rainy season. Fruits are up to 40 cm long, similar to a large cucumber, with juicy, edible, slightly sour flesh, in which a large number of seeds are embedded (Plate 24). Wood is soft, light, porous, without annual rings. Usually large hollows are formed in the tree due to fire because of the burning of pith, but the tree continues to live. The roots spread out to hundreds of meters from the tree. Baobab is a typical tree of the African savanna. It is becoming rare because of its multiple uses, because of which it is recommended to be placed under protection in several countries of southern Africa. The oldest specimens are preserved in the Kruger National Park.

FAMILY BURSERACEAE

The family Burseraceae includes up to 20 genera and about 600 species, which are widespread in the tropical countries, especially in America and northeast Africa. The incense or olibanus tree genus (*Boswellia*) includes more than 20 species. These are small trees or shrubs with shapeless, curved trunk and drooping branches. Their leaves are pinnate, hairy. The **Carter's incense tree** (*B. carteri* Birdw.) is found in all the elevated areas of Somalia, but its population is declining every year due to collection of benzoin resin (Plate 25).

FAMILY CAPPARACEAE*

The members of the family Capparaceae are most frequently found in the arid regions of Africa, although they grow in the tropical, subtropical, and partly temperate countries of other continents also. It comprises 45 genera and about 900 species. They are trees, shrubs, and herbs, usually covered with glandular hairs. Leaves are alternate, simple or palmate. Flowers are irregular [*sic*],** usually tetramerous. Fruit is a pod or capsule, rarely a berry or nut.

All species of the genus *Boscia* have been placed under protection in the Orange Free State province of South Africa. These are trees or shrubs with simple leathery leaves, opposite or whorled. Flowers are arranged in racemes, terminal or axillary, small, usually with 5 sepals and often without petals; stamens are few or many. Fruit is a roundish berry of the sweet cherry size, fleshy or hard, with 1–2 seeds. The species are typical of arid regions.

FAMILY CAESALPINIACEAE

The family Caesalpiniaceae is close to the family Fabaceae.

---

*Same as Capparidaceae—General Editor.
**In this family the flowers are mostly regular—General Editor.

Unijugate eurypetalum (*Eurypetalum unijugum* Harms) is a tree with leaves consisting of a pair of opposite leaflets short-petioled. Leaflets are oblong-lanceolate, slightly bent. Sepals number 4, are unequal in size. One petal is large, sessile, with distinct veins, wrinkled and naked, 6 × 10 mm in size; others are very small. The plant is found only in border areas of Cameroon and Gabon.

Two species of the Paleotropical genus *Gigasiphon* are found in Africa, one in Madagascar, and others in New Guinea and the Philippines.

Microsiphon gigasiphon [*G. macrosiphon* (Harms) Brenan] is a tree up to 20 mm tall with roundish crown. Bark is whitish or pinkish-white. Branches are initially covered with short rusty hairs, bearing papery, broadly ovate or almost cordate leaves, 8–17 cm long and broad. Flowers are like those of magnolia, in short erect racemes. Petals are 9–13 cm long, pure white, only one of the petals has a yellow spot. Gigasiphon is known from only 4 habitats in Kenya and Tanzania. A part of the Rhondo plateau, where the plant is found, is included in the forest reserve; other habitats are not protected. The species is included in the Red Data Book of IUCN.

Kololo [*Gillediodendron glandulosum* (Port.) J. Leonard] is an endemic of Mali. The local population is proud of the fact that the remnants of ancient forests with this unique plant, attracting tourists to the region, have been preserved on their land.

The genus *Microberlinia* is represented by trees with bipinnate leaves and flowers in terminal racemes. Petals number 5; they are equal or the upper petal is larger than others; stamens number 10; upper stamen is free and the remaining 9 are fused at the base. Pod is oblong, woody.

Bisulcate microberlinia (*M. bisulcata* Chevalier) is endemic to Cameroon. It differs from the Gabon species, **Brazzaville microberlinia** (*M. brazzavillensis* A. Chev.), in having 12–18 pairs of leaflets. Both species provide valuable timber and are rare.

The monotypic genus *Paraberlinia* is represented by **bifoliate paraberlinia** (*P. bifoliata* Pellegrin). It is a tree with glabrous, bilobed leaves. Inflorescence is terminal; flowers are small, sessile or on very short pedicels. The species is quite common in dense humid forests of Gabon and is also distributed in Nigeria and south of Gabon. In Cameroon, it is found only in the Mbanga and Kribi regions, where it is extremely restricted, forming groups of several trees. Protection of the species in Cameroon is essential.

Lemon zenkerella (*Zenkerella citrina* Taubert.) is a tree with simple, glabrous, oblong leaves. Racemes of flowers are up to 2 cm long, axillary, each consisting of 10 flowers. This species is distributed from northern Nigeria to Cameroon, in Fernando Po (Luba Bioko) and in northern Gabon.

FAMILY CISTACEAE

The family Cistaceae includes **round-calyx sun rose** (*Helianthemum sphaerocalyx* Gauba et Jahchen.), an olive-green shrub with a height of 1 m and glabrous, slightly fleshy, flat, ovate or ovate-lanceolate leaves. Flowers are fragrant, up to 3 cm in diameter; petals are spreading, broadly obovate, serrated at the apex, yellowish; calyx is swollen at fruiting, spherical, with reddish veins and white hairs. Capsule is ovoid-roundish. Distribution is restricted to the Mediterranean coast, over an area about 150 km long from El Ameria to Ras El Hapman, west of Alexandria (Egypt). This territory is being brought under cultivation, which is disastrous for the sun rose. About 10 habitats are known where a few hundred plants are found. The sun rose grows on limestone rocks among white limestone dunes, at a distance of 1–2 km from the sea. It is included in the Red Data Book of IUCN as endangered species.

FAMILY COMBRETACEAE

The family Combretaceae includes trees or shrubs with entire leaves and flowers in terminal and axillary inflorescence. Flowers are small, regular; fruit is leathery or drupaceous. The family comprises 20 genera and 500 species, distributed in tropics.

Lead tree or **elephant-tusk tree** (*Combretum imberbe* Wawra) reaches a height of 21 m and a diameter of 1 m. Its wood is heavy; a dead tree remains standing for a long time along with its branches. Trunk is pale gray, sometimes almost white, bark characteristically fissured into small squares or rectangles. The main branches, almost white, are called 'elephant tusks', and young branches often terminate in hard spines. Foliage is drooping. Leaves are small, simple, opposite, petiolate, silvery-gray, pale grayish-green, or yellowish-green, covered with silvery, golden, or reddish scales on the lower surface and sometimes on the upper surface. Small yellow or cream-colored flowers are borne in loose cylindrical spikes, in leaf axils or at the tips of branches. Fruit is solid, roundish, up to 1.9 cm in diameter, 4-winged, yellowish-green. The plant is distinguished by slow growth, living more than 1000 years.

This tree grows along rivers, in shrubby belts in Zululand, Swaziland, Transvaal, and southwestern Africa.

The leaves serve as fodder for many animals, its juice is used in food, and its wood, which burns very slowly and gives a lot of heat, is an excellent fuel. The Africans consider the lead tree a holy plant, an ancestor of humans and domestic and wild animals.

FAMILY CUPRESSACEAE

Duprez's cypress tree (*Cupressus dupreziana* A. Camus) is endangered. This coniferous tree grows to a height of 20 m and more with reddish-brown, deeply cracking bark. The branches are densely covered with scaly, minutely dotted, dull green leaves with glaucous tinge, usually appressed to the stems; young leaves are needle-shaped, 2–3 mm long, branching in 2 rows. Male cones are conical, yellowish, oblong, 6 × 3 mm in size; female cones are 18–24 mm long, grayish chocolate-brown, ovoid, with 10–12 scales. Seeds are reddish brown, ovoid, flat, and winged.

The area of distribution of the cypress tree is restricted to about 200 square km in Algeria, in Edehi (or Temrat), on the Algerian Tassili plateau in the central Sahara, in the extreme southwest of the country. It grows on alluvial gravel or sands along the bottoms of wadis, at altitudes of 1000–1800 m. Only 2 seedlings were reported and not a single tree younger than 100 years has been found (only 5 trees have diameter less than 50 cm, and the diameter of the smallest tree is 13 cm). It is difficult to judge the natural living form of the trees, as they are all strongly deformed. The species is very long-lived: possibly the oldest tree is about 2000 years old.

At present only 153 living and many dead trees are known. In the 19th century, the trees were the main source of wood for local needs. The pollen analysis shows that this cypress tree was widespread earlier in the Sahara.

Measures to conserve the species have not been taken as yet. Because of scattered distribution (there are only a few trees over an area of 1 km^2), the conservation of cypress tree is difficult under natural conditions. This is one of the most drought-resistant species of Algeria, with high frost resistance. It is cultivated from seeds in many botanical gardens of Algeria and France. The species is included in the Red Data Book of IUCN.

Abyssinian juniper (*Juniperus procera* Hochst.) is a large shrub or small tree with scaly needles, similar to Greek juniper, which is protected in Europe. Earlier the Abyssinian juniper together with podocarpus formed a belt in the mountains of Sudan, but now only isolated trees have survived on the rocks and along riverbanks. Frequent fires and agricultural use of the land has so much reduced its population that its outgrowths are not capable of regeneration. All the trees are protected in Sudan and in northern Somalia.

Cape cedar tree (*Widdringtonia cedarbergensis* Marsch.) grows up to 18.2 m tall, with trunk diameter 0.9–1.2 m and widespreading branches. Bark is thin, reddish gray, and cracking. Young leaves are needle-shaped, old leaves scaly and flat. Male cones are small; female cones are up to

PLATE 25. 1—Carter's incense tree; 2—Altenstein's bread tree; 3—caffer bread; 4—Dreg's tree fern.

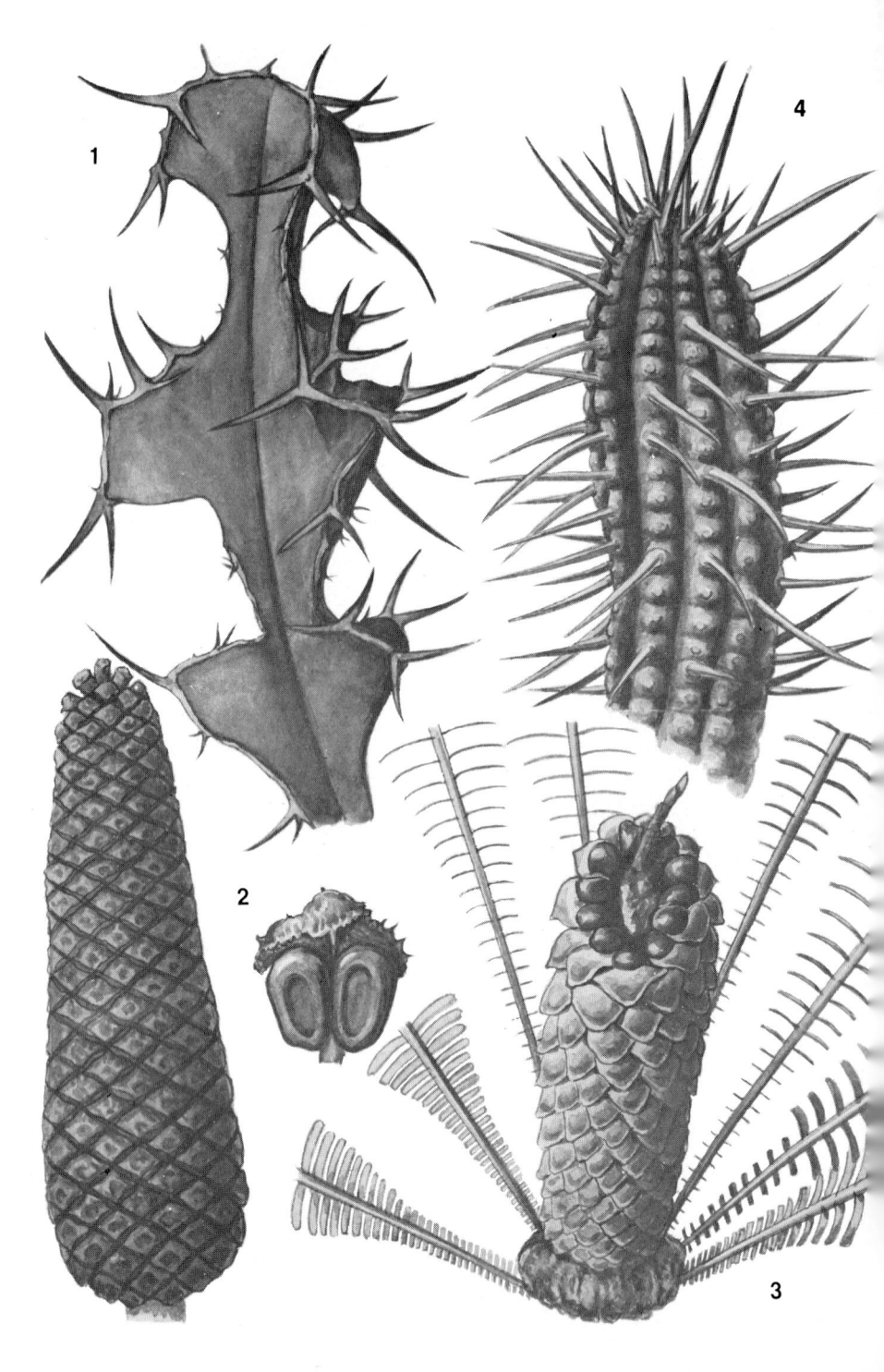

PLATE 26. 1—Large-horned spurge; 2—umbeluzian bread tree; 3—shaggy bread tree; 4—horrid spurge.

3.8 cm in diameter, woody when mature; cones are often in clusters. Its wood is very beautiful, heavy, white or yellowish, giving a good surface and easily carved.

The distribution of this species is restricted to Cedarburg in the Clanwilliam region, about 241 km north of Cape Town. The cape cedar grows at altitudes of 915–1980 m in the mountains, along rocks and cliffs, without forming a dense forest. Now it is difficult to establish the causes for destruction of the cape cedar forest. Possibly, there was a forest fire. A good number of trees have been cut to make furniture, telegraph poles, and carvings. Its resin was collected for medicinal purposes.

At present, the forests of cape cedar are protected.

FAMILY CYATHEACEAE

The family Cyatheaceae includes more than 1000 species, mainly distributed in the tropics; half of them are tree-like ferns.

The tree fern genus (*Cyathea*) comprises about 600 species, distributed in the tropical regions of the Old and New Worlds. Many tree ferns are endemic to the humid mountainous tropical forests (in the lower tier of trees). Their trunks sometimes grow to more than 20 m and have typical structure for the tree-like ferns; hard adventitous roots are developed, which completely cover the stem forming a thick layer particularly in the lower part. Leaves are pinnate to quadripinnate, up to 6 m in length, with long verrucose or spiny petioles; the reduced, sometimes almost filiform basal segments are a unique structure on the leaves (together they appear to be forming an additional open crown at the apex of the stem). The wilted leaves sometimes persist on the stem for a while, bending down and forming a unique 'frock' below the crown of functional leaves. Large scars are visible on the stem after falling of the leaves. The tree fern is characterized by a dense cover of scales, hairs, and spines. Externally these plants are similar to palms. They have all been included in Appendix 2 of CITES, being endangered.

Dreg's tree fern (*C. dregei* Kunze) grows to a height of 5.5 m and has a strong, thick, unbranched trunk and an arcuate crown of very long, elegant, tripinnate leaves, which are dark green on the upper side and lighter beneath (Plate 25). It is found in South Africa (towards the eastern Cape of Good Hope region, Natal, eastern and central Transvaal) up to the tropical zone. It usually grows on the plateau at an altitude of 350 m, in velds and grassy slopes along streams. It is preserved in the Natal National Park, but its protection in Transvaal is also essential.

Cameroon tree fern (*C. camerooniana* Hooker) has a trunk height of 0.4–0.8 m. The leaves are bipinnate, lanceolate, 0.6–2 m long. This tree-like fern is a rarity and requires special measures for conservation in Cameroon. It is also found in Guinea and Gabon. It usually grows along rivers and streams.

Mann's tree fern (*C. manniana* Hooker) has a trunk height of 5–6 m or more. Leaves are tripinnate, 2.5–3 m long; pinnules are 40–50 cm long and 10–15 cm broad; petioles are up to 2 m long. The fern grows in tropical Africa, in humid lowlands. It is still common at some places, but has already disappeared from many areas.

Cape tree fern (*C. capensis*) is a fairly tall plant; its stem bends under the weight of large, elegant, very soft leaves which cannot stand dry air, reaching more than 3.5 m in length. It is found in constantly humid, dense, low-lying, and shady forests of Swaziland (quite rarely), as well as in the forests of the coastal belt of the Cape of Good Hope (in the west up to the Cape peninsula).

FAMILY CYCADACEAE

Woolly Hottentot's head [*Stangeria eriopus* (Kunze) Nash.] has a carrot-like main root, which is gradually converted into underground rapid stem (caudex), reaching 10 cm in diameter. Large, fern-like leaves, rarely more than 1–2, develop above the soil; they are pinnate, with long and broad (up to 6 cm) opposite or almost opposite segments, 2 m long together with petiole. This species is endemic to the coastal zone of southeastern Africa (Cape and Natal provinces, South Africa), where it is found in the velds as well as in the shade of forest shrubs. This unique plant of Africa is endangered; as it is being replaced by sugarcane plantation at the southeastern coast and by pineapple plantations in the east of the Cape province.

Most interesting among cycads is the bread tree genus (*Encephalartos*) because of its ancient origin. It includes about 40 species. These are mostly short, palm-like plants with a height of stem 1–4 m, rarely up to 8–15 m; acaulescent species are also found. Trunk is simple or branched from the base. Leaves are rigid with spiny, sharply pointed leaflets usually dentate on one or both margins. Its local name is bread tree: earlier the trunks were split to collect hardened pieces of resin, which were eaten. The seeds resemble dates, with a hard center, a layer of juicy flesh, and hard, often bright-colored coat. The stone of all species is poisonous, but the flesh is edible. In the past, many cycads were exported from Africa or transferred to botanical gardens, parks, and private collections. Now all the bread tree species found in Africa are endangered.

Caffer bread [*E. caffer* (Thunb.) Lehm.] is distinguished by very slow growth; it can live up to 500 years. Its distribution area is small, spreading along the Indian Ocean coast in the Cape province of the South African Republic from Port Elizabeth in the south almost to Durban in the north. Caffer bread has been used since ancient times as a food plant, but at present its reserves have ominously declined because its habitats have been occupied by maize crops (Plate 25).

Altenstein's bread tree (*E. altensteinii* Lehm.) is a plant usually with a height of 2–4 m, rarely 7 m; full-grown specimens are found among low, almost acaulescent bushes. During pollination, the yellowish female and male strobili emit a unique, strong smell which attracts a mass of insects, especially beetles; the megastrobili of Altenstein's bread tree have a length of 40–50 cm, diameter up to 30 cm and weight up to 40 kg (Plate 25). The plant grows in southeastern Africa and is widely distributed in botanical gardens.

Natal bread tree (*E. natalensis* R.A. Dyer et Verdoorn), up to 5 m tall, is found in northern Natal on slopes with sandy soil, in plantations. Its trunk, especially in the upper part, has a dense crown of 2 m long leaves and is covered with remnants of dead leaves occupied by numerous epiphytes (ferns, orchids, succulent stone crops).

Transverse-veined bread tree (*E. transvenosus* Stapf et Burtt Davy) is the tallest species of the genus: the height of its trunks reaches 13 m, and diameter more than 0.5 m. Sometimes the trunk branches in the upper part on being damaged, forming 2 crowns. The strobili, particularly female, are very large, up to 80 cm long and weighing 50 kg. In contrast to other species it is spread more toward the coast, being distributed in the shady mountain forests of Transvaal (toward northeast Pretoria).

Hildebrand's bread tree (*E. hildebrandtii* A. Br. et Bouche) grows up to a height of 6 m, and at some places up to 8 m. The trunks have a crown of large, up to 6 m long leaves. Two habitats of this species are known in Uganda northwest of Lake Victoria, on both sides of the equator. The main part of the distribution area is the coastal belt in Tanzania and Kenya, south of equator. Here it is common in evergreen dry and mesophytic forests and coastal shrubby growth. In the forests, it is included in the lower tier.

Umbeluzian bread tree (*E. umbeluziensis* R.A. Dyer) is much smaller than Hildebrand's bread tree. It does not have a trunk above the soil surface: because of the contractile action of its roots the trunk sinks into the soil during growth, so that only the crown of leaves is visible above the soil, and even the old leaf bases are underground (Plate 26). The tree grows in dense forests along the valleys of tributaries of the Mgulizi River, from where it spreads into Mozambique through mountains.

The **shaggy bread tree** (*E. villosus* Lehm.) is distributed from the eastern part of the Cape province and Natal to Swaziland. This is also an acaulescent plant and differs from Umbeluzian bread tree by longer and fleshy leaves and much longer and thinner male cones (Plate 26).

Paucidentate bread tree (*E. paucidentatus* Stapf et Burtt Davy) is found in the mountains of Swaziland, as well as in the neighbouring regions of Barberton (Transvaal). This is a very beautiful plant with leaves up to 3 m long and large female cones.

Smooth-leaved bread tree (*E. laevifolius* Stapf et Burtt Davy) was recently found in the steppe rocky slopes in the mountains of Swaziland, and was earlier known from a single habitat in Transvaal. The trunks are a few meters tall and grow in clusters. The leaves differ significantly from other plants of the same genus: they consist of numerous sharply acute leaflets.

FAMILY CYPERACEAE

The family Cyperaceae includes about 95 genera and up to 3800 species, which are widely distributed all over the world, especially in temperate and cold regions. One nutgrass (*Cyperus*) species is endangered. **Papyrus galingale** (*C. papirus* L. ssp. *hadidii* Chrtek et Slavikova) is a giant perennial grass with short woody rhizomes and triquetrous to cylindrical stems, 2-5 m tall, with a crown of slender arcuate branches at the apex; spikelets are short and ellipsoidal. This subspecies is found only in Egypt (in the Vadi En Natrun depression, in the western desert, west of the southern part of the Nile delta at a small area of freshwater marshes around 3 salty lakes, in a few small patches around the freshwater springs).

In Egypt, it was known even during the time of the pharaohs. The habitat of papyrus galingale in the Vadi En Natrun has possibly been preserved since ancient times when a system of drainage was developed in the delta. The lakes are now turning saline and are drying out. Lotos, another Egyptian species on the verge of extinction, found only at one point in Upper Egypt, is also found here. Its habitat must be protected. Papyrus and lotos are the most famous plants of ancient Egypt. Papyrus galingale was a symbol of the king's house of Lower Egypt, lotos was the symbol of Upper Egypt. In ancient Egypt, papyrus galingale was widely used as a food and medicinal plant as well as for preparing mats, shoes, boxes, burial garlands or bouquets and, most important, material for writing. The word 'paper' comes from the word 'papyrus'. It is cultivated together with other subspecies in the Papyrus Institute in Giza (Egypt) and is included in the Red Data Book of IUCN.

FAMILY DAVALLIACEAE

The family Davalliaceae includes epiphytic tropical ferns, mostly distributed in the Old World. The family includes 12 genera and about 230 species.

The oleander genus (*Oleandra*) comprises 40 species, 3 of which are distributed in Africa. These are typical epiphytes with long rhizomes. Their leaves are borne on the rhizomes in a cluster, but the rhizome continues to grow and new clusters of leaves are formed at some distance. The leaves are simple, entire, smooth and glossy, very rigid. **Annet's oleander** (*O. anneti* Tardieu) is protected in Cameroon.

FAMILY DICHAPETALACEAE

The **African tapura** (*Tapura africana* Oliv.) from this pantropical, predominantly African family of 4 genera and 150 species is protected in Cameroon. This is a small tree or shrub with simple alternate leaves and flowers in cymose inflorescence.

FAMILY DROSERACEAE

The family Droseraceae includes 4 genera and about 100 species of herbaceous plants distinguished by great variability. The members of the family are very widely distributed. Their characteristic property is the ability to trap insects and feed on their juices. For this purpose the plants have leaves with glandular hairs, setaceous processes, and palps with glandular heads which are sensitive to stimulation. Under such stimulation, the leaves move. The insects captured are digested by the enzymes released by the leaves, which are similar to pepsin. The largest genus, sundew (*Drosera*), is typical of the temperate and tropical regions of both hemispheres. It includes perennial, usually marshy herbs. The sundews are fairly widespread in southern Africa. The **royal sundew** (*D. regia* Stephens) deserves protection. Its height reaches more than 0.5 m; the plant is capable of digesting even snails and frogs.

FAMILY EBENACEAE

Cape ebony (*Euclea pseudebenes* E. Mey ex A. DC.) is a 4.5–9 m tall tree with drooping branches and a single trunk, 30 cm in diameter. Bark is reddish or dark gray. Leaves are slightly leathery, slightly bent, narrowed toward the end. Flowers are small, male and female flowers borne on different trees; male flowers are in small clusters, female flowers are solitary. The fruit is roundish, pisiform, black, with a single, edible seed. The valuable timber of this tree has been known since ancient times. The species is ornamental. It grows along the banks of dry rivers in the Namib desert and provides a good shade. Its population is declining due to excessive exploitation.

FAMILY ERICACEAE

The heath genus (*Erica*) consists of 500 species, which are typical for southern Africa, and only a few are found in the Mediterranean region and in Europe.

Chrysocodon heath and jasmine-flowered heath are endangered in southern Africa. **Chrysocodon heath** (*E. chrysocodon* Guthrie et Bolus) is an erect shrub up to 45 cm tall with slender branches. The leaves are borne in whorls of 4, linear, incurved along the margin and fimbriate due to the presence of hairs. One or two terminal flowers are at the tips of numerous short lateral branches, which are clustered at a distance of 8–15 cm from the tip; corolla is golden-yellow, funnel-shaped, with 4 short, spreading or sometimes recurved lobes. The only habitat of the species is in the east of Cape Town, in Franschhoek, in an area of about 1 ha, in a marsh at an altitude of 300 m; it grows on acidic sandy soils. The population is very small and can easily perish due to fires (only a few adult specimens have survived after the last fire). Collection of flowers and introduction of other plants are also harmful for the species and has been prohibited since 1974. It is unknown in cultivation.

Jasmine-flowered heath (*E. jasminiflora* Salisb.) is an erect, sparsely branched shrub up to 60 cm tall with slender and densely leafy branches. Leaves are acicular. Flowers are in terminal racemes in groups of 1–3. Corolla is white or pale pink with reddish veins, broadly stellate at the apex (the sticky tube of corolla scares away the bees feeding on the nectar after they penetrate the flower). Flies from families Nemestrinidae and Tabanidae could be the pollinators of this plant as they reach up to the nectar using the nonsticky lobes as passage. The only habitat of this heath species is near Caledon in the northwest of the Cape province. The area is 1 ha, which includes about 150 specimens. The area is surrounded by fields, burnt vegetation, and roads, which may serve as a cause for isolation of heath from pollinators. It is proposed to create a buffer belt for the population. The jasmine-flowered heath is cultivated in McGregor and Caledon. Both heath species are included in the Red Data Book of IUCN.

FAMILY EUPHORBIACEAE

The spurge genus (*Euphorbia*) is the largest in the family Euphorbiaceae (about 2000 species) and is well represented in Africa. Particularly interesting among the plants of this genus are the arboreal succulents, imparting a unique appearance to the South African landscape. All succulent spurges are included in Appendix 2 of CITES, as they are endangered.

Cameroon spurge (*E. cameronii* N.E. Brown) is endangered. This succulent, spineless shrub, up to 3 m tall and 3.5 m across, branches from the base and bears a dense, conical crown. Branches are cylindrical, 1.5–3 cm thick, with spirally arranged leaf scars. Leaves are terminal, fleshy, obovate. Flowers are small, yellowish-cream, borne at the apices of branches. This spurge is endemic to Somalia, known from 4–5 locations in the region of the Golis mountains (however, it may have died during the drought of 1970). The species is disappearing due to overgrazing and as a result of changes in its environmental conditions. As a succulent shrub with juicy and fleshy branches, it is relished by domestic cattle, mainly camels, but also sheep and goats. It is used as a source of water in the arid regions during periods of drought. It grows mainly on stony hills, but one habitat is known from the sandy alluvial plain.

Wakefield's spurge (*E. wakefieldii* N.E. Brown) is also endangered. This tree, resembling a palm in external appearance, has a slender and erect trunk with a height of 7–15 m. Its small crown consists of slender but nonsucculent tri-quadrangular branches; ridges (ribs) dentate, each tooth being surrounded by a pair of thin, needle-like hooks up to 1 cm long. Leaves are small, found only on young plants. Flowers are small, arranged in terminal clusters. Scattered habitats are known in the coastal forest near and north of Mombasa (Coast province). At present, it has completely disappeared from the main locations as a result of destruction of the coastal forests. It is still known in a few habitats further north, but is also disappearing gradually from them. The population is the only confirmed center that had more than 100 trees.

This spurge grows on exposed Jurassic limestone deposits, in dense outgrowths of shrubs and forests. Its vegetative reproduction is interesting: clusters of roots form on the lower branches, which give rise to a dense cluster of branches if they come in contact with the soil. It is cultivated in the Kew Botanical Garden (England) and a few botanical gardens in Kenya. Both spurge species have been included in the Red Data Book of IUCN.

Waterberg spurge (*E. waterbergensis* R.A. Dyer) is known only from isolated locations (a few small, possibly relict populations) in Watterberg in northwestern Transvaal. It is a fairly densely branched shrub from the base, up to 1.5 m tall; branches are usually simple, 4–6 angled, with spines up to 5 mm long.

Proball's spurge (*E. proballyana* Leach.) is also quite a rare species, restricted to relatively small regions in the ravine of the great Ruaha River (Tanzania). This highly branched shrub, less than 1 m tall, bears up to 8–10 mm long quadrangular branches with conspicuous, up to 1 mm long hooks (spines).

Curved spurge (*E. curvirama* Dyer) is a prickly, arboreal succulent, 4–6 m tall, with simple or branched trunk. Branches are dark green, quadrangular, distinctly segmented with paired spines along the edges, bent, forming a roundish crown. The plant is found in the Peddi and Fort Bofort region of the Cape province.

Milk-bush (*E. tirucallii* L.) is a spineless, arboreal succulent up to 9 m tall with slender, terete branches which may be opposite, alternate, or clustered. The fruits are eaten by monkeys and other animals; the juice has commercial value as a source of rubber. It is used for medicinal purposes, as an insecticide, and for obtaining fish poison. It is distributed from the east of the Cape province to central Africa and is cultivated in cities. The Cape province of South Africa has a great variety of spurges. The most rare species are **large-dentate spurge** (*E. grandidens* Haw.), triangular spurge (*E. triangularis* Desf.), **mamillate spurge** (*E. mammillaris* L.) (Plate 27), **large-horned spurge** (*E. grandicornis* Goebel.) (Plate 26), **round spurge** (*E. obesa* Hook. f.) (Plate 27), **poisonous spurge** (*E. virosa* Willd.), **angular spurge** (*E. angularis* Klotz), **horrid spurge** (*E. horrhida* Boiss.) (Plate 26), and **melon spurge** (*E. meloformis*) (Plate 27).

FAMILY FABACEAE

Protection of a fairly large number of plants of the family Fabaceae is essential in Africa.

Cordeauxia edulis Hemsley is a shrub up to 2 m tall, with numerous ascending, highly branched shoots. Leaves are pinnate, usually consisting of 4 pairs of leathery, ovate leaves. Flowers are bright yellow, about 25 mm in cross-section, in terminal racemes. Pod is flat, ovoid, and curved like a horn. The seeds when eaten fresh, boiled, and sometimes fried have a pleasant, sweet taste. Cordeauxia is found in Ethiopia and Somalia only at 2 places in the arid region. It grows in open shrubby savanna on very poor red sandy soils.

The population of this species has declined sharply mainly because of change in the vegetational cover as a result of overgrazing by cattle. Earlier the shrub had great importance as a source of food; however, because of large-scale collections of its fruit as well as destruction of seedlings, young plants are altogether absent now, and this means that generation is delayed by many years. Wild populations are not protected. It is cultivated in Somalia and Kenya as an edible plant. The species is included in the Red Data Book of IUCN.

African braziletto wood (*Peltophorum africanum* Sond.) is a widely spreading tree up to 9 m tall. Trunk is often twisted or branched almost right down to the base. Leaves are alternate, silvery gray, bipinnate. Flower has bright yellow, drooping petals and pubescent calyx. Flowers are located in bunches, in terminal inflorescences or in leaf axils (Plate 28). Wood is reddish, medium heavy, easily polished and processed, and used widely.

This acacia is found in Zululand, Natal, Swaziland, in the northern and eastern parts of Transvaal, central tropical Africa, Botswana, southwestern Africa, and Angola. It grows on sandy soils in dry shrubby outgrowths and open savannas; it is a common plant of the Transvaal bushland. Its population is declining as a result of its territory being covered under agriculture and use of the pods for cattle feeding. This is one of the raintrees of Africa; drops of water appear on the branches in late spring and fall like raindrops on the land under the crown. The reason for this phenomenon is still not explained; possibly, it is caused by insects. It is a good garden tree, its seeds germinate synchronously, and the tree is resistant to cold.

African pararubber (*Burkea africana* Hook.) is a tree 4.5–8 m, sometimes up to 21 m tall, highly branched almost from the base. Crown is flat; branches are rough, bark resembling crocodile skin, dark red. Leaves are drooping, borne in clusters at the apices of branches, bipinnate or tripinnate, young leaves silvery, later turning dark greenish-light blue; flowers are pale yellow (Plate 28). The plant grows in tropical western Africa, usually on sands, in acacia savanna, dry open shrubby velds at an altitude of 600–1370 m. The species is recommended for protection in South Africa. The present status of African pararubber in Nigeria is unknown.

Colored-seeded guibourtia [*Guibourtia coleosperma* (Benth.) J. Léon] is a slender, sometimes deciduous or almost evergreen tree up to 19 m tall with a broad crown of numerous drooping branches. Bark is gray to reddish, chocolate-brown or black. Leaves are in pairs, 10 cm long and 4 cm broad. Wood is heavy, reddish-pink with dark- and light-colored veins and fine texture. This tree grows in the Kalahari desert on sand dunes and is widely used by the local population.

Lister's umtisa (*Umtisa listerana* Sim.) is one of the most rare species in the world. It is a shrub or a small, slender, evergreen, prickly, highly branched tree, up to 8 m tall with supports, especially at the base. The trunks are usually very twisted, wavy. Bark is dark, chocolate-brown or red; branches are often modified into 2.5–10 cm long thorns, serrated.

Leaves are alternate, compound. Flowers are small, white or yellowish-green, stellate, abundant at the tips of branches. Wood is fine-layered, red or purple-black, very heavy and oily; juice is pale yellow. This umtisa has a very restricted distribution in southern Africa; it is known only from a forest-covered ravine near East London and in a few isolated places in the eastern part of the Cape province, in the Kektani region. This is the only species in the genus. The Africans consider it a holy plant.

Rhodesian teak (*Baikiaea plurijuga* Harms.) is an evergreen, spineless tree, 8–18 m in height and 0.6–1 m in diameter; crown is broad, comprising many branches providing shade. Bark is smooth, cream-colored or grayish dove-blue in young plants and reddish-gray or dark chocolate-brown in old trees. Flowers are white, pink, or purple-red with dark chocolate-brown, velvety calyx. Wood has a broad pith, chocolate-brown when fresh and dark red on drying, fine-layered, heavy, heat-resistant, and difficult to carve. The distribution of this species is limited to the Kalahari-Botswana region, north and northeast of southwestern Africa, Zimbabwe, and Zambia. In the Kalahari desert, Rhodesian teak is the main source of wood; as a result, its population is declining. It is protected in the Orange Free State of South Africa.

Cape lonchocarpus or **raintree** (*Lonchocarpus capassa* Rolfe) is a small tree, 4.5–12 m tall. A good part of the trunk is without twigs. Bark is smooth, white or gray, sometimes fissured, and then its inner cream-yellow part is exposed; juice is red. Leaves are compound, grayish-green. Flowers are small, aromatic, similar to the flowers of pea, dove-blue or violet, with velvety calyx, on large, unbranched peduncles at the apex (Plate 28). Wood is yellowish, used by local people for making pots and canoes and for medicinal purposes. The bark and roots are highly poisonous, used as fish poison. The tree is found in Bush velds and lower velds of Zululand and Swaziland, and in the eastern and northern parts of Transvaal. It also grows in the forests of northeastern parts of southwestern Africa, in Botswana, and towards northern tropical Africa. It is a good garden tree, one of the raintrees or weeping trees of Africa. The species is recommended for protection in Transvaal.

Zambian xanthocercis [*Xanthocercis zambesiaca* (Bak.) Dumaz-le-Grand] is an evergreen tree up to 18 m tall with many thick trunks from 0.46 to 2.4 m in diameter (sometimes there is only one trunk). The branches 'weep' at the tips. Leaves are alternate, consisting of 5–12 alternate or opposite leaflets with a larger terminal leaflet. Flowers are small, white, with grayish velvety calyx, in small racemes at the tips of branches (Plate 28). Fruit is unusual for Fabaceae; 2.5 cm long and 1.3 cm across, with smooth chocolate-brown pericarp and a black seed in thin and juicy flesh. Wood is white and heavy. The tree grows in forests in the plains, on deep sands along rivers, and in the hot dry regions between South Pansberg and Limpopo River, in the northern part of the Kruger National Park, northern Botswana, Zimbabwe, and Zambia. It is easily raised from seeds and is not much known in cultivation. The tree is protected in the South Pansberg region of Transvaal.

Angolan barwood (*Pterocarpus angolensis* DC.) is a graceful tree up to 12 m tall (in eastern Africa up to 36 m) with diffuse crown and a single trunk up to 0.6 m in diameter; dark red bark fissures into pieces. Leaves are alternate, drooping, with 4–12 pairs of leaflets. Flowers are pisiform, numerous, orange-yellow. The tree contains blood-red juice, the so-called false dragon blood which was widely used in cosmetics, pharmacology, and dyeing. The seeds germinate with difficulty. The tree is found in southern, southwestern, eastern, and tropical Africa, in forests and shrubby velds, on sands, and in plains.

FAMILY IRIDACEAE

Many ornamental plants of the family Iridaceae have been placed under protection in Africa. The genus *Acidanthera* is represented by corm-producing perennials with 5–7 green linear or narrow sword-shaped leaves and spicate inflorescence with 3–6 flowers. Flowers are aromatic, white, pinkish to light purple. Corm is roundish, milky white, with reticulate, dense, light chocolate-brown scale. **Divine acidanthera** (*A. divina*) is protected in Cameroon. **Golden sword lily** (*Gladiolus aureus* Baker) is endangered. It is a slender, herbaceous plant with 15–85 cm tall stems, arising from an ovoid corm, with a diameter of 2–2.5 cm and flattened at the base. Leaves (numbering 3) are erect, linear, very finely pubescent, up to 45 cm long, covering the stem in the lower part; spike consists of 6 golden-yellow flowers. The sword lily grows in southern Africa where it is known from several habitats in a belt about 5 km long along the western coast of the Cape peninsula (Cape province of South Africa). All the known populations except one are already destroyed as a result of the land being used for agriculture. At the places where in 1975 as many as 50–70 plants were known from an area of 20 × 70 m, only 18 remained in 1977. This is explained by the fact that the sword lily is being replaced by introduced plants; besides, a picnic spot has been developed where a children's playground and a footpath were laid (the sword lily was uprooted at the time of flowering). Gravel mining at a short distance from this place has a fatal effect on the population (drainage of the area is chocked). This extremely ornamental plant is cultivated by amateurs from bulbs. It is included in the Red Data Book of IUCN.

Loubser butterfly iris (*Moraea loubseri* Goldblatt) is endangered and possibly has already disappeared. This is a tender herb, 15–20 cm tall, with corms up to 1 cm in diameter. The solitary leaf is vertical, linear, adjacent to the stem at the base and usually longer than the latter. Flowers are 4–5 cm in diameter, solitary, and terminal; outer perianth lobes are bluish-lilac, glossy, with blackish hairs in the center of the flower; inner lobes are very narrow, also glossy, with a pair of short lateral lobes each. The plant is known only in southern Africa, on the hills near Langeban, in the southwest of the Cape province. It was first discovered on the granite hills in 1973, but now these hills are partly destroyed by mining; in 1977 not a single plant was found here. Possibly, the rhizomes or seeds are preserved in the soil. The butterfly iris grew on fine turfy, fairly heavy and dry sandy soils, on the coastal hills up to a height of 70 m. A small number of seeds are being preserved in the seed bank. It is easily cultivated from seeds and has been conserved under cultivation by amateur florists and traders in southern Africa, California, France, New Zealand, and Portugal. It has been included in the Red Data Book of IUCN.

Two of the 54 species in southern Africa may already be extinct, 3 are endangered, and 3 have a highly reduced population.

FAMILY LAURACEAE

The **black stinkwood** [*Ocotea bullata* Burch.) Baill.] belongs to the family Lauraceae. It is a large evergreen tree, 9–27 m tall, with a trunk of diameter 0.3–1.8 m. The bark is light gray, smooth, often with a pinkish tinge when young, and wrinkled, scaly, dark chocolate-brown when old. Leaves are simple, opposite, dark green, oblong, entire; many leaves have small outgrowths at the base on each side of midrib (hence it is sometimes called 'blistered laurel'), usually infested with small insects. The flowers are small, creamish, in small clusters in the axils of upper leaves. The tree has dark red, heavy golden, chocolate-brown, or black wood with fine texture. A freshly cut tree emits a very unpleasant smell. It grows in the forest of southern Africa from the ravines of Table mountain to northern Transvaal. The wood of black stinkwood is the most valuable among all the South African trees. It has been extensively exploited, and the best trees were already cut by 1812. At present, a few specimens have survived in Transvaal, and it is possibly destroyed in Natal. Its cultivation is difficult, as almost all the flowers are infected with fungi and are sterile.

FAMILY LILIACEAE

The genus *Aloe* is widespread throughout the entire African continent, but the tropical regions are particularly rich. Aloe is a deciduous succulent, often arboreal, with strongly branched trunks at the apex, bearing rosettes of fleshy leaves at the ends of juicy green branches. The trunks are often woody in the lower part and are covered with brownish bark like the large branches. The genus includes about 240 species.

All the species are included in Appendix 2 of CITES. It is proposed to protect them by law in several African countries or individual regions.

Eel-shaped aloe [*A. allooides* (H. Bol.) Van Druten.] is a plant up to 2.1 m tall with long and narrow leaves. The upper half of the leaves is grooved, green, sometimes with reddish veins; the lower half is green, with reddish margins. Old leaves are recurved, touching the stem and almost parallel to it. Inflorescence consists of a few long and slender racemes. Flowers are campanulate, yellow. The plant grows in mountains, in regions of heavy rainfall and on dolomites in eastern Transvaal.

Angelica aloe (*A. angelica* Pole Evans) has restricted distribution: it was collected only from a small region in the northern slopes of South Pansberg in Transvaal and in the northern part of the Kruger National Park, among outgrowths of shrubs. Plants are up to 4 m tall with simple or branched trunk. Leaves are in a dense rosette, narrow, initially slightly spreading, later recurved, green, notched, with reddish chocolate-brown margins and teeth. Inflorescence is paniculate, branched. Flowers are yellow or greenish-yellow. It is a highly ornamental plant.

Woody aloe (*A. arborescens* Mill.) is a beautiful plant with a height not more than 3.3 m. Stems branch from the base, each of them terminating in a rosette of leaves. Leaves are long, fairly fleshy, green or grayish-green, with serrated margins. Inflorescence is usually unbranched. Flowers are light scarlet, tubular, densely covering the peduncles (Plate 29). Flesh of leaves is used for medicine. The distribution of this species is fairly wide; the plant is found in South Africa, Swaziland, Mozambique, Zimbabwe, Malawi. Woody aloe is one of the few aloe species having a significantly high amplitude: it is distributed from sea level to mountain peaks at an altitude of 1829 m. The plant grows in the coastal shrubby outgrowths as well as on naked mountain slopes among stones. It is cultivated widely.

Baines' aloe (*A. bainesii* T. Dyer) is the largest and tallest among tree aloes: it has a height up to 18.2 m with a trunk diameter of 0.6–0.9 m, sometimes up to 1.8 m. It is repeatedly branched dichotomously. Leaves are in dense rosettes, long, narrow, deeply grooved, and recurved. Inflorescence is also branched dichotomously [*sic*], with 3 vertical cylindrical racemes. Flowers are usually pink, with greenish tips, rarely orange. The tree grows in dense shrubby outgrowths and forests along mountain slopes in southern Africa (South Africa, Swaziland). It is often cultivated in parks as an ornamental plant.

Chestnut aloe (*A. castanea* Schonl.) is a tree-like plant up to 4.5 m tall with a thick branched trunk. Leaves are grooved on the upper side, light blue or grayish-green on both sides, with reddish margins and minute reddish teeth. Inflorescence is simple, consisting of thin cylindrical racemes. Flowers are campanulate; chestnut is chocolate-brown. The plant is found in a few regions of eastern Transvaal (South Africa) along forest-covered slopes or in open places.

151 **Dolomite aloe** (*A. dolomitica* Groene-Wald) has a simple stem, 2.1 m tall, usually surrounded with remnants of old leaves. Leaves are thick, broader at the base and narrowed toward apex, twisted, forming a dense rosette, green (reddish in winter), with reddish margins and reddish chocolate-brown teeth. Inflorescence consists of a few cylindrical racemes. Flowers are yellow or creamish-yellow. The plant is adapted to high peaks and slopes in the Dragon mountains and other mountain systems of Transvaal. It is usually found on dolomites.

Fibrous aloe (*A. fibrosa* Lavranos et Newton) is a shrubby plant with branches up to 2.5 m long and 3 cm thick. Leaves are lanceolate, acute, sometimes recurved at the apex, bright green (turning brownish in sun), sometimes with spots. Inflorescence is simple or with 1–2 branches, about 100 cm long, conical. Perianth is orange-red with yellow margins (Plate 29). The plant grows in Kenya on sandy soils and among gneissic rocks in woody savanna.

Coastal aloe (*A. littoralis* Bak.) reaches up to 3 m in height. Trunk is thick, surrounded with old and dead leaves, sometimes appearing acaulescent. The trunk ends with a rosette of long leaves, among which the upper leaves are erect, lower leaves divergent. Leaves are pointed, grayish or light bluish-green, sometimes with reddish veins or white spots, with dentate margins. Flowers are in highly branched racemose inflorescence, red or pinkish-red. The tree grows in velds and scrub velds in South Africa and southwestern Africa.

Alpine (Marlot's) aloe (*A. marlothii* Berger) is one of the most famous aloes with strong, unbranched stem up to 4 m tall, sometimes more than 6 m tall. Lower fleshy leaves are clustered in dense rosettes, grayish or light bluish-green, grooved on the upper surface, with reddish chocolate-brown pointed spines scattered along both surfaces, and with reddish chocolate-brown teeth along margin. Inflorescence differs significantly from that of other species: it consists of simple panicles, originating from a rosette of leaves, long racemes (usually 20–30 in number) often almost horizontal. Flowers are yellow or yellowish-orange, borne only in the upper part of peduncle. The alpine aloe is widespread in Transvaal, Swaziland, Zululand, and northern Natal.

Lesotho aloe (*A. polyphylla* Schonl ex Pillans) is a succulent perennial with roundish rosette of 75–150 usually erect leaves, up to 80 cm in diameter, arranged spirally. Leaves are highly fleshy, oblong-ovate. Peduncles are 50–60 cm long, branched almost from the base, with flowers at the apices of branches. Flowers are pale red or pink, rarely yellow (Plate 29). The plant is endemic to Lesotho (southern Africa), found on the Thaba Putsoa range and Maseru in the Dragon mountains. This rare species has great significance for horticulture; however, its reserves are shrinking due to digging of plants for sale to gardeners. At present, only about 3500 specimens are known in about 50 places, most of which have an area less than 1 ha. The plant disappeared from 12 earlier known places and its population has declined in many others.

The Lesotho aloe grows on basalt slopes, rocks, at the altitude of 2230–2720 m. The species is unique in strictly spiral arrangement of leaves, which is much valued in ornamental gardening. At present, it is protected by law, and its export is prohibited. It is cultivated in several botanical gardens. The species is included in the Red Data Book of IUCN.

Sessile-flowered aloe (*A. sessiliflora* Pole Evans) is a plant with simple, very short trunk (0.9–2.1 m), surrounded with dead leaves. Leaves are in dense rosettes, green or red above, green beneath with reddish margins and small reddish teeth. Inflorescence is unbranched, consisting of 1–5 long cylindrical racemes, yellow. The plant grows in Mozambique, Swaziland, Zululand, and eastern and northern Transvaal on stony slopes of mountains.

Striped aloe (*A. striatula* Haw.) is a perennial succulent herb with a height 0.5–0.7 m and commonly verticillate branching. Leaves are 20–25 cm long and 3–3.5 cm broad, at the tips of branches. Flowers are in racemes up to 40 cm long. The plant is found in the Cape province on dry stony, sandy, and dry clayey slopes in well-illuminated places. The plant is drought resistant. It is cultivated as an ornamental plant in many countries, including the USSR.

Varicolored aloe (*A. variegata* L.) is a very rare species of the Cape province (Plate 27), whose population has reduced as a result of destruction of its habitats. It is widely cultivated. The **Pillans' aloe** (*A. pillansii* Guthrie) and **small-flowered aloe** (*A. parviflora* Baker) are also endangered (Plate 29).

The dragon plant genus (*Dracaena*) includes very ancient plants. The red juice of the members of this genus has been called dragon blood since ancient times. On drying, it was used in the Mediterranean countries as a medicinal and dyeing material, and in India for religious ceremonies. The genus includes about 50 species. (Some authors include this genus under the family Agavaceae.)

Nubian dragon plant (*D. ombet* Kotschy et Peyr.) is on the verge of extinction and is included in the Red Data Book of IUCN. This 3–4 m tall tree has umbellate crown, with strong branches, branching repeatedly after flowering and bearing dense rosettes of thick, sword-shaped,

40–70 cm long leaves at their apices, and broadly ovate at the base. Numerous flowers are aggregated in cylindrical racemes. Flowers have 6 white or pale pink, narrow, oblong-lanceolate perianth leaves. Fruit is a berry, spherical, yellow (Plate 30). The tree is an ancient relict and one of the most marvellous plants of Djibouti, Ethiopia, and Sudan (Erythrea and Red Sea hills); possibly, it grows also along the northern coast of Somalia.

The Nubian dragon tree is found among shrubs on dry mounds along sandstone and quartzite outcrops at the altitude of 750–1200 m together with the arboreal **Abyssinian spurge** (*Euphorbia abyssinica* J.E. Gmelin) and different acacias. The vegetation of these hills is suffering due to overgrazing of cattle and is on its way to extinction. Only a few specimens of dragon plant have survived on naked rocks; possibly, earlier it was subdominant here. The population of this species has reduced not only due to overgrazing and recent drought, but also because of exploitation (juice extraction, cutting for wood, collection of fibrous leaves for weaving). In the past, this species was protected in the region of the Erkowit oasis and on the Red Sea hills (Sudan) but now the vegetation there is changed so much that in 1961 only dead trees were found. Most probably, now the Nubian dragon plant can be preserved only by its cultivation in botanical gardens.

Usambar dragon plant (*D. usambarensis* Engl.) is known from only one point in northeast Tongaland, but it is still widespread in tropical Africa.

Batten's aloe (*Haworthia batteniae* C.L. Scott) has a simple stem up to 10 cm [sic] tall and 13 cm in diameter. Leaves are light bluish-green, about 120; young leaves are straight, with bent tips, old leaves spreading, twisted, with darker green stripes, acute margins with numerous transparent teeth. Perianth is greenish-white in the beginning, later turning pinkish-green.

The plant is found on a very small territory near Agter Skiber in the Cape province in an area of intensive agriculture. Here one population of about 50 plants has been found.

Umbrin torch lily (*Kniphofia umbrina* Codd) is a herbaceous perennial up to 1.8 m tall with thick rhizome. It has 6–8 leaves, sword-shaped, initially straight, subsequently divergent, sometimes drooping, 45–70 cm long. Flowering shoot is longer than the leaves, with dense cylindrical inflorescence. Flowers have a faint smell and are usually drooping. Perianth is reddish to dark chocolate-brown, tubular. This plant is found only in a small area north of Mbabana in Swaziland, where it grows at an altitude of 1400 m in well-drained, fairly humid stony meadows. One population was destroyed in 1977 (cultivated with maize crop), other habitats are still protected, but they are already identified for agricultural use. The total population consists of 3000–4000 plants over an area of 4 ha. It

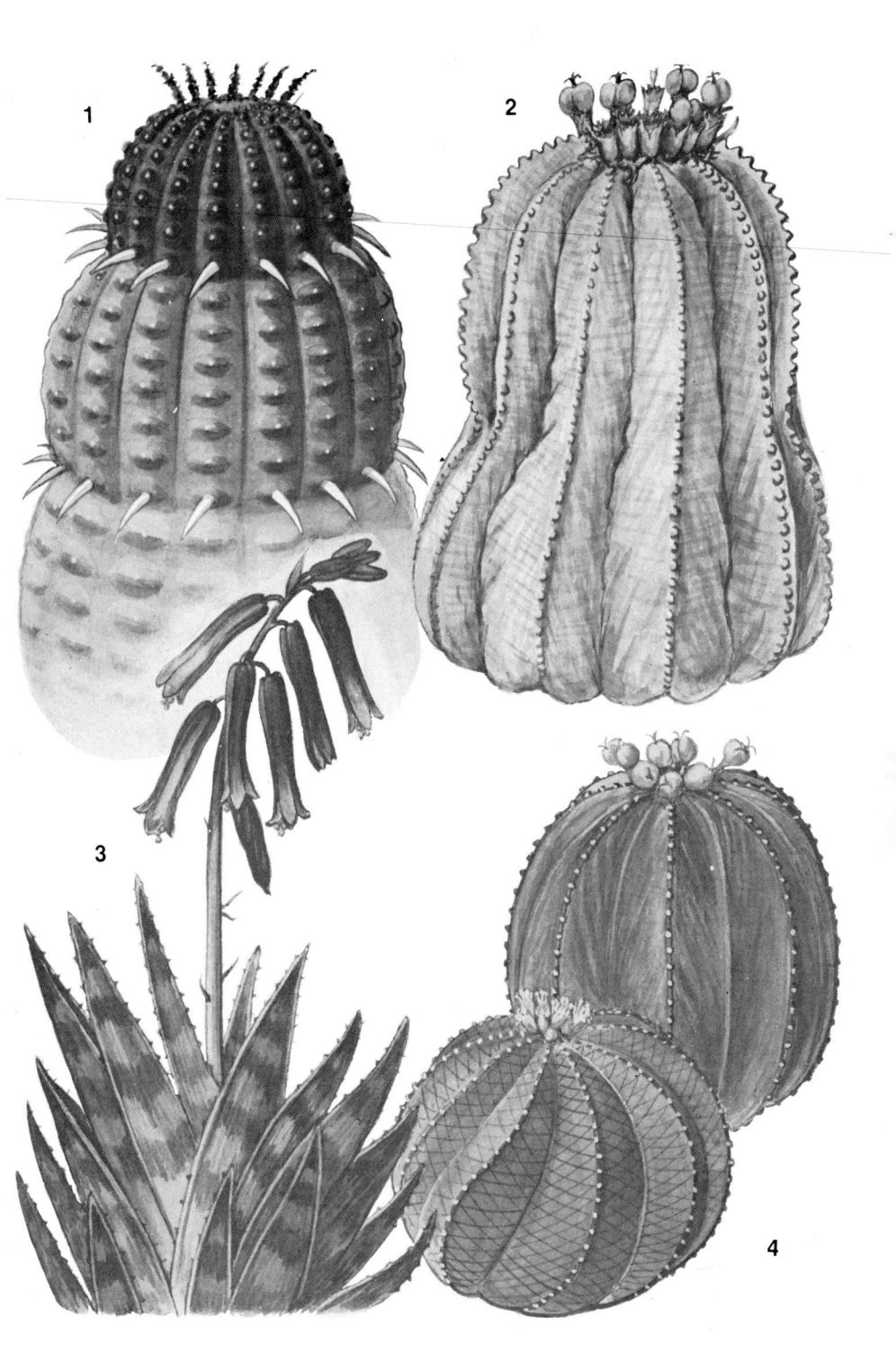

PLATE 27. 1—Mamillate spurge; 2—melon spurge; 3—varicolored aloe; 4—round spurge.

PLATE 28. 1—African pararubber; 2—African braziletto wood; 3—rain tree; 4—Zambian xanthocercis.

PLATE 29. 1—Woody aloe; 2—fibrous aloe; 3—small-flowered aloe; 4—Lesotho aloe.

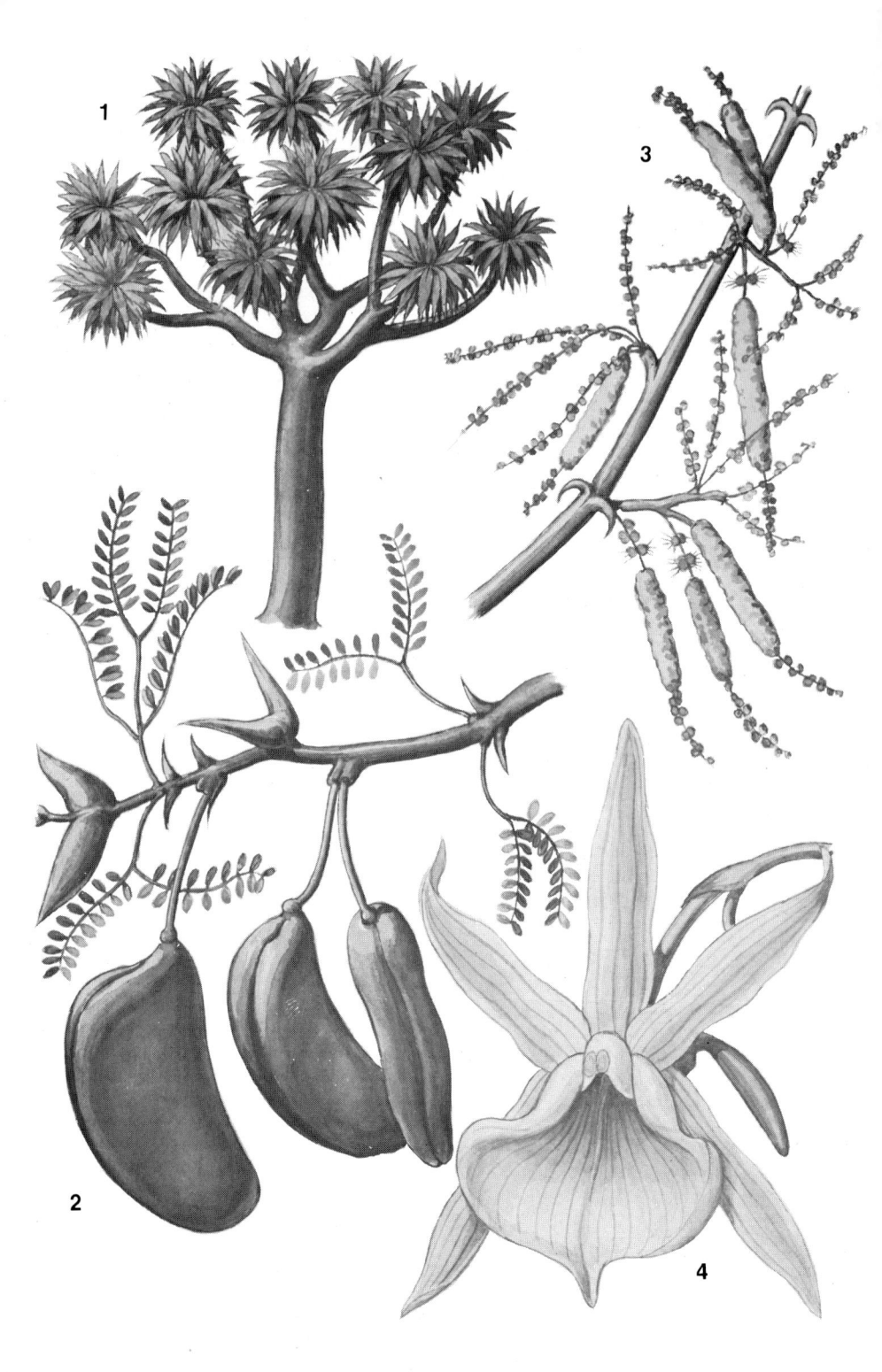

PLATE 30. 1—Nubian dragon plant; 2—camel thorn; 3—Galpini's acacia; 4—two-rowed angrec.

is already proposed to organize a national reserve for conserving this species. The plant is promising for floriculture. As a vulnerable species it has been included in the Red Data Book of IUCN. The **splendid torch lily** (*K. splendida* E.A. Bruce) is protected in Transvaal. It is a rough plant with a very thick spike of numerous yellowish-green flowers and grows on stony hills.

FAMILY MELIACEAE

The family Meliaceae includes 50 genera and about 1400 species distributed in the tropics, partly in the subtropics, and very rarely in the temperate zone.

Forest redwood [*Entandophragma spicatum* (DC.) Sprague] is a very elegant tree with 18 m tall stem and a crown of light green foliage, grayish trunk, with smooth bark cracking in roundish pieces. Leaves are at the ends of branches, with 3–5 pairs of linear oblong to almost roundish leaflets. Flowers are green in long and loose heads in leaf axils. The tree grows in mixed forests and the population is being depleted for its valuable wood.

FAMILY MELIANTHACEAE

The family Melianthaceae includes 2 genera and 35 species distributed in tropical and southern Africa.

Sutherland greyia (*Greyia sutherlandii* Hook. et Harv.) is a shrub or tree up to 11 m tall. Young trees have a compact crown and old trees a spreading crown. Leaves are simple, alternate, leathery, finely glandular, deciduous. Wood is pale pink, light-colored. Flowers are bright red, rich in nectar, and in racemes at the tips of branches. Fruit is a cylindrical capsule. The tree is particularly beautiful in flower. It grows on slopes and stony ranges of the Dragon mountains at an altitude of 1800 m (Orange Free State, Cape, Natal, Swaziland, and eastern Transvaal provinces). It is sometimes cultivated.

FAMILY MIMOSACEAE

The family Mimosaceae is very close to Fabaceae and some taxonomists consider it a subfamily of the latter. The *Acacia* species are the most numerous and are typical for the tropics and most of the subtropics of the southern hemisphere.

The tropical acacias are mainly small trees with broad umbellate crowns. Leaves are compound, pinnate, with tender, minute leaflets, falling at the time of drought, and flat, lanceolate-broadened petioles, the phyllodes, covered with thick cuticle, performing the photosynthetic functions. Many acacia species need special conservation measures because of environmental changes in the savanna due to human activity.

Sadra-beida (*A. albida* Delile) is a tree up to a height of 20 m, in southwestern Africa up to 31 m. Its crown is open, flat, umbellate. Branches are white; leaves are thin, drooping, light green, sometimes gray. Thorns are about 4 cm long, paired, white at the base and reddish chocolate-brown at the ends. Flowers are cream-yellow, borne in long spikes. The tree is found in Mozambique, northern Tanzania, and northern Transvaal in southwestern Africa, and in tropical Africa in the north. It grows in savannas, along river shoals, on alluvial deposits, mainly on forest soils, but often also on deep pure sand. It needs to be protected in southern and southwestern Africa.

Galpini's acacia (*A. galpinii* Burtt Davy) is a tree up to 82 m tall, including the underground portion of the trunk (usually the lower part is covered with silt for centuries), possibly, trees up to a height of 120 m were also known. The circumference of the trunk at the height of 1 m above the soil surface reaches 23.2 m, and diameter of crown 55 m; the height above the ground is usually 25 m. Branches are widely spreading, with long bent thorns on trunk and branches; leaves are light green (Plate 30). Wood is heavy, dense, with darker pith. Earlier the Galpini's acacia grew along the banks of the Magalakvena river, a tributary of Limpopo, in northwestern Transvaal. At present, almost all the old trees are destroyed by fires and storms, but trees up to 25 m tall have still survived at isolated places in Transvaal.

Camel thorn (*A. giraffae* Willd.) is a tree up to 9 m tall in South Africa and up to 2 m in Botswana. Usually it has a broad crown of dense tender leaves and erect trunk with dark red bark. Young branches carry 2 small thorns at the base. Inflorescence is in the form of roundish yellow heads borne in clusters in leaf axils (Plate 30). It is found in deserts and savannas of south Africa: central and western Transvaal, the west of the Orange Free State province, Zimbabwe, Angola, Namibia, southern Africa, southwestern Africa, and Botswana. It grows very slowly; the old trees may be over 100 years old. The seeds germinate poorly, seedlings are photophilous. The Katu reserve (65 km from Kuruman) has been established in South Africa for conservation of camel thorn.

The rare species **Radde's acacia** (*A. raddeana* Savi), a tree up to 10–12 m tall with blackish bark and dirty white flowers, is found in Chad and Chadian Sahali. It grows at the bottom of dried rivers.

FAMILY MONIMIACEAE

The family Monimiaceae includes 38 genera and about 450 species distributed in the tropical regions of America, Asia, Polynesia, and very rarely Africa. The **short-pedicel glossocalyx** (*Glossocalyx brevipes* Benth.) is a shrub or 3–8 m tall tree. Leaves are up to 18 cm long, entire. The male flowers have campanulate calyx of 5 unequal lobes, the upper portion of which is strongly stretched out (basis for the generic name), and others sometimes highly reduced or about half as long as the lobe. Female flowers are similar to the male, with 5–8 free carpels with one ovule each. It usually grows in secondary forests near rivers in Cameroon and Gabon.

FAMILY OCTOKNEMACEAE

The family Octoknemaceae comprises 2 genera and about 10 species distributed in the rainforests of tropical western Africa. The **Aubrevill's okoubaka** (*Okoubaka aubrevillei* Pellegrin et Normand) is a monoecious* tree, up to 25 m tall, with horizontally spreading branches. Inflorescence is dense, hemispherical, many-flowered, at the tips of branches. Petals are triangular or broadly lanceolate, and densely pubescent; fruit is yellow. The tree is found in isolated groups in Gabon, Ivory Coast, and Cameroon. It is a rare species.

FAMILY OLEACEAE

The **laperrian olive** (*Olea laperrinei* Battand. et Trabut.), a tree with a height up to 12 m and trunk diameter 2 m, has been included in the Red Data Book of IUCN. Its leaves are linear-lanceolate, silvery on the lower surface. Flowers are small, borne in racemes in leaf axils; corolla is white. Fruit is similar to a small olive fruit, dark purple, elliptical. The distribution of this species covers Algeria, Niger, and Sudan. It grows on stony grounds under precipices and near water sources at altitudes of 1000–3000 m. This is a small tree, close to the cultivated olive, with discontinuous distribution in the mountains of Sahara and Sahali. It is being rejuvenated in Algeria and Niger due to changes in vegetation and soil moisture. The young branches are cut for cattle fodder. Cattle grazing and cutting during the drought of the 1970s increased particularly in the habitat of this species. In Sudan, the trees are in good condition; one population with 1000 trees was found between 1953 and 1955, and another in 1967 over

---

*Error in the original Russian text. Should read dioecious—General Editor.

an area of at least 5 km². This species is an important genetic resource, as it is resistant to diseases and drought. It lives up to 3000 years.

South African olive (*O. capensis* L.) is a multibranched evergreen tree with a height up to 30 m. Leaves are narrow or broadly elliptical, with acute apex, leathery, almost glabrous on both sides. Racemes are at the tips of branches, with numerous small white flowers. The tree is found in the forests of Zambia.

FAMILY ORCHIDACEAE

In comparison with the tropics of Central America and southeast Asia, Africa is poor in orchids; however, hundreds of species are found even here in different countries. It must be noted that many African orchids have restricted distribution. The biology of many species has been studied incompletely. The reduction in their reserves in several regions is mainly associated with human activity. The epiphytes are faced with maximum danger due to cutting of the host trees serving as substrate for them, and also due to increasing aridity of climate as a result of deforestation. The terrestrial orchids are more tolerant, some species can even withstand fire clearings. About 400 species have been listed in western Africa, of which 77 (19.2 percent of all the west African orchids) require urgent protection. It is proposed to organize a few reserves for them in Nigeria and Cameroon. Orchids are also protected in southern Africa. Unfortunately, we do not have information from other African countries.

The angrec genus (*Angraecum*) with 206 species is very typical of Africa and Madagascar. Most of them are epiphytes with leafy stems and highly branched swollen roots. Leaves are in 2 rows, lyrate; flowers are solitary or in inflorescence. Many species have white flowers with a spur, emitting a strong smell during the night, so that they are pollinated by noctuid moths, whose proboscis is as long as the spur. Due to their smell, the flowers of some species are used for preparing tea.

The **two-rowed angrec** (*A. distichum* Ldl.) (Plate 30) and **Eichler's angrec** (*A. eichlerianum* Kränzl.) are found in western tropical Africa.

Anselia confusa N.E. Br. is a plant with cylindrical stems covered with transparent whitish sheaths in the lower part, found in the forests of western Africa. Its leaves are lanceolate, with acute apex. Inflorescence is a many-flowered raceme; the lip of the flowers is trilobed, middle lobe directed forward, slightly wavy, and lateral lobes strongly incised, light greenish-yellow with longitudinal reddish chocolate-brown stripes from the base and similar spots along the margin.

The *Bulbophyllum* genus includes up to 900 species distributed in the tropics of the Old and New Worlds. All of the species are epiphytes, distinguished by their size: from 5 mm to a few meters. The unique, highly mobile lip of their corolla is interesting.

Cocos-like bulbophyllum (*B. cocoinum* Batem.) and **bearded bulbophyllum** (*B. barbigerum* Ldl.) are found in western Africa. The former species has a horizontal aerial rhizome with slender greenish roots and densely borne stems. The inflorescence is a many-flowered raceme; flowers are small, whitish. Bearded bulbophyllum produces 3 cm long, broadly ovoid tubers with a solitary leaf. The peduncle, up to 15 cm long, bears 8–14 chocolate-brown to purple flowers (Plate 31).

Carreyanian bulbophyllum (*B. carreyanum* Ldl.), with broadly ovoid tubers and a solitary leaf, is found in South Africa. Its 15 cm long peduncles bear 6–8 flowers; perianth leaves are yellowish-green with reddish spots.

The *Disa* genus has about 80 species distributed in Africa, Madagascar, and the Mascarene Islands. The Cape region is particularly rich in them. They grow in humid and marshy places as well as in meadows, in more dry conditions. The species are extremely variable in external morphology and size, some of them bear a rosette of leaves, while in others the leaves are similar to those of cereals; the peduncle may arise from the center of the leaf rosette or develop from buds away from the leaves. The flowers are usually borne in inflorescences, and their color is extremely variable: white, yellow, pale lilac, red. Like other orchids of southern Africa, disas are under usual protection and their collection without special permission is prohibited. The **table mountain orchid** (*D. uniflora* Berg) is particularly protected (Plate 31).

FAMILY PALMAE

One of the centers of origin of palms is located in Africa. Two species of African palms are endangered and many other species are rare.

Argun medemia [*Medemia argun* (Mart.) Württemberg ex H. Wendl.] is a palm up to 10 m height with naked unbranched trunk, with a crown of up to 1.25 m long, fan-shaped leaves on long petioles. Lobes of leaves are stiff, sword-shaped; lateral lobes are much shorter and narrower than the middle ones. Male and female flowers grow on different trees. Male flowers are small, with 3 spreading 3–4 mm long petals, covered by tomentose bracts and borne in dense spikes about 15–28 cm long and 1 cm thick. Female flowers are 5 mm across, roundish, on 1 cm long, strong pedicels. Fruits are ellipsoidal, 2–5 cm long, with lustrous chocolate-brown to violet surface (Plate 32).

This palm is found in only a few places in Egypt and Sudan. Three habitats are known in Egypt: an uninhabited oasis 220 km southwest of

Aswan where one tree and a few seedlings were found in 1963; 200 km south of Aswan where one tree was found in 1964 (here this palm grows together with another palm species, *Hyphaene thebaica*, as well as date palm *Phoenix dactylifera*); and on the eastern bank of the Nile (in the south). In Sudan, argun medemia is known from a single habitat, about 200 km southeast of Wadi Kalif. It grows in the river shoals, in wadis and oases. Perhaps it had much wider distribution in the past: there is information about some of its habitats till 1910, which were not conserved. It was widespread in ancient Egypt (numerous sketches of the tree and its fruits are present in the ancient Egyptian pyramids). In general, the populations of this palm are at a critically low level because of excessive exploitation (its fruits being edible, and leaves used for preparing mats). Its natural habitats are destroyed due to irrigation works in the Nile valley. Complete conservation of this excellent plant in all its habitats is essential.

Keeled wissmania [*Wissmannia carinensis* (Chiov.) Burret] is endangered. It is a palm with solitary grayish-brown trunk, 15–20 m tall and 40 cm in diameter. Its crown consists of 40 fan-shaped leaves with 120 cm long petioles bearing recurved spines along the margin and yellowish-green on the lower surface. Laminas are up to 95 cm long, green on both surfaces. The inflorescence is axillary, each branch bearing a yellow bisexual flower. The mature fruit is round (Plate 32) and the plant is very similar to some species of the widely cultivated genus *Livistona*.

This wissmania species is known from Somalia, Jabuti, and Southern Yemen. It grows along river banks, in valleys and oases, and in Jabuti near swati waterbodies. Its population has significantly declined. In Somalia it was known from 3 points in the northeast of the country; one of them is already destroyed, and only 25 trees have been preserved in the others. In the neighboring regions of Jabuti, 7 locations are known in the Goda mountains, but even there its population is small, the total strength does not exceed 97 trees, and their regeneration is not taking place. It is not known whether it has been preserved in Southern Yemen; earlier it was reported from one location, where, most probably, it is now endangered. Reduction in the population of this palm is related to its cutting (its wood is valued as construction material) and sheep and cattle grazing, which prevent its regeneration. It is being cultivated in Crimea and also grows in botanical gardens in England. Both these palms have been included in the Red Data Book of IUCN.

The doum or fan palm genus *Hyphaene* comprises about 11 species, distributed in tropical Africa, Arabia, and the Mascarene Islands; one species is found in southern Africa and one more in southwestern Africa and Botswana.

In distinction from all other palms, doum has a branched crown. The leaves are fan-shaped with sword-shaped lobes; fruits are covered with scales. The juice of fruits is used for preparing palm wine.

Doum palm [*H. thebaica* (L.) Mart.] has almost disappeared from the world. It is a tree up to 10 m in height with 3–4 branches, each terminating in a crown of fan-shaped leaves, among which flowers arise. The flowers in female plants develop into large bunches of reddish, glossy, yellow to chocolate-brown fruits (one bunch has up to 200 fruits). The fruits are edible—their fibrous and floury peel reminds one of finger powder in taste—but very dry. In Upper Egypt, this plant is called 'doum palm'. On oases it grows with other rare plants. Its population has declined due to irrigation works in the Nile valley. The doum palm is also known in Chad and Chadian Sahali.

Another doum species, *H. ventricosa* Kirk., is found in the north of southwestern Africa and in Botswana, and towards the north penetrates into tropical Africa. This is a tree with 15–18 m height and simple trunk, often having a characteristic swelling in the middle or lower part. The crown consists of large fan-like leaves, arranged at the apex of trunk (Plate 31). The plant is dioecious bearing hard chocolate-brown fruits with a layer of edible flesh, in which a hard seed is embedded. The juice of young fruits reminds one of coconut milk. Often the upper part of the trunk is cut for obtaining juice, which is used for preparing palm wine. Population of this palm has also decreased significantly.

Southern raffia or **Kozi palm** (*Raphia australis* Olem. et Strey) is an unbranched tree up to 24 m tall. The trunk base is surrounded by small respiratory roots and the rest is covered with remnants of the bases of old leaves. Numerous roots appear from the corners of lower leaves. The leaves of this palm are the longest (up to 9 m) among all the plants of the world. They are light bluish-gray, often with a thin reddish midrib. Inflorescence is 3 m long, branched, with a chocolate-brown pinnate form at the tree apex. Male and female flowers are borne in the same inflorescence; fruit is up to 9 cm long, with a chocolate-brown peel enclosing a thin layer of porous orange flesh in which a single seed is embedded.

The Kozi palm lives only for 30 years and dies after fruiting. Its fruits mature in 2 years and germinate easily. This palm is found in South Africa, around Kosi Bay, at the border with Mozambique, growing as isolated specimens or in groups in freshwater marshes. It is also found in tropical Africa, where it is a source of raffia fiber. In South Africa, its leaves (midrib) are used. However, this may lead to its complete disappearance because of limited distribution of the palm here.

Caffr jubaeopsis (*Jubaeopsis caffra* Becc.) is a rare plant of South Africa, is included in the Red Data Book of IUCN. This palm is of medium size with a cluster of stumps, usually root suckers. Its height, as a rule, does not exceed 6 m; leaves are pinnate, up to 2 m long, bent, with numerous stiff leaflets arranged in one plane; inflorescence has male and female flowers borne in leaf axils; fruit is 2.5 cm in diameter, with thin peel covering the hard nut. The adult plant produces about 200 seeds in a year, but 80 percent of them die due to insects. This palm grows in Transkei along the northern banks of the Mtentu and Msikaba rivers (eastern part of the Cape of Good Hope). Along the Mtentu River, the palms are found in groups at large distances from each other; the total number of palms here does not exceed 5000. Because of destruction of seeds by insects and collection of fruits, seed regeneration is almost completely absent. Its population is still less along the Msikaba River, 200–300 trees, concentrated at one place. It is found on steep, rocky banks. Its fruits are similar to coconut, but smaller. The local population collects its fruits, for which they often cut a fruiting tree. At present, this palm has survived only in inaccessible places, on cliffs. Since 1973, it has been protected by law.

FAMILY PASSIFLORACEAE

The family Passifloraceae includes about 20 genera and 650 species, distributed in the tropics and subtropics, mainly in America and Africa, and rarely in Asia, New Guinea, Australia, and New Zealand. Adult plants of **Pechuel's adenia** (*Adenia pechuelii* Harms) have a thick uneven trunk with grayish bark, often more than a meter in diameter, with pulvinate or short grayish-green branches. This adenia grows in South Africa on stony places, fixing itself in fine crevices. It is damaged by goats and wild animals.

FAMILY POACEAE

Members of 2 genera of the family Poaceae need protection in Africa. The genus *Guaduella* consists of herbaceous plants in which the leaves are present only in the upper part of stem, the bases being surrounded with leafless sheaths; leaves are oblong to linear-lanceolate; inflorescence is conical, raceme or panicle, spikelets sparse, with a few flowers, 1–3 inner flowers male, others bisexual. The genus includes about 8 species, distributed mainly in Cameroon and Gabon. Several species are endemic to Cameroon; among them **Bedermann's guaduella** (*G. bedermannii* Pilg.) and **large-spiked guaduella** [*G. macrostachus* (K. Schum) Pilg.] need protection.

The genus *Puelia* comprises herbaceous plants with leaves at stem apices and short, conical racemes. The leaves are ovate-lanceolate, 25–30 × 10–20 mm, with ciliate margins. Five species of this genus are found in Gabon, Cameroon, Uganda, and Kongo. **Acuminata puelia** (*P. acuminata* Pilg.) is protected in Cameroon.

FAMILY PODOCARPACEAE

Sickle-shaped podocarp [*Podocarpus falcatus* (Thunb.) R. Br. Ex Mirb.] is recommended for protection in Mozambique. This is a large, beautiful tree with gray, chocolate-colored, or purple bark, red in mature trees, with roundish or rectangular cracks. Leaves are narrow, alternate; male cones are 1.3 cm long, solitary or in small groups, mature cones turning yellow. This podocarp grows in southern Africa from Swaziland to northern Transvaal and in Mozambique. It is found from the coast to the mountains; the species is still common in the Mozi swamps in the north of Zululand. Its wood is widely used; therefore, the best trees have already been cut and the population of this species continues to decline. It can be used for greenery.

Milan-Jian podocarp (*P. milanjianus* Rende) is a highly branched evergreen tree with a height up to 30 m and smooth or finely cracked pale chocolate-brown bark. Leaves are borne on the flowering shoots, linear-lanceolate or linear. This podocarp grows in small isolated groups in the mountains of tropical Africa. It is quite tolerant to fires; it is widely used in construction and as a result its population has sharply declined everywhere. For example, its felling was so intensive near Djilo that all the medium age trees were cut in a radius of 3–4 km from the city from 1940 to 1956. It is protected in Sudan, Cameroon, Angola, and Zambia.

FAMILY PROTEACEAE

The family Proteaceae includes more than 70 genera and about 1400 species. It is mainly distributed in the southern hemisphere, and South Africa is particularly rich. The silver tree genus (*Protea*) comprises more than 100 species. These are trees and shrubs with long, thin, evergreen leaves, usually arranged spirally. Inflorescence is brightly colored large heads. Flowers are small and hidden in a disk of petaloid bracts. Earlier, the leaves and bark were used as source of tannin, which reduced the population of this genus. It is recommended to place all its species under protection in Natal. They are fairly widely cultivated.

The **Angolan silver tree** (*P. angolensis* Welw.) is 1.5–2.2 m tall, with very large white terminal flowers. It must be protected in Guinea. It is found as isolated trees among the grasses of savanna and belongs to the relicts of the ancient Guinean flora.

Marsh rose (*Orothamnus zeyheri* Pappe ex Hook.) is a vulnerable species. It is a 1–4 m tall tree or a slightly branched shrub; leaves are densely arranged on twigs, elliptical, leathery, and hairy, particularly along the margins; floral heads (1–3 at the apex of each branch) are subulate, 5–7 cm long, with pink-red, pubescent, 4–6 cm long bracts, surrounding lemon-yellow flowers. Fruit is oblong, about 6 mm long (Plate 32). The marsh rose is found only in South Africa. Nine populations and a few small groups of this plant are known from the Kogelberg mountains in the southwest of the Cape province and one population near Germanus, 25 km east of the former. It grows on steep slopes of southern exposure, at altitudes of 500–850 m, but may disappear due to uncontrolled burning of vegetation, fungal diseases, and damage. This is the most beautiful species in the family Proteaceae. The rose-like bracts of the heads are highly ornamental and have long been attracting the attention of flower traders. The plant has been protected by law since 1974. Although its population has increased during the last 10 years as a result of the measures taken, the plant is endangered. In 1968, 90 plants were found but they were threatened due to collections of the floral heads. By 1977, 1940 plants of marsh rose had already been found as a result of further search as well as stimulation of seed germination by controlled burning of vegetation. A reserve has been created in the Kogelberg mountains; it is planned to restrict its vegetation and take measures for rat control. A reserve has also been created in Germanus over an area of 20 ha to study the biology and develop cultivation methods for this plant. Seeds have also been sown in neighboring reserves. This member of a monotypic genus has a great scientific value. An attempt has been made to cultivate it, in view of which it is necessary that the seeds of this plant are kept in the seed bank. It has been included in the Red Data Book of IUCN.

Roxburgh's serruria (*Serruria roxburghii* R. Br.) is a small, slender shrub with erect up to 50 cm tall stem; leaves are 1–2 cm long, twice divided. Floral heads are borne in dense racemes at the tips of stems, many-flowered, bright pink. The shrub is found only in South Africa, where its numerous locations were known earlier north of Wellington, in the southwest of the Cape province. This plant was seen for the last time in 1850, after which it was not found for almost 100 years and was considered to have become extinct. In 1976, 2 isolated populations were discovered northwest of Wellington. One occupies an area of 25 ha, where about 3000 plants are growing. The other in an area of 35 ha is much smaller, as it has been destroyed due to overgrazing of cattle. Both populations are located on private lands. The species has been included in the Red Data Book of IUCN.

Ciliate serruria (*S. ciliata* R. Br.) is endangered with extinction. It is a small branched shrub up to 60 cm tall with erect stems and thin, usually ternate leaves. Floral head is solitary, terminal, with numerous flowers; outer bracts are 6–8 mm long, narrow lanceolate, pointed, with long hairs; bracts are 5–9 mm long. The plant is found in 3 locations in the Cape peninsula on deep white sands. Earlier it was fairly widely distributed on the sandy plains, but the number of its habitats has decreased due to urbanization as well as replacement of this species by acacia introduced from Australia. Now it is represented by 3 small relict populations, each of which has less than 200 plants over an area of 2–20 ha. The plant is protected by law in the Cape province since 1974. It has been included in the Red Data Book of IUCN. The **blooming serruria** (*S. florida* R. Br.) is protected in South Africa.

Buek's diastella [*Diastella buekii* (Gaudoger) Rourke] is a prostrate shrub with numerous branches forming a dense pad with a height of 15 cm and diameter up to 1 m. The floral heads are pink, with 15–20 flowers. The biology of this species is not properly understood. Two populations were found in 1976 in the Franschhoek valley, east of Cape Town in the northwestern part of the Cape province. For 40 years before this the species was considered extinct. Diastella grows in the fertile mountainous valley on sandy soils along the slopes at an altitude of 150–300 m. The valley where it has been conserved is intensively used for agriculture. One population is scattered over several square kilometers, another occupies an area of 2–3 ha. In both cases, diastella is threatened by the pine plantations. Although this plant can survive under the crowns of pine trees, the thick litter of the pine needles inhabits the germination of its seeds. The species is included in the Red Data Book of IUCN.

Whorled silver tree [*Leucadendron verticillatum* (Thunb.) Meissner] is a slender and erect shrub up to 2 m tall. The plant is dioecious; leaves are silvery gray, narrowly oblanceolate. Floral heads are small, located among leaves at the tips of branches. Male heads are hemispherical, 7 mm long and 11 mm in diameter, with 6 linear, sessile, 3–4 mm long bracts and minute bracteoles. Female heads are similar to the male heads, but smaller, up to 8 mm in diameter, with 12 flowers and 10.5 mm long bracts. Its distribution is limited to 4 relict areas of natural vegetation between Cape Town and Paarl in the southwest of the Cape province. These oases of natural vegetation (their area does not exceed 30 km^2) have been preserved in cultivated land. The silver tree grows at an altitude of 1000 m in the zone of winter rains in the communities of sclerophyllous heath vegetation, typical for this region. At the same time, other rare and endangered species of South Africa are also found: *Cliffortia acockii* Weimark; *Restio duthieae* Pillans; *Tritoniopsis elongata* (L. Bolus) G.J. Lewis; and *Watsonia strictiflora* Ker-Gawl. Earlier such vegetation was widely distributed, but due to agricultural use of the land only a few patches have survived. Three of the known populations of silver tree have only a few specimens each, and may soon disappear. The fourth population over an area of 4 ha has up to 2.5 m tall adult plants; a large part of this population is on private lands, and a few specimens are on the outskirts of a natural reserve. The plant is protected by law since 1974. It is proposed to include the main population in the reserve, remove the introduced acacia, and create a buffer zone for reducing the pressure of agriculture on it. The species is included in the Red Data Book of IUCN.

Cape silver tree [*Leucadendron argenteum* (L.) R. Br.] is one of the most beautiful species of the genus. It is a slender, elegant tree with terminal crown in young plants and broadly scattered crown in adult trees. Its height reaches 15 m, main trunk bearing long, leafy, horizontal branches. Male and female flowers are borne in heads on different trees. The male heads are apricot-colored, terminal on the branches; female heads modified into beautiful silvery cones with a length of 7.8–10 cm protruding from the tree. Fruits mature in autumn and the small, black, suspended seeds are carried by wind. Its leaves are silvery white. The plant is found only in the Cape peninsula on the Table mountain, growing in dry and windy slopes, preferring acidic, well-drained soils. It is used for bouquets and the soft and porus wood is used for carving. The silver tree is threatened by small gray squirrels which were introduced into this region, as they eat its young cones, and thus may reduce its regeneration.

FAMILY PSILOTACEAE

The family Psilotaceae includes 2 genera with primitive characteristics, indicating their very ancient origin. They grow either on trunks or at the base of tree trunks, sometimes in rock crevices. They have no roots and their underground organs are long, dichotomously branched, rhizome-like structures, with mycorrhizal fungi. Therefore, psilotes mostly live as saprophytes. The stems are dichotomously branched with scaly leaves. The plants form spores.

Naked psilotum [*Psilotum nudum* (L.) Beauv.] is one of the most rare plants of Swaziland, found on a single location on rocks near a river. It is also known from a few places in South Africa.

FAMILY RESTIONACEAE

The family Restionaceae is an ancient family of the Cape province and practically not found outside that area. The plants are herbaceous perennials, similar to sedges or rushes, but some of them grow up to the height of a human. The **Acock's chondropetalum** (*Chondropetalum acockii* Pillans) is endangered. It is a rush-like perennial with creeping rhizomes and very slender, straight, unbranched, 70 cm tall stems. Male inflorescences are loose panicles 5–10 cm long; female inflorescences are similar to the male, but smaller (Plate 31). A few habitats of this species are known in South Africa, in a very thickly populated area between Cape Town and Mair over a distance of 45 km. The search for other populations preserved in the areas of relict vegetation was unsuccessful. It grows in poorly drained clayey soils conserving moisture at an altitude of 100–300 m. Four natural populations are known which have been preserved in this reclaimed region. The area of each population is less than 2 ha, and the scattered turfs have a few hundred plants. The plant may disappear within a few years as a result of development of farms and cities. All the populations are already mixed with acacia.

Rope grass (*Restio acockii* Pillans) is also endangered. This is an erect, rush-like perennial with numerous simple or sparsely branched, very slender stems up to 1 m height. It differs from other species by very compressed stems. For 50 years the species was considered extinct. Presently, a small population has been found as a result of survey in an isolated area of natural vegetation at a distance of 45 km north of Cape Town, which has about 1000 grass patches over an area of about 6 ha. Unfortunately, a recent fire destroyed a larger part of this area. Near this population are found outgrowths of introduced acacia which are gradually spreading, and cultivation of the lands (the population is surrounded by cultivated fields) may lead to complete disappearance of the species. Besides, it is proposed to build a city at a distance of about 1500 m from the population. The rope grass is so rare that its biology is not understood. It has been protected by a special law since 1974. Both rope grass and Acock's chondropetalum have been included in the Red Data Book of IUCN.

FAMILY RUBIACEAE

A few species of the small genus *Alberta* are distributed mainly in Madagascar and only one species is found in South Africa. The **large alberta** (*A. magna* E. Mey) is a tree up to 13 m tall with gray or red bark and green or chocolate-brown branches. Leaves are evergreen, simple, opposite, oblong, dark green on the upper surface and pale green on the lower. Flowers are arranged in large panicles at the tips of branches, tubular, up to 2.5 cm long, bright red; 2 lobes of the calyx are modified into bright red wings after flowering, which protect the small dry fruit. The fruits remain suspended on the tree for months and make the tree very ornamental. It is rare in nature and is found in forests, ravines, and valleys from Transkei to Natal and Zululand, from the coast to an altitude of 1800 m. It is an ornamental plant, cultivated in humid regions.

The members of the genus *Calochone* are twining plants with conical umbel or paniculate inflorescence. The **acuminate calochone** (*C. acuminata* Keay) is a creeper or small tree. Its inflorescence is pink, turning black on drying, terminal umbellate and fluffy. The plant is found in dense forest. It is a less known and rare species of equatorial western Africa (Gabon, Cameroon).

The genus *Corynanthe* includes small trees. Inflorescence is terminal or axillary, usually subterminal. Flowers are numerous, small, on short pedicels or sessile, 3–6-merous.

In Cameroon, it is proposed to protect the **long-fruited corinanthe** (*C. dolichocarpa*), a small tree with terminal or axillary (usually subterminal) inflorescence of numerous small 3–5 merous flowers.

FAMILY RUTACEAE

The family Rutaceae is widely distributed in Africa, especially in the south. The glandular vepris [*Vepris glandulosa* (Hoyle et Leskey) Kokwaro], a prickly evergreen tree up to 6 m tall with smooth gray bark, is endangered. Leaves are opposite, with 3 leaflets each; flowers are small, in axillary and terminal inflorescences, densely tomentose. This vepris grows only in Kenya, in the Muguga forest, northwest of Nairobi; it is known from an area of 15.5 ha in a forest protected as a natural reserve. Here 8 adult trees of about 4 m height were found; a few seedlings up to 20 cm tall are also present near the parent trees, but they are very sensitive to bright light, and only a few survive during the drought period. This plant was first noted here in 1967; earlier it was also known from other places, but the plants have not been preserved there.

FAMILY SAPOTACEAE

The family Sapotaceae includes more than 60 genera and about 800 species with pantropical distribution. They are trees or shrubs with milky juice, often characterized by cauliflory. Among the members of the monotypic tree genus *Argania*, endemic to Morocco, the population of **Morocco ironwood** (*A. sideroxylon* Roem. et Schult) is declining. These are trees or shrubs with hard wood and, therefore, also called 'iron tree'. Oil is obtained from the fruits. The tree grows in northern Africa along the Mediterranean coast at an altitude of 1500 m. It is being destroyed for fuel. It is cultivated in Morocco.

FAMILY URTICACEAE

The family Urticaceae comprises 45 genera and more than 700 species, mainly distributed in the tropical countries; however, some species are found in the temperate or cold latitudes of both hemispheres. These are herbs or semishrubs, rarely small trees with very minute flowers. Their fruit is small, a nut or stone.

The **odum tree** [*Chlorophora excelsa* (Welw.) Benth. et Hook. f.] is large with brown, heavy wood, resembling the wood of oak, and is protected in Ghana. The wood is used for underwater structures, construction, and furniture. It is a typical plant of dry forests and savannas.

FAMILY VITACEAE

The family Vitaceae includes up to 12 genera and more than 700 species in hot and temperate regions of the world. The tree-vine genus (*Cissus*) has pantropical distribution. Members of this genus are creepers with annual branches, herbaceous or woody, sometimes slightly succulent. The root systems have nodules. Leaves are simple, entire or lobed, rarely 3–5 times divided. Inflorescence is always opposite. Flowers are tetramerous, bisexual; corolla is usually oval. Berry is single-seeded. At one time this was a very widespread genus, but it is now reduced to the ancient section *Cissus*, having about 300 species in the tropical and equatorial regions.

In Angola, it is proposed to protect the rare species of this genus—**uter tree-vine** (*C. uter* Exell et Mendonca).

FAMILY VITTARIACEAE

The family Vittariaceae includes very specialized, predominantly epiphytic ferns, small herbaceous plants with simple, entire (rarely lobed) leaves and with creeping rhizomes covered with reticulate scales. The family comprises 7–9 genera, distributed in the tropics and subtropics.

Out of the 40 species of the genus *Antrophyum*, distributed in the tropics and subtropics of the Old World, it is essential to protect **Annet's antrophyum** [*A. annetii* (Jeanpert) Tardieu] in Cameroon—an epiphyte with short rhizome, covered with deltoid-lanceolate, chocolate-brown to red scales. Laminas are lanceolate-ovate, sessile, 18–22 cm long and 2.5–3 cm broad, pointed toward apex. The plant grows on tree trunks and in shady places on rocks where atmospheric humidity is high.

The genus *Vittaria* is represented by ferns with narrow-linear leaves, similar to the leaves of cereals, with unique venation: the lateral veins converge with each other, forming one marginal vein on each side of the midrib. Leaves are sometimes scattered on the creeping rhizome, but usually clustered in small numbers and drooping in adult epiphytes, which distinguishes these ferns from others.

Schefer's vittaria (*V. schaeferi* Hieronymus) is a rare plant of Cameroon. It is a fern with rhizomes covered with deltoid-lanceolate, curly, dark scales. The linear-lanceolate leaves are closely placed.

FAMILY WELWITSCHIACEAE

The family Welwitschiaceae has only a single genus and species, and is a transition between the typical gymnosperms and angiosperms.

Tumboa [*Welwitschia bainesii* (Hook. f.) Carr (*W. mirabilis* Hook. f.)] is a unique dwarf tree of the desert, which produces only 2 leaves through its entire life. The trunk usually grows up to a height of 30 cm, very rarely 1.5 m, but it may be up to 3 m long under the ground; trunk diameter is more than 1 m. The wood is as heavy and dense as in sequoia. The leaves, when they appear, are small but become broad, thick, leathery, and prostrate during growth, up to 3.7 m long, greenish chocolate-brown. They never drop and continue to grow piling on the sand. Male and female cones appear near the leaf base on different plants. Male cones are pink and female cones initially grayish-green, but red on maturity. The seeds are winged, light, carried by wind (Plate 32). All the plant organs discharge transparent resin. Tumboa is found in southwestern Africa only along the western coast, from southern Angola towards the south, reaching up to the southern tropic in a branch of the Keiseb River in the Namibian desert. Its distribution is restricted to the area of oceanic fogs; the maximum distance from the sea is 80 km. It grows scattered in sandy deserts, as individual isolated specimens, never in groups. The plant is protected by law.

In Africa, other plant species also need protection. In Guinea, for example, *Medusandra richardsina* Brenan is already protected, it is a rare endemic, the only species in the genus of a very small family. *Pitcairnia feliciana* (A. Chev.) Harms et Mildbr., the only member of the family Bromeliaceae in Africa is also protected. In Guinea and Nigeria, *Blaeria mannii* Engl. has been placed under protection. Isolated specimens are known from 3–4 locations in the Namib range. The list of plants which require special conservation measures in Cameroon includes a few dozen names.

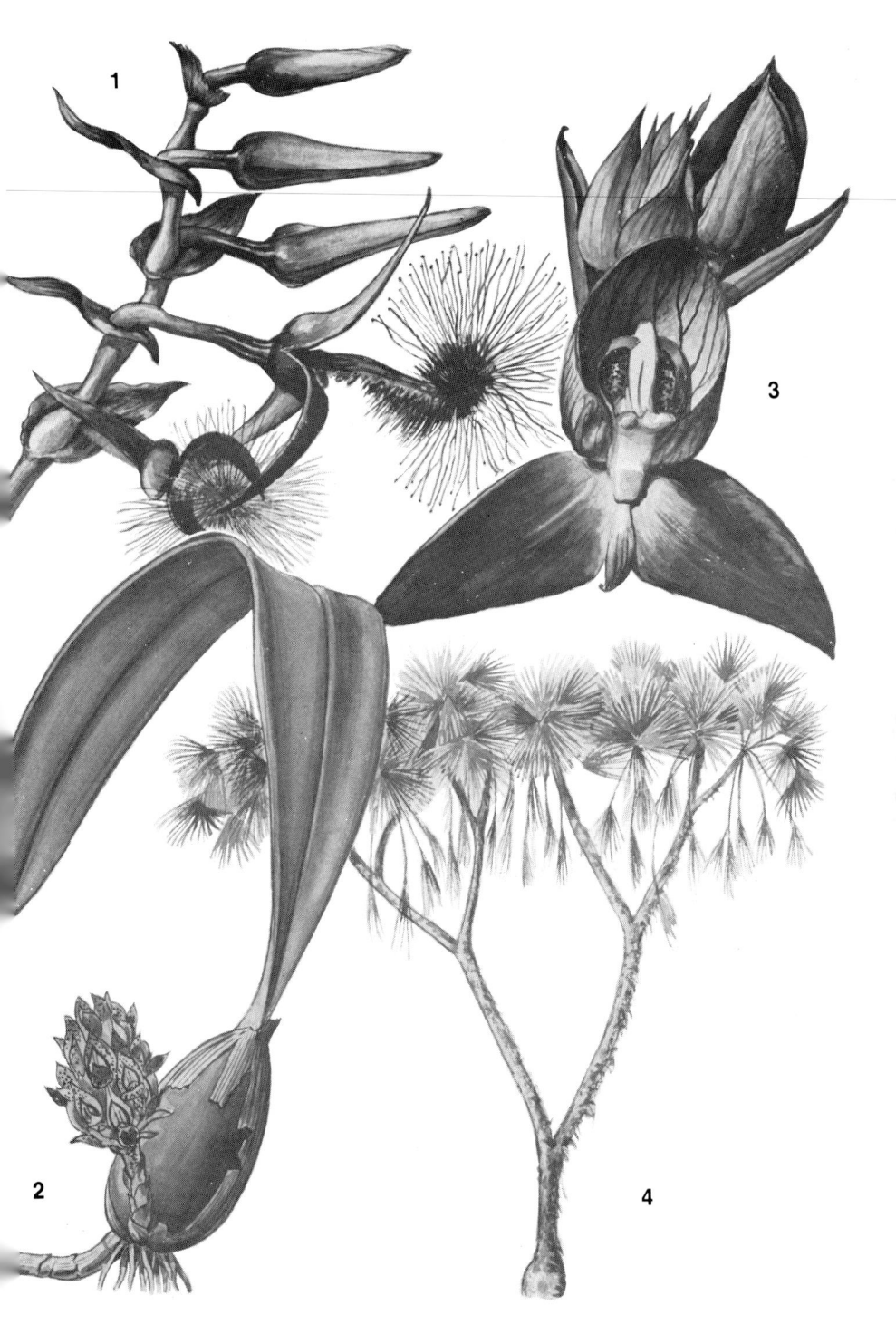

PLATE 31. 1—Acock's chondropotalum; 2—bearded bulbophyllum; 3—table mountain orchid; 4—doum.

PLATE 32. 1—Argun medemia; 2—keeled wissmania; 3—marsh rose; 4—tumboa.

PLATE 33. 1—Two-colored pachycormus; 2—Huron tansy; 3—golden barren-wort; 4—American ginseng.

PLATE 34. 1—Goat-horned astrophytum; 2—boat-nose ariocarpus; 3—Reichenbach's echinocereus; 4—hedgehog cactus.

4

North America

The flora of the North American continent mainly belongs to the Holarctic floristic kingdom and only the southern part of the Mexican plateau and the southern Florida peninsula are included in the Neotropical kingdom. The northern part of the continent, Canada and Alaska, belongs to the Circumboreal region, whose vegetation is similar in many respects to that of northeastern Asia. A large part of the territory of the United States of America is in the Atlantic-North American region, whose flora is distinguished by its richness and high endemism (almost 100 endemic genera). The Rocky Mountain stretch from Alaska in the north to New Mexico in south is considered an independent region.

At present, the flora of North America, especially Canada and the United States, has been studied fairly completely. The higher flora of Canada includes about 3500 species belonging to 734 genera and 174 families; more than 10 per cent of the flora consists of rare species. Identification of the rare and disappearing vascular plants has been completed in almost all the provinces of Canada.

The flora of the continental United States (excluding the Hawaiian Islands) includes 20,000 species, of which about 2140 species or 10.7 percent of the total flora require urgent measures for conservation. The list of such plants was prepared by the Smithsonian Institute on the basis of the 1973 law on endangered species. Inclusion of a plant in the list means that it is protected by the federal government. The plants in the list are divided into 3 categories: extinct or most probably extinct plants (90 species); endangered or disappearing plants (818 taxa); and plants that may be endangered in future over the entire area or a larger part, i.e., the vulnerable plants (1342 species and subspecies). The preliminary list of rare and disappearing plants in Mexico includes 212 names, of which 8 species are listed as extinct, and 72 as disappearing. The family Cactaceae is in the most endangered condition: 44 per cent of the total plants require protection.

FAMILY AGAVACEAE

The members of Agavaceae are distributed in the southern part of North America, Central America, and the northern part of South America. They are evergreen perennials with giant leaves and a single scapigerous inflorescence, produced once, after which the plant dies. More than 100 agave species are known.

Arizona agave (*Agave arizonica* Gentry and J.H. Weber) is on the verge of extinction. The plant bears a rosette of linear-lanceolate, dark green leaves reaching 24 cm in length and 4 cm in breadth. The floral axis (arrow) is 3–4 m long, terminating in an inflorescence of 35–50 pale yellow flowers. Fruits are thick-walled, developing after pollination. A total of 12–14 habitats are known, each with one or a few specimens in an area with a radius of 3–5 km. The area is located in the central part of Arizona state at the border of Maricopa and Yavapai counties in the new river mountains at an altitude of 910–1830 m. The scrub communities with Arizona agave—chaparali—consist of evergreen broad-leaved plants: *Quercus turbinella*, *Arctostaphylos* spp., *Cercocarpos*, and *Ceanothus greigii*.

In spite of the conservation measures for this species, its present condition remains threatened. This agave is poorly reproduced by seeds and by lateral branches and, therefore, its natural population is continuously declining. It has also been established that its flowering shoots are often eaten by animals.

The Arizona agave is cultivated in the Tucson desert botanical gardens (Sonora desert). In 1976, one specimen produced flowers, the capsules from which had less than 1 percent mature seeds with very low germination. On the one hand, this can be explained by non-correspondence of conditions between the natural area and the botanical garden and, on the other, by the need for cross-pollination. It is proposed to transfer some of the plants from the botanical garden to the natural habitat and also distribute plants to other botanical gardens and institutes. The Arizona agave is included in the Red Data Book of IUCN.

FAMILY ANACARDIACEAE

A member of the family Anacardiaceae, the **two-colored pachycormus** [*Pachycormus discolor* (Benth.) Coville], is a characteristic and, at the same time, unique plant of Baja California (Mexico). Translated from the Greek, the word 'pachycormus' means thick stems. This plant is known, as is the elephant tree, for its short and twisted trunk which is divided into several strong, horizontally spreading branches up to 6 m long. The branches terminate in a few short inflorescences with a mass of red flowers (Plate 33). The chocolate-brown bark of this plant, resembling skin, cracks every year and peels off so as not to disturb the growth process. The height of the tree is 4.5–6 m, reaching up to 9 m, and a diameter at the base 90 cm. Pachycormus is a succulent. It can store a huge quantity of water in the trunk and the thick roots and can tolerate the long periods of drought typical to the desert regions. The leaves are small, produced immediately after the rainy season in April. They turn yellow by May and drop. Pale pink or pinkish-red paniculate inflorescences appear from June to August, and the tree is noticeable from a long distance during this period. Pachycormus grows on the western coast of the Pacific Ocean, occupying stony slopes of mountains up to an altitude of 1500 m. The areas where it grows are permanently exposed to strong, cold, moist and salty winds blowing from the ocean. That is why the trunks and branches of the tree are curved and deformed.

Pachycormus is endemic to Mexico, with a very limited distribution, and it requires protection and control of its population. It has its own means of protection. Like many other desert plants, it has practically no wood, all the tissues remaining soft, so that neither the trunk nor the branches can be used as construction material or fuel. Latex, a fast drying thick sap, oozes from the injured areas, thereby protecting the tissues from infection and excessive water loss.

The sumach genus (*Rhus*) is represented by shrubs with poisonous roots.

The mature fruits of **Kerny's sumach** (*R. kearneyi* Barkley) are red, glandular-pubescent; leaves are broadly ovate, pointed, dark green, on up to 1 cm long petioles. The shrub is 5.5 m tall. It grows on dry overhanging cliffs at an altitude of 300–450 m. One habitat of this species is known in the Arizona-Tinajas Atlas. It belongs to the category of endagered species in the United States.

FAMILY ANNONACEAE

Of the family Annonaceae, only American papaw (*Asimina*) is found in the temperate zone of North America. All other members of the family are distributed in the tropics of both hemispheres.

Rugel's papaw (*A. rugelii* B.L. Robinson) is a short shrub, 10–20 cm tall, with prostrate, ascending, or erect stems, covered with reddish, straight hairs. It is very attractive at the time of flowering, when it is covered with numerous lemon-yellow axillary flowers. This is a rare plant of North America. Its condition is such that without special conservation measures, it may soon come under the category of disappearing plants. Only 2 locations are known. The first was discovered in 1848 in the outskirts of New Smyrna beach (Florida). Its existence at the same place could be confirmed only in 1927. At the same time, a second location was also discovered in Seminole county (Florida). This papaw species grows on poor but well-drained sandy soils under the cover of pine plantations in the tier of shrubs together with lyonia, holly, bepharia, and Vacciniaceae species, among which it is difficult to locate it. Fires are the most important ecological factor affecting the existence of papaw. It has been noted that a huge mass of flowering plants appears after burning of vegetation at the end of summer or during winter. Detailed mapping of the natural populations and study of the ecological properties are proposed so as to decide on conservation measures. This plant is not cultivated in any of the botanical gardens of America. It has been included in the Red Data Book of IUCN and in the list of vulnerable plants of the United States.

Tetramerous papaw (*A. tetramera* Small) is endangered. This endemic of Florida was initially found in the coastal belt on ancient coastal sand dunes covered with bushy outgrowths, from Stuart in the north to West Palm Beach in the south. The total length of this belt is about 60 km. At present, this plant has been preserved at 2 places because of destruction of its habitats. The first is near the estuary of Santa Lucia (here only 4 plants are growing), and the second on the territory of the Jonathan-Dickinson Park of Florida state (25–30 specimens). This papaw is a 1–3 m tall shrub with naked virgate branches. Young branches are reddish due to pubescence. The species name was given because the number of perianth lobes is a multiple of 4 (4 or 8), but, as noted by the author of the first description, this is not an essential condition, as trimerous as well as tetramerous flowers can be found in the same population. The flowers are white up to the middle part and reddish-purple on the lower side. Pruning of old branches increases its height to 2 m in a single season. The plant has been included in the Red Data Book of IUCN and the list of disappearing plants of the United States.

FAMILY ARALIACEAE

The ginseng genus (*Panax*), which belongs to the family Araliaceae, includes 2 species in the United States: **American ginseng** (*P. quinquefolium* L.) (Plate 33) and **three-leaved ginseng** (*P. trifolium*). They differ in the number of leaves in a rosette as well as by the color of berries: red in American and yellow in 3-leaved ginseng. The differences are not so sharp when American ginseng is compared with **Asiatic ginseng** (*P. ginseng* C.A. Mey.). The similarity of these species is manifested in the structure of the aerial and underground organs, life span, and duration of life cycle. Both species are used in medicine and also in making creams, shampoos, tea, cigarettes, and other preparations. Both species are widely cultivated. The American ginseng is included in CITES.

FAMILY ARISTOLOCHIACEAE

The genus *Hexastylis* of the family Aristolochiaceae is distinguished from the wild ginger genus (*Asarum*) by the number of styles, being 6. The plants of hexastylis are perennial rhizomatous herbs with short hairs; leaves are on long petioles, roundish, and lustrous; flowers are solitary, on short drooping peduncles. The perianth is campanulate, greenish-purple outside and dirty dark purple inside. The **small-flowered hexastylis** (*H. naniflora* Blomq) is the rarest endangered species. The length of its calyx is 10 mm. A few habitats of this plant have been recorded in North Carolina, South Carolina, and Virginia states. Another disappearing plant, **showy hexastylis** (*H. speciosa* Harper), has widely separated calyx lobes. It has been found only in Alabama state.

FAMILY ASTERACEAE

Newcombe's groundsel (*Senecio newcombei* Greene) is a very rare, most probably disappearing, species of Canada. Its distribution is restricted to the mountainous regions of the Graham and Moresby islands from the group of Queen Charlotte Islands (British Colombia province). It is a small herbaceous perennial with solitary capitula having yellow or orange-yellow peripheral florets. In origin, this groundsel belongs to more southern elements of the North American flora, and its existence at these islands together with other endemic island species confirms the hypothesis about the existence here of a refugium during the last glaciation. The plant was found in good condition in open rocky and marshy mountain slopes along the western coast. Here it is almost a common species of heathlands in the alpine zone on the slopes of Lake Takakia. An ecological reserve has been organized at the western coast of Graham Island to conserve and study the botanical-geographical associations of this and other endemic species from the northern islands of Canada. The plant has been included in the Red Data Book of IUCN.

FAMILY BERBERIDACEAE

The plants of barberry genus (*Berberis*) are shrubs, in which the leaves of long branches are modified into simple or branched spines. Many species of this genus are distributed fairly widely, with the exception of tropical regions.

Sonne's barberry [*B. sonnei* (Abrams) McMinn] is a species on the verge of extinction. It is an evergreen shrub, 25–60 cm tall with a well-developed system of underground shoots. The leaves are 5-lobed with serrate margins. Yellow flowers are clustered in dense inflorescence. The habitats of this plant are confined to the zone of winter-spring warming at the bank of the Tracy river (Nevada county of California state), where it grows under canopy of floodland forests of California poplar, small-leaved alder, Virginian plum, numerous willows, Wood's brier, and other trees. Two populations of this species are known from a very restricted territory of 50 × 5 m. One population had 8–9 small bushes, and the other about 40. In spite of the fact that Sonne's barberry produces aerial shoots from rhizomes and produces flowers and fruits regularly, its condition in nature is critical. Its spread by vegetative means does not extend beyond a small territory, and the fruits formed on the plants often do not have seeds. The plant has excellent ornamental and soil-binding qualities. Therefore, it can be used in cultivation for preventing water erosion. It has been planned to create a special natural reserve on a small territory at its natural habitat to protect this endangered plant. This species has been included in the National List of Plants of the United States and in the Red Data Book of IUCN.

The endemic barren-wort genus (*Vancouveria*) of North America comprises 3 species, which are distributed along the Pacific coast from Washington state to California. It is a herbaceous plant up to 45 cm tall with bilobed or trilobed leaves. The inflorescence is paniculate, flowers freely suspended on long pedicels, outer sepals numbering 6–9 and soon dropping off, inner sepals similar to petals; corolla forms the innermost whorl.

American barren-wort (*V. hexandra* C. Morr. et Decne.) has white, 1.3 cm long flowers with yellow nectaries. The leaves are thin, dropping toward the end of the vegetative growth of the plant. The plant flowers from May to July. It grows in humid shady forests in northwest California and western Washington state. **Smooth-petalled barren-wort** (*V. planipetala* Calloni) also has white but shorter (8 mm long) flowers. The leaves are thick, lustrous, almost leathery, retained on the plant throughout the winter. The plant flowers during May–June. It grows in shady forests of the coastal ranges (California) and in the southwestern parts of Oregon. The **golden barren-wort** (*V. chrysantha* Greene) with 1.3 cm long yellow flowers (Plate 33) is the rarest species of the genus. Its leaves are retained over the winter. It flowers in June in open areas and in the shrubby outgrowths in northwestern California and southwestern Oregon. Golden barren-wort belongs to the category of vulnerable plants.

FAMILY BETULACEAE

The **fruiting birch** [*Betula uber* (Ashe) Fernald], endemic to the Appalachian mountains, is among the species on the verge of extinction. It is assumed that this species is of hybrid origin (*B. pumila* L. var. *glandulifera* Rogel and *B. lenta* L. or *B. alleghaniensis* Britton). It was discovered in 1914 in the southern part of the Appalachian mountains (Smith county, Virginia state). In 1953, its collections were not confirmed, and it was resolved that the specimens preserved in herbaria belong to an entirely different birch species with different characters and could not be considered an independent taxon. In 1975, a population of 12 adult trees and 21 seedlings was found in another place on private land. Some young plants were removed for transplantation and scientific observations and a few adult and young specimens died at different times for various reasons, but 2 trees were soon found at a short distance on the territory of the Jefferson National Park. The natural habitats of the fruiting birch are confined to mixed deciduous forests in the southern parts of the Appalachian mountains, which are rich in broad-leaved trees. A fence has been constructed on the private lands as well as in the national forest to protect the only surviving population of this species. Attempts to cultivate this plant have been successful. About 30 plants raised from seedlings are growing. Seedlings have also been obtained from collected seeds, of which about 1 percent are similar to the parent plants, and the others have characters of *B. lenta*. The fruiting birch has been included in the Red Data Book of IUCN and the National List of Plants of the United States.

FAMILY BRASSICACEAE

The **torulose laevenworthia** (*Laevenworthia torulosa* A. Gray) belongs to the family Brassicaceae. It is an annual plant, whose entire life cycle is completed during the winter. The seeds mature and disperse in May and germinate only during autumn when the soil is highly moist and becomes cold. If the autumn is warm, seed germination is delayed. The sensitivity of seeds to moisture and low temperatures is considered an ecological adaptation to drought conditions not only for this species but the entire *Laevenworthia* genus (11 species). The communities with laevenworthia are characterized by the presence of areas with open calcareous or gravel deposits. Torulose laevenworthia often grows on shallow soils with a depth not exceeding 5 cm, and their humidity depends only on atmospheric precipitation. Till recently, this plant was the most widespread in the genus. However, it is now included in the category of vulnerable plants because of destruction of the habitats over its entire area of distribution. Its distribution covers the central regions of Tennessee state with a total area of

980,000 ha; 2 small populations, each with an area of 10 m², are known in Kentucky and one in northern Alabama. The disappearance of laevenworthia is accompanied by reduction in the population of other endemic plants adapted to the same habitats.

Bladder-pod (*Lesquerella densipila* Rollins) is a small, 10–40 cm tall, annual plant, covered with simple and branched hairs. Leaves at the base of stem are pinnate, petiolate; cauline leaves are lobed or toothed, sessile. The paniculate inflorescence consists of a large number of female flowers on pedicels. The *Lesquerella* genus includes 69 species distributed in the North American continent from the arctic to the southern border of Mexico. The southwestern part of the continent has the largest number of rare, vulnerable, and disappearing 3 species of the genus, of which 21 taxa, including the hairy bladder-pod, are included in the National List of Plants of the United States. Even recently it was found in fairly large numbers in a majority of habitats in Tennessee in the main area of its distribution, in the zone of spring floods of the Duck, Harpeth, and Stones rivers. All the areas with bladder-pod are within a radius of 40 km. A few isolated habitats are known in the northern part of Alabama. The decrease in the number of plants is mainly associated with destruction of its habitats. Bladder-pod is included in the Red Data Book of IUCN.

FAMILY BURMANNIACEAE

The family Burmanniaceae includes herbaceous autotrophic or saprophyte plants distributed in the tropical regions. The areas of distribution of only 2 *Thismia* species are located beyond the tropics. One of them, **American thismia** (*T. americana* N.E. Pfeiffer), is from North America, and the other, **Rodway's thismia** (*T. rodwayi* F. Muell.), from Tasmania and New Zealand. At present, the former species has disappeared in nature. It was seen for the last time in 1913 at the border of the meadow on the eastern side of Lake Calumet, between Torrens Avenue and the Nickel Plate railway line. Now this area is occupied by the industrial establishments of Chicago; all the attempts to find this plant in the earlier known habitats or under similar conditions were unsuccessful. It has also not been preserved under cultivation. The American thismia is a very small, white, saprophytic plant with an erect, up to 6 mm tall stem, ending in a whitish-green flower. The roots are short and the leaves are reduced into scales. At one time, moist meadows with a continuous cover of dormant mosses were its typical habitats. This environment is completely opposite to the one in which the tropical species of thismia grow (the latter inhabit fallen trunks or leaves under the canopy of rain forest). The loss of this species from the world flora is mentioned in the Red Data Book of IUCN.

FAMILY CACTACEAE

The family Cactaceae consists of xerophytes with succulent cylindrical, spherical, columnar, or round stems and leaves modified into spines. The cacti are distributed most unevenly in the American continents: they are almost completely absent in the tropical rainforests of Central and South America, but are represented by a huge variety of forms in Mexico and hilly deserts of Peru, Chile, and Bolivia. Most cacti are adapted to desert life, some species grow in dry, thorny, sparse forests; rarely these plants can be found in the tropical evergreen forests of the Amazon River basin. Many species are ornamental, cultivated in public and private collections, and are objects of trade. For the conservation of all cactus species, the entire family has been included in CITES. Some particularly rare species have been included as independent items. Cacti have been used as medicinal plants since ancient times.

Classification of cacti is difficult due to large variability in nature, ease of interspecific and intergeneric hybridization, and absence of herbarium material in several cases. Therefore, the estimates of different authors differ widely: number of genera from 80 to 225, and species from 800 to 2000.

Tobusch's ancistrocactus [*Ancistrocactus tobuschii* (W.T. Marshall) Backeb. ex L. Benson] is an extremely rare plant with limited distribution and is endangered due to excessive collections. Two small populations of this dwarf cactus, located at a short distance from each other, are known in Texas state, one at the border of Bandera and Kerr counties, and another in the Big Bend National Park. This ancistrocactus grows on limestone deposits at an altitude of 450 m in the canyons facing the Edwards Plateau among the outgrowths of junipers, oaks, and grasses. A few bright yellow flowers develop during the blooming period of this spherical thorny plant, which is up to 5 cm tall and slightly flattened at the top. The small size of the cactus and its bright flowers attract the attention of collectors. This plant is found in several private collections. Under natural conditions, it is being conserved in the Big Bend National Park. It has been included in the National List of Endangered Plants of the United States and in the Red Data Book of IUCN.

Goat-horned astrophytum [*Astrophytum capricorne* (Dietrich) Br. et R.] is a member of a small genus of star-shaped cacti (4–6 species). It grows in the northern part of Mexico (Coahuila state) on open and brightly illuminated areas. The only source of shade in these areas are its long, gray, flat, bitwisted needles, for which it has been called 'goat-horned'. The cactus itself is columnar, its large yellow flowers with red center persist through the entire summer (Plate 34). It is characterized by slow growth; the young seedlings are sensitive to moisture deficiency.

Agave-like ariocarpus [*Ariocarpus agavoides* (Castan.) E.F. Anders] is a small plant with a rosette of up to 4 cm long, leaf-like organs, semicircular in cross-section. It produces pink flower in November–December, with reddish chocolate-brown fruits developing later. It is successfully multiplied by seeds, but grows very slowly. It prefers dry stony calciferous soils. Its only location is the area between Monterrey and San Luis Potosi (Tamaulipas state, Mexico), at an altitude of 1200 m. This region has a great variety of cacti, including rare endemics and endangered species.

Boat-nose ariocarpus (*A. scapharostrus* Bödeker) is a rare species. Known from somewhat closely located habitats in Nuevo Leon state (Mexico) with a small number of specimens, this cactus has a short stem, consisting of numerous papillae, each of which is similar to the nose of a boat. The flowers are pink-purple (Plate 34). The cactus grows in limestone mountains at an altitude of 820 m. All members of the Ariocarpus genus are in great demand in the world market. It is difficult to raise them from seeds, therefore, collectors prefer adult plants. The only source of such cacti is the natural habitat. At present, the Mexican government has banned the export of all wild cacti. The genus is included in the CITES.

The monotypic genus *Aztekium* is represented by an endangered species. This is **Ritter's aztekium** (*A. ritterii* Bödeker), a unique plant. It is a minute, roundish plant, slightly flattened on the top, its light bluish stem covered with numerous furrows alternating with ridges. Its apical, white, polypetalous flowers differ from the flowers of other cacti. Aztekium grows on limestone rocks in Nuevo Leon state. A small population of this species near Los Raiones grows on the walls of the canyon together with *Ariocarpus scapharostrus*. This population has 2000–3000 plants. Aztekium was also found in the outskirts of Cadereyta (Queretaro state), but this population was destroyed by fire in 1970. The critical condition of this species in nature is associated with commercial trade (the plants were dug out by the thousands to meet the demand in the world market, and the local people used them as decorations at Christmas). The plant has been included in Appendix 1 of CITES so as to impose strict control on the export and import of aztekium and protect its natural population.

The hedgehog cactus genus (*Echinocactus*) has about 1000 species. They have a roundish stem with longitudinal ridges which are modified into tubercles. Most species are ornamental, which is the cause of their fast reduction in nature.

Gruson's hedgehog cactus (*E. grusonii* Hildm.) is endangered. Its young and adult plants have the almost regular shape of a ball. The largest plants are 1.3 m tall and 1 m in diameter, with the stem slightly flattened at the top. The entire plant is covered with clusters of 8–10 whitish, about 3 cm long spines. Each cluster carries up to 4 cm long, curved, pale or golden yellow spines in the center; the flowers are light chocolate-brown (Plate 34). The mature seeds have a good germination. Large-scale collections of adult plants lead to a decline in the already depleted populations, whose distribution is restricted to the territory extending from San Luis Potosi state to Hidalgo state of Mexico. Gruson's hedgehog cactus is cultivated in the United States and in southern France.

Engelmann's echinocereus (*Echinocereus engelmannii* Lem.) is a cactus with cylindrical stem, almost uniformly covered with spines, some of which are long and reddish, others gray and much shorter than the former and arranged in clusters around the longer spines. The flowers are pinkish-red. Three varieties are distinguished on the basis of shape and color of spines, of which var. *purpureus* is distributed in Utah state, and var. *howei* and var. *munzii* in California. All the varieties have limited distribution and are included in the list of endangered and vulnerable plants of the United States. The category of endangered plants in this list has a total of 7 taxa, and the category of vulnerable plants has 5 taxa. **Lindsay's echinocereus** (*E. lindsayi* Meyran.), growing on rocks together with other cacti and succulents including *E. engelmannii, Ferocactus tortulispinus, Idria columnaris, Yucca whipplei,* and *Agave deserti,* is included in the second category. The only known habitat of Lindsay's echinocereus is in Baja California, Mexico. This plant may soon disappear if measures are not taken for its protection. The catastrophic reduction in its population is a result of collection for commercial trade. One of its habitats was destroyed during construction of a highway. The transfer of Lindsay's echinocereus from Appendix 2 to Appendix 1 of CITES imposed strict control on the trade of this species.

The list of endangered plants includes a variety of **Reichenbach's echinocereus** [*E. reichenbachii* (Anus) Engelm. ex Haage var. *albertii*], the only habitat of which is in Texas. Its population strength is not determined because of difficulty in access to its habitat, which is surrounded by dense outgrowths of scrub vegetation (Plate 34).

The variety of **prickly barrel cactus** *Ferocactus acanthodes* Br. et R. var. *eastwoodiae* is a beautiful and very rare cactus from Arizona. It is known from 2 distant desert habitats. The plants are adapted to rocks, where they grow in fissures and scraps, practically inaccessible to collectors. The adult specimens reach 2 m in height and 50–60 cm in diameter, with their surface densely covered with golden yellow, 5–7.5 cm long spines (Plate 35). The weight of such a plant is about 150 kg. This variety has been included in the National List of Endangered Plants of the United States.

Glaucescent barrel-cactus [*F. glaucescens* (DC.) Br. et R.] is characterized by bluish-green stem. Its spines and flowers are yellow, fruits are white with pinkish tinge (Plate 35). This is one of the most beautiful barrel cacti, growing in mountainous limestone rocks at an altitude of 1500 m in the eastern regions of central Mexico.

De-Negri's obregonia (*Obregonia denegrii* Frič.) has numerous grayish-green, fleshy papillae up to 1.5 cm long, giving an appearance of a raised rosette with a diameter of 20–25 cm. It produces white flowers and fruits (Plate 35). It grows on dry limestone slopes at an altitude of 760 m in the communities of xerophytes together with *Mamillaria, Opuntia,* and various other cacti. It is on the verge of extinction. At one time it was found along the slopes of 2 valleys in Tamaulipas state, Mexico, but now not a single specimen of obregonia survives there. The earlier known population has been found west of San Antonio and must be protected.

De-Negri's obregonia is known to many collectors and cactus admirers in many countries of the world, although its first description was published in 1925. Because of the difficulty in raising plants from seeds, the collectors prefer to buy adult plants. To enhance international control on export and import, this species has been shifted from Appendix 2 to Appendix 1 of CITES. The species is included in the Red Data Book of IUCN.

Knowlton's snowball cactus (*Pediocactus knowltonii* L. Benson) is an endangered species in the United States. It produces small, succulent stems up to 25 mm tall and 19 mm in diameter, which are in clusters. The flowers are white-pink, with a diameter of about 2 cm (Plate 35). Only 2 populations of this species have survived, the total number of plants in which is difficult to estimate, as this cactus is small and often grows among other plants. It grows along the boundary of New Mexico and Colorado, near the Los Pinos River at an altitude of 1800 m in pine-juniper communities. Future survival of this pediocactus is threatened by flooding of the area during heavy rains due to construction of a dam on the Los Pinos and San Juan rivers. The heavy recreational stress is already making the survival of this plant doubtful, and collection of the plant for commercial trade would lead to complete extinction. Unfortunately, no effective measures are yet being taken to protect this plant in nature. It is only proposed to reduce the recreational stress in one of the areas. The plant is included in the Red Data Book of IUCN.

Winkler's snowball cactus (*P. winkleri* Heil) was discovered in 1960, but initially included in another species. Its independent species status was confirmed only in 1979. Taxonomically, it is closest to Knowlton's snowball cactus and differs from the latter by the number of stems in the cluster and their size (height 3.9–6.8 cm and diameter 2–2.6 cm), strong hairy growth, the number of spines on the outgrowths (9–11 as compared to 18–26 in Knowlton's pediocactus), and flowers of peach color (Plate 36). The distribution of this species is limited to southeastern Utah state, where it grows on mountain peaks, slopes, and foothills on gravelly and coarse sands in the Navaho desert, rising up to an altitude of 1600 m. It is extremely rare.

Brady's snowball cactus (*P. bradyi* L. Benson) and **Siler's snowball cactus** [*P. sileri* (Engelm. ex Coult.) L. Benson] have also been included in the category of endangered plants. Both cacti grow in Arizona. Brady's snowball cactus is particularly attractive during flowering, but it is not easy to locate it, as it flowers only for 2–3 days in a year. For the rest of the year, it is almost mixed with soil, because of its small size and a thick layer of dust covering the stem. Till now, very little was known about the biology of this species as well as about the importance of its dry fruit which is used by the local Indians.

The population of Siler's snowball cactus has dropped to critical level due to excessive collections. This species can be preserved only by extending the boundary of the nearby national monument.

The **Peeblesian snowball cactus** [*P. peeblesianus* (Croizat) L. Benson var. *peeblesianus*], discovered in 1938, is included among the endangered plants. It is a small plant, the main part of the huge stem is underground. It is well adapted to severe growth conditions. The limited area of its distribution is in the gravelly uplands to the north and northwest of Holbrook (Arizona state); it is distributed over a length of 8–11 km.

The plants of the genus *Pelecyphora* are endangered. The **sowbug-shaped pelecyphora** (*P. aselliformis* Ehrenberg) has round and mildly flattened stem with chocolate-brown to gray papillae and numerous thin and flattened hooks, which resemble the body segments of the sowbug in shape. Its flowers are pink-red (Plate 36). It grows on limestone rocks at an altitude of 1900 m together with *Prosopis* sp., *Jatropha* sp., and *Opuntia* sp. The habitat of this cactus is known only in San Luis Potosi state (Mexico). The population strength of the species is unknown.

False comb-like pelecyphora (*P. pseudopectinata* Backbg.) is the rarest plant of Mexico. It has a globose shape, reaching 6 cm in height and 4.5 cm in diameter; spines in the form of small comb-teeth adjoin the slightly extended and flattened papillae; flowers are white to pink, 2.3 cm in diameter, with a reddish stripe in the middle of petals (Plate 36). It is found in northern Mexico (Coahuila and Nuevo Leon states) on the scraps of rocks, at the places of humus accumulation. The plant is highly ornamental, much valued for collection.

The **cone-shaped pelecyphora** [*P. strobiliformis* (Werdermann) Frič] has a stem resembling coniferous cones. The papillae are flat, imbricately arranged, rusty to chocolate-brown; flowers are pink-red. A few habitats of this species are known in Tamaulipas and Nuevo Leon states (Mexico). It grows on limestone rocks at an altitude of 1500 m. The disappearance of this pelecyphora is related to uncontrolled collection of plants and commercial sale. Including the genus in Appendix 2 of CITES was found to be insufficient and it was therefore shifted to Appendix 1.

Twisted sclerocactus (*Sclerocactus contortus* Heil) was discovered in 1979, and immediately the question was raised about the need to protect this species. This is a plant 3–9.5 cm tall and 2–8 cm across, with 13 costate, spirally twisted outgrowths. The papillae are 12–15 mm long, terminating in long, up to 7 cm, white, twisted spines. The flowers are dark pink, up to 4 cm in diameter (Plate 36). Its distribution is limited to the territory of the Canyonlands National Parks, Utah state. It grows on hills and mountains in the Navaho desert in pin-juniper outgrowths at an altitude of 1500–1600 m. The population of this species is small.

Terre-Canyon sclerocactus (*S. terrae-canyonae* Heil) has a small distribution. It covers southeastern Utah, southwestern Colorado, and northwestern New Mexico. It grows on the peaks, slopes, and rock base, on sandstone rocks, in pine-juniper communities at an altitude of 1800–2100 m. It is morphologically similar to twisted sclerocactus, but slightly taller, up to 11.5 cm high, and broader, 8–11 cm across, and its spines are shorter, 5 cm long, and slightly bent. The flowers of this plant are light yellow, 5.5 cm in diameter. This sclerocactus does not form large groups, and its total population size is unknown.

Multihooked sclerocactus (*S. polyancistrus* Br. et R.) was found in the Mohave desert. The young plant is usually spherical, growing uniformly in height and width, but the adult plant becomes ovoid. The stem reaches 30 cm in height and is covered with up to 12.5 cm long, white flat and red cylindrical hard spines with a hooked tip. The flowers are pink, 7.5 cm in diameter, and fruits are red (Plate 37). The cactus flowers from April to June. Its distribution is limited to the Mohave desert in California and western Nevada. It grows on the southern and western slopes of rocky elevations at an altitude of 750–1800 m together with *Echinocactus polycephalus*, *Opuntia basilaris*, and *Larrea tridentata*. This sclerocactus was adapted to the bottoms of ancient lakes and riverbeds of the Pleistocene. Its habitats have alkaline soils (pH 8.2) with high calcium and calcium carbonate content and low humus content, as well as dry air and soils over a larger part of the year. At present, the population of this species is declining due to absence of natural pollinators; overgrazing of cattle in its habitats; increase in recreational stress; and collection of plants. It has been established that loss of even a single population of sclerocactus decreases the possibility of outcrossing (which is essential for the species), thereby reducing its seed setting. At present, a part of its habitat is protected in the Valley of Death national monument in California.

Mesa-Verde sclerocactus [*S. mesae-verdae* (Boissev. et C. Davidson) L. Benson] is distributed in southwestern Colorado and northwestern

New Mexico. It grows on rocks where the soil layer is almost completely absent. The ecology of this plant has been studied insufficiently.

It has been reported that the sprouts develop and survive only in small rock crevices. The plants collected from nature do not respond to cultivation and die.

Glaucous sclerocactus [*S. glaucus* (K. Schum.) L. Benson] is an endangered plant of the United States. Its distribution is restricted to the areas famous for their crude oil reserves in western Colorado and Utah. The sclerocacti are included in Appendix 1 of CITES.

A few species of the genus *Thelocactus* need protection. **Yellow-spine bicolored thelocactus** [*T. bicolor* (Galeottii) Br. et R. var. *flavidispinus* Backbg.] has short, cylindrical stems compressed on the sides with 13 distinct green ridges and spines in clusters of 20. The spines are yellowish, 1.8–2.4 cm long. The flowers are 5–6 cm in diameter, purple or pink, red at the base, filaments white, anthers yellow (Plate 37). It grows near the Rio Grande (Texas). It has been included in the National List of Endangered Plants of the United States.

The **bolanian** variety of **bicolored thelocactus** (*T. bicolor* var. *bolansis* F.M. Knuth) from the environs of Sierra Bol (Chihuahua state, Mexico) is a large columnar plant up to 25 cm tall, covered with snow-white, long and hard spines (red spines appear at the apex in spring). It produces large, pale lilac flowers in summer.

Rinconada thelocactus [*T. rinconensis* (Pos.) Br. et R.] grows in the small area Rinconada, northeast of Saltito in the Chihuahua desert. It has an ovoid stem, 12–15 (20) cm in diameter and 8 cm in height, light bluish to grayish-green, with strongly projecting ridges (13–21) and large spines. The flowers are 4 cm in diameter, variable in color from pure white to pale yellow and pink.

Tula thelocactus [*T. tulensis* (Pos.) Br. et R.] was found near Tula (Tamaulipas state). Its habitat is near San Luis Potosi. The plant forms turfs. Its stems are ovoid, short cylindrical, with 8–13 ridges; the spines are 10–15 mm long, whitish. Flowers are small, red or pink with red streaks and broad wide or yellow margins.

Multicolored thelocactus [*T. heterochromus* (Web.) van Oost.] grows in northeastern Mexico (Coahuila state). Its stems are globose, up to 14 cm in diameter, hairy. The glaucous ridges support long white spines, 9 in each cluster, some of them bent. The flowers are large, 7.5–10 cm in diameter, reddish-purple with yellow stamens.

FAMILY CISTACEAE

Mountain American heath (*Hudsonia montana* Nutt.) is a relict plant, which differs significantly from other *Hudsonia* species by the primitive flower structure. It is assumed that the species evolved as a result of hybridization between 2 subspecies of American heath and further isolation of hybrid population during the postglacial period. At present, only 3 habitats are known. They are located in a hilly region in Burke county, North Carolina, which is being used intensively for recreation. The total population of this species here does not exceed 100 plants (including the young ones). The mountain American heath is a small shrub, 30–40 cm tall, morphologically resembling heath, but with yellow flowers clustered in terminal cymes (Plate 37). At present, all the areas with American heath are included in the territory of the Pisgah National Forest, but its population is gradually declining under recreational stress. Measures are being developed for restricting the factors of disturbance and maintenance of population level. The species is included in the Red Data Book of IUCN and in the national list of endangered and vulnerable plants of the United States.

FAMILY CRASSULACEAE

The **Traskia's dudleya** [*Dudleya traskiae* (Rose) Moran] belongs to the family Crassulaceae. Till 1975, this plant was considered extinct: its repeated search on the Santa Barbara Islands in California did not yield results. A few regenerating specimens were, however, found in 1975, and a population of a few hundred plants was discovered a year later on a vertical rock. This dudleya is a herbaceous perennial with short branched stem, capitate inflorescence, and rosette leaves. It grows on rocks facing the sea, at an altitude up to 100 m. Its distribution is limited to the Santa Barbara Islands over an area of 259 ha. Development of farms, sheep grazing, and introduction of European hares are responsible for the disappearance of dudleya and the entire natural vegetation of the islands. In 1938 in connection with the establishment of the Channel Islands National Memorial, the Santa Barbara Islands were included in the system of protected territories. The farms were closed, and cattle grazing stopped. Further survival of dudleya depends on population control of hares, which is being undertaken by the US National Parks' Administration. The species was included in the National List of Endangered Plants of the United States and in the Red Data Book of IUCN.

FAMILY CUPRESSACEAE

The **Monterey cypress** (*Cupressus macrocarpa* Hartweg) of the family Cupressaceae is distinguished by an umbellate crown broadening upward,

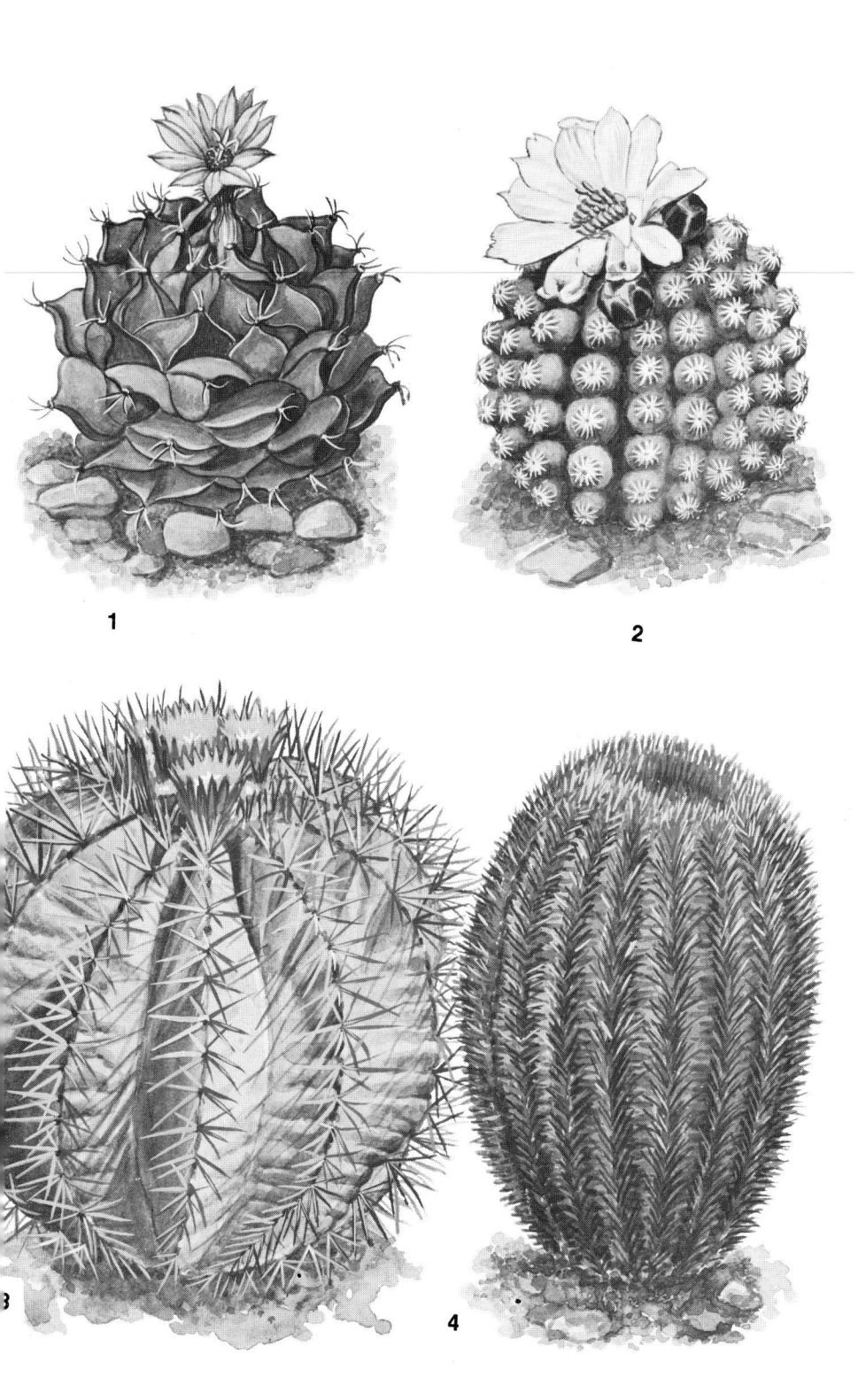

PLATE 35. 1—De-Negri's obregonia; 2—Knowlton's snowball cactus; 3—glaucescent barrel cactus; 4—prickly barrel cactus

PLATE 36. 1—Twisted sclerocactus; 2—Winkler's snowball cactus; 3—false comb-like pelecyphora; 4—sowbug-shaped pelecyphora.

PLATE 37. 1—Mountain American heath; 2—Monterey cypress; 3—bicolored thelocactus; 4—multihooked sclerocactus.

PLATE 38. 1—Lily-like fritillary; 2—Chiapa tiger flower; 3—Venus fly-trap; 4—American wych-hazel.

growing to a height of 20–25 m (Plate 37). The leaves have the smell of lemon. The symmetry of the crown is disturbed with age and the color of bark gradually changes from dark chocolate-brown to gray. The tree grows up to 300 years. This rare species forms 2 plantations on the Pacific coast of the United States in central California: the Point Lobos plantation (length about 4 km, width 180 m) near Monterey, and the Point Cupress plantation in the environs of Carmel. These coastal regions have an annual precipitation not exceeding 430 mm, are frost-free, and the temperature does not rise above 32°C. Cloudy and foggy days are frequent, and strong sea winds maintain constant humidity during a large part of the year. The soil contains high proportion of lime. At present, the Point Lobos plantation is included in the reserve of the same name and forms a part of the state park since 1933; Point Cupress is located on the territory of the protected Del Mont Forest.

The Monterey cypress is widely used for city plantation, as it has a beautiful crown, is easily reproduced by seeds, and grows rapidly. A few specimens of this tree grow in the Soviet Union in the area of Sochi, Adler, and Sukhumi, as well as on the southern coast of Crimea. Monterey cypress is included in the Red Data Book of IUCN; it has been included in the category of vulnerable plants in the List of Protected Plants of the United States.

Arizona cypress (*C. arizonica* Greene) is distributed in the more southern regions of the United States. Five geographically isolated varieties of this cypress are distinguished, which are considered by some botanists as independent species. Two varieties—var. *nevadensis* and var. *stephensonii* are included in the category of vulnerable plants in the National List of the United States. Their distribution is limited to California.

The **Abrams** variety of **Goven cypress** (*C. goveniana* Gord. var. *abramsiana*) is included in the category of endangered plants. This is a shrub up to 1.5 m tall. Its bark is smooth, branches thin, leaves flat, and cones semiconical or ovoid. Its only location is known at the Pacific coast in the environs of Monterey, California.

FAMILY DROSERACEAE

The fly-trap genus (*Dionaea*), with only one species, **Venus fly-trap** (*D. muscipula* Ellis), belongs to the family Droseraceae, whose area of distribution covers the Atlantic coast of South America. This endemic plant is found over a distance of 320 km from Beaufort county (North Carolina) to Charleston county (South Carolina). The Venus fly-trap grows in marshes among pine forests. It prefers open peaty areas, mainly pads of sphagnum moss, where other insectivorous plants are also found: sundew, butterwort, and *Sarracenia*. This plant bears a rosette of leaves at the base, and the stem terminates in an inflorescence of 4–10 large white flowers (Plate 38). Flowering begins at the end of May and continues till mid-June; young seedlings appear at the end of July. The leaves of fly-trap are unique: each has a winged petiole and bilobed blade; long teeth are located along the margins of leaf blades, and 3 sensitive hairs in the middle of each lobe. When an insect attracted by the smell of nectar touches these hairs, the lobes of the leaf close 10–30 seconds, the marginal teeth are entangled with each other, and the insect cannot escape from the leaf. After the assimilation of the digested parts of the insect, the leaf blade opens again and the chitinous residues are carried away by rain or wind. Each leaf performs the role of a trap not more than 2–3 times, after which it turns black and dies, and is replaced by a new leaf.

Like other members of this family, the Venus fly-trap has not lost its usual method of nutrition, typical for all green plants, but growing on poor peaty soil with a high nutrient deficiency, it supplements nutrition through live food. The plant does not die for want of it, but it becomes very weak.

The survival of Venus fly-trap is threatened by the drying up of the marshy territories with subsequent agricultural use or for establishing productive forest plantations which completely disturb the conditions of moisture and light, change the soil composition, and facilitate the spread of more competitive species better adapted to the new conditions. The other reason for the reduction in the population of fly-trap is digging of living plants for cultivation. The fly-trap survives well when transplanted to appropriate conditions. The plants raised from seeds are more viable under cultivation.

At present, about 100 habitats of Venus fly-trap are known in nature. It can grow on open areas along roads and in young pine plantations, as well as in the mountains where there is no shade or competition from other species, and the level of ground water is higher than in the forest plantations.

In North Carolina, the fly-trap is protected by law, but its populations are continuously declining. With increase in the demand for this plant in the world market, the annual sale increased from 1.4 to 4.5 million plants (it is sold in Europe, Australia, Japan, and South Africa). The species has been included in the National List of Endangered Plants of the United States and in the Red Data Book of IUCN.

FAMILY ERICACEAE

The family Ericaceae is fairly widespread in North America and is distinguished by a great variety of species. In the United States, 36 species and subspecies of this family are included in the category of endangered and vulnerable plants, 24 of which belong to the bearberry genus (*Arctostaphylos*).

The **dense-flowered bearberry** (*A. densiflora* M.S. Baker) is among the endangered plants. It is an evergreen shrub with spreading branches, covered with dense racemes of white-pink flowers. At present only 2 populations have been preserved: one of them includes 19 adult and 26 young plants, and the other 10 adult and 3 young plants. The plant was found in the western part of Sonoma county (California) on acidic sandy soils, covered by a compact crust during summer, which prevents excessive water loss and converted into thin slush during winter. The seeds germinate in this slush. Californian coffee-tree, 2 species of *Ceanothus* (family Rhamnaceae), and *Xerophyllum* (family Liliaceae) grow together with bearberry. Most probably, in the past bearberry occupied a fairly large territory, which was subsequently brought under orchards and vineyards. Abundant flowering and fruiting, high seed germination, and easy maintenance of the plants have earned bearberry fame as an excellent plant for cultivation. At present, this species is included in the List of Endangered Plants of the United States and in the Red Data Book of IUCN.

Bunchy elliotia (*Elliotia racemosa* Muhlenberg ex Elliot) is endangered. Initially it was a fairly common plant in the valley of the Savannah River (Georgia state). Now only a few isolated areas are left in the central part of the state, where this elliotia grows in low-lying, wet but well-drained soils in mixed forests together with other ericaceous plants and Virginian magnolia, as well as in pine forests and pine-oak plantations. A few habitats of elliotia were also known in South Carolina state, but by now they have disappeared. The disappearance of this plant from the natural habitats and universal reduction in its populations is related to forest felling, sale of timber, and further clearing of the areas for agriculture. To conserve this species, the Department of Natural Resources of Georgia has placed under protection the Big Hammock territory (the largest number of elliotia populations has been found in this 300 ha area of Pleistocene sands). The plant is also protected in the 29 ha Charles Harold Natural Reserve.

Observation of elliotia in the protected territories showed that its seed productivity is very low (the reasons are not yet clear). It has been assumed that the plants are self-sterile and each population consists of only one clone. That is why there is practically no rejuvenation under natural conditions. Under cultivation, elliotia is propagated by rhizome cuttings. The plant is cultivated in a few botanical gardens and its snow-white inflorescence attracts the attention of many horticulturists. It has been included in the Red Data Book of IUCN.

FAMILY FABACEAE

The milk-vetch genus (*Astragalus*) of the family Fabaceae is distinguished in North America by the richness and variability of species. This is indirectly indicated by the number of species and varieties included in the List of Protected Plants in the United States. At present their number is about 100.

Beatley's milk-vetch (*A. beatleyae* Burneby) is a perennial, fast-growing herbaceous plant with bluish-violet flowers and pale green fruits. At present, it has survived in only 2 small areas (Nevada state) located at a distance of 13 km from each other. One area extends over 5 km, and the other much less. Beatley's milk-vetch grows on volcanic soils under the canopy of pine and juniper, in community with shrubby wormwood, other milk-vetch species, phacelia, and other plants. A road has been constructed through its habitat. It is proposed to be fenced so as to eliminate human movement along the roadside areas where this plant grows.

Purple red milk-vetch (*A. phoenix* Barneby) is adapted to the Ash Meadows plains, located at the northern tip of the Mohave desert. Springs with low salt content exist in the plains, and the ground water is shallow. All this makes vegetative growth of the plants possible in the plains even during summer in spite of their proximity to the desert. The vegetational cover is predominated by bunchy-leaf wormwood and haplopappus. Five more plant species, endemic only to Ash Meadows and included in the National List of Endangered Plants of the United States, grow together with milk-vetch: *Grindelia fraxino-pratensis*, *Mentzelia leucophylla* Brandegee, *Ivesia eremica* Rydberg, and *Centaurium namophilum* Reveal, Broome et Beatley. All the plants have been found on a small area of 20 ha.

The population of purple red milk-vetch, forming compact hemispherical pads 40–50 cm in diameter and carrying pink-purple flowers, is declining due to overgrazing of cattle. The only possibility of preventing its extinction is to ban cultivation of the land and cattle grazing. Both the milk-vetch species described above are included in the Red Data Book of IUCN and the National List of Endangered Plants of the United States.

FAMILY FAGACEAE

The oak genus (*Quercus*) comprises about 450 species, the largest number of which are typical to North and Central America. They are distributed from southeast Canada to the Columbian Andes. **Shumard's oak** (*Q. shumardii* Buckl.), a tall tree with deeply cracking bark, is a fairly rare species. Its leaves are glabrous, only slightly pubescent along the veins on the lower surface, with 5–7 lobes deeply cut and wide apart, and toothed apex. Diameter of cupule is 1.5–3 cm, tightly sticking to the acorn. The oak grows in low-lying and moist soils. The **maple-leaved Shumard's oak** (*Q. shumardii* var. *acerifolia*) is included in the list of plants that may be endangered in the near future if special conservation measures are not adopted. It has 5-lobed leaves, resembling the leaves of maple, and is found in Arkansas state.

Quercus georgiana M.A. Curt, *Q. oglethorpensis* W.H. Duncan, *Q. parvula* Greene, and *Q. tomentella* Engelm. are also included in the category of vulnerable plants. *Quercus hinckleyi* C.H. Mueller, *Q. graciliformis* C.H. Muell., and *Q. tardifolia* C.H. Mueller are endangered. These 3 species are distributed in Texas.

Ozark's chestnut (*Castanea ozarkensis* Ashe) is a disappearing plant of the Ozark plateau (Missouri, Arkansas, and Oklahoma states). It is a 20 m tall tree. Its leaves are narrow, oblong-elliptical, 15–20 cm long, acute, with serrate margins and with cusps. It grows on dry hillocky elevations.

FAMILY GENTIANACEAE

The family Gentianaceae includes **Centaurium** (*Centaurium namophilum* Reveal, Broome et Beatley), an annual plant up to 45 cm tall with linear-lanceolate leaves and long panicles (more than half the length of the entire stem) consisting of deep pink flowers with white base. Its main habitats are the mid-mountain meadows at an altitude of 600–700 m around the springs, near rivulets. The areas with this centaurium are known in southwestern Nye county (Nevada) and Inyo county (California). As a result of cattle grazing, hay cutting, and regulation of runoff, the population of this plant has sharply declined, and has even disappeared completely at some places. To conserve this species, included in the National List of Endangered and Vulnerable Plants of the United States, it is proposed to develop a reserve over an area of 20 ha in the Ash Meadows region (Nye county), where 4 more endemic species grow, one of which has been included in the Red Data Book of IUCN, with this centaurium.

FAMILY HAMAMELIDACEAE

The family Hamamelidaceae is represented in North America by the monotypic American wych genus (*Fothergilla*). The **American wych-hazel** (*F. gardeni* Murr.) is a deciduous shrub with entire pinnately veined leaves, reappearing after flowering. Its bisexual, apetalous, white, and fragrant flowers are arranged in terminal capitate or spicate inflorescence

(Plate 38). This plant is found in Georgia and Alabama states. It is included in the National List of Endangered Plants of the United States.

FAMILY IRIDACEAE

The tiger-flower genus (*Tigridia*) is represented in Mexico by a number of species. The members of this genus are highly ornamental plants, used in floriculture.

Chiapa tiger-flower (*T. chiapensis* Molseed ex Cruden) is a bulbous perennial. The stem is up to 36 cm tall with basal leaves up to 40 cm long, cauline leaves 13–15 cm long; bracts are greenish, sometimes purple. Flowers are 4–5 cm in diameter, lobes of the outer whorl white with purple spots up to the throat, limb bright yellow; lobes of inner whorl are also yellow with spots along the margins (Plate 38). It flowers from mid-May to mid-July. The seeds mature in August. It grows in wet meadows in the outskirts of San Cristobal de las Casas.

Hinton's tiger-flower (*T. hintonii* Molseed) has 30 cm tall stem with 2–3 leaves at the base (about 40 cm long) and a few cauline leaves. Flowers are 2–3 cm in diameter white with purple spots in the centre. The only habitat of this species is the Fesnos pine forest (Mina region) at an altitude of 2100 m.

Purpus' tiger-flower (*T. purpusii* Molseed) is a bulbous plant up to 70 cm tall. Stem is branched, leaves are linear-lanceolate, 45 cm long. Flowers are violet, 5–9 cm in diameter. The plant was found near Tehuacan (Pueblo state) in a dry valley surrounded by flat desert areas.

FAMILY ISOETACEAE

One of the species of the family Isoetaceae—**Louisiana quill-wort** (*Isoetes louisianensis* Thieret)—was discovered only in 1973. Soon afterwards, the area of river where it was growing was destroyed during bridge construction. Since then, the plant was considered extinct; however, in 1978 it was again found upstream along the river. The new location has only 30 plants, and 20 more specimens were found in a small stream near the main river course. This extremely restricted area of distribution of Louisiana quill-wort is located in Louisiana state, in the environs of Washington-Perish. The quill-wort is a water plant with bilobed rhizophore and a large number of 15–40 cm long leaves. The leaves are flat on one side and roundish on the other. The sporangia with microspore (male) or macrospore (female) are axillary, protected by the leaf base. Other water plants are also found together with quill-wort, which grows in calm or flowing water along the banks with sandy or gravelly bottom:

American bur-reed, slender pondweed, and floating arum. Louisiana quillwort is included in the List of Endangered Plants in the United States. This list also includes *I. lithophylla*, distributed in Texas.

FAMILY LAMIACEAE

The family Lamiaceae includes **whorled wild rosemary** (*Conradina verticillata* Jennison), one of the 4 species included in the National List of Endangered Plants of the United States (there are totally 5 species in the genus). The whorled wild rosemary is the only member of the genus found in the Camberland plain (Kentucky and Tennessee); the remaining 4 species grow in Florida. Most probably, the habitats of whorled wild rosemary are relict remains of once very wide area of ancient ancestral type. The large-flowered wild rosemary is morphologically closest to this species.

Whorled wild rosemary is an evergreen semishrub, not exceeding 0.5 m in height, with opposite leaves and flowers arranged in verticillasters, giving a strong smell. Its distribution is limited to isolated areas in the Camberland plain. The plant is found on sandy river banks, shoals, ridges, and deposits, exposed after winter rains. Its population is shrinking as a result of destruction of the habitats and their use for recreation. Some habitats have been included in the protected natural territories of various regimes. It is proposed to list all the populations, determine their size, and also restrict recreational activity in the habitats.

FAMILY LAURACEAE

The candle-wood genus (*Persea*) of the family Lauraceae comprises about 50 species, distributed in tropical and subtropical America. The North American plants belong to several species, including **Borbon candle-wood** [*P. borbonia* (L.) Spreng.] also known as red laurel. This is a tree up to 20 m tall, with leathery, narrow, 10–20 cm long and 2–5 cm broad leaves, narrowed towards both ends. The fruits are black, dry, hemispherical, 10–12 mm long. The plant flowers in May–June. It grows in places water logged for several weeks or months along the borders of marshes.

The **dwarf Borbon candle-wood** (*P. borbonia* var. *humilis*) is included in the list of vulnerable plants. It grows in Florida. Its population is declining due to drying of the marshy areas.

FAMILY LILIACEAE

Not all the fritillaries are speckled. There are monotone species also, to which **lily-like fritillary** (*Fritillaria liliacea* Lindl.) belongs. This plant, found only in California, bears pure white, wide open flowers resembling a lily (Plate 38). Initially it was found in 9 counties, but during recent years its distribution has been sharply reduced, as also the population strength, because of change in its habitat due to clearing of for construction. Collection of plants has also reduced the population. This fritillary grows on heavy soils in areas with 650–2000 mm annual rainfall, with frequent fogs and strong winds, on open elevations, without raising above 150 m. It is proposed to organize conservation of all its known habitats and include this species in the National List of Endangered Plants of the United States. The plant is included in the Red Data Book of IUCN.

Interrate nolina (*Nolina interrata* Gentry), a North American plant, is an unusual species: its underground succulent rhizome, up to 3.5 m long, forms numerous aerial rosettes of 10–20 or more glaucous leaves. The leaves are 70–90 cm long and 8–15 mm broad. The peduncle, up to 2 m long, terminates in a dense panicle. Most probably, such a rhizome makes it possible for nolina to withstand fires and droughts, which are common in the areas of its distribution. Four habitats of this nolina are known in California, each of which contains a few populations; their total strength is a little more than 1000 plants. Nolina mostly propagates vegetatively, not flowering every year. Its rhizome is sold for its attractive shape. It grows on dry slopes of granite rocks and on rocky debris. It is a component of arid shrubby thicket, the chaparrals.

The interrate nolina, a Californian species, together with 2 Florida species of nolina, is included in the National List of Endangered Plants of the United States. Nolina is cultivated in the Huntington desert park of San Marino, California.

San Clemente island is located near southern California. It was never connected with the mainland; at the end of the Pleistocene, when the sea level rose, its area was significantly reduced.

San Clemente and the Mexican island of Guadalupe, 400 km farther south, have a unique flora, which differs from that of all the Californian islands and at the same time is a bitter example of human destruction of the island biota.

FAMILY MALVACEAE

The **Clement malacothamnus** [*Malacothamnus clementinus* (Munz et E.M. Johnston) Kearney], a highly branched, 1 m tall semishrub with 3–5-lobed leaves and numerous pink flowers, belongs to the endemic plants of San Clemente Island. Only 2 habitats of this species are known (each with 10 plants) on the walls of deep canyons below the chaparral belt at an altitude of 920 m. Together with wormwoods and four o'clock, malacothamnus form open plant communities. Most probably, in the past it was much more widely distributed over the island territory and occupied areas in the lower parts of slopes. Later, it was displaced by more tolerant species under the influence of cattle grazing and was preserved only in inaccessible places. The other rare species also had a similar fate. At present, a programme has been initiated on the island to eliminate grazing by goats and other domestic animals, and the status of rare plant species is being studied to determine the means for conservation and maintenance of their population levels. The Clement malacothamnus is cultivated on San Clemente Island in the nursery of natural flora as well as in the botanical gardens of California.

FAMILY NYCTAGINACEAE

The members of the family Nyctaginaceae are distributed in the tropical and subtropical regions of all the continents, but most of them are found in North and South America. The family includes 30 genera and about 300 species of trees, shrubs, and herbs; creepers are also found. Many species are ornamental. The four o'clock genus (*Mirabilis*) is represented by the largest number of species, almost 50 species in North and South America. The name in Latin language means 'wonderful'.

The **MacFarlane's four o'clock flower** (*M. macfarlanei* Constance et Rollins) is among the endangered plants of North America. It is a herbaceous perennial with thick, short-petiolate leaves and spirals of 4–7 bright pink-purple flowers. It blooms in May and grows on rocks along the Snake River and its tributaries in northwestern Oregon state and adjoining regions of Idaho. The **pudic four o'clock** (*M. pudica* Barneby) is a small plant, hardly reaching 20 cm; the leaves are short-petiolate, and the spirals consist of 6 white flowers. It grows in Nevada. Both species are included in the category of endangered plants in the United States. The decline in their population strength is a result of unrestricted collection of plants.

FAMILY ONAGRACEAE

The family Onagraceae includes the evening primrose genus (*Oenothera*). Some species of the genus are cultivated as ornamental plants. In the United States, 4 species and subspecies of the genus are included in the category of endangered plants.

Howell's evening primrose [*O. deltoides* Torrey et Frémont ssp. *howellii* (Munz) Klein] is a perennial herbaceous plant, 40–80 cm tall, with large white flowers up to 7.5 cm in diameter, in racemes. The flowers open in the late afternoon. The blooming period is fairly long, from March to May, and a brief flowering is also recorded in September. The plant is insect-pollinated. It grows on open, sunny, well-drained, shifting coastal sand dunes up to an altitude of 1050 m. Another endemic plant found with evening primrose is the narrow-leaved **capitate hedge mustard** [*Erysimum capitatum* (Douglas) Greene var. *angustifolium* (Greene) G.B. Rossbach] from the family Brassicaceae, also endangered.

The initial area under evening primrose was 200 ha. Its habitats were located along the Sacramento and San Joaquin rivers (California). The total length of the areas was 5 km and the width 400 m. As a result of agricultural and industrial exploitation of the lands, and construction of residences, the area under evening primrose has been reduced to 6 ha, and its population to a few hundred specimens. At present, it is proposed to protect one of the areas of coastal dunes as a preserve and also introduce evening primrose on an isolated dune located along the East Bay. The plant is cultivated in botanical gardens and private collections. Under cultivation, it lives for a few years, i.e., a perennial plant is changed into biennial. It is included in the Red Data Book of IUCN and the list of endangered plants of California.

FAMILY ORCHIDACEAE

The family Orchidaceae includes **medeo-like isotria** [*Isotria medeoloides* (Pursh) Raf.], an endemic of North America. This orchid is distributed sporadically, without forming large populations. It has a relatively large area of distribution and flowers rarely (interval between bloomings is 10–20 years). The population of this species is declining as a result of destruction of its habitats. At present, only 2 populations of 4 plants each have survived in Canada (southwestern parts of Ontario province). A total of about 10 habitats are known in the United States, from Maine to North Carolina, southern Illinois, Missouri, and Michigan. This isotria is found in dry, somewhat closed deciduous forests including maple, birch, mocjernut, beech, and hemlock, under the canopy of which Virginian medeola, Canadian may-lily, honeysuckle, Solomon's seal and other plants grow. This small ground orchid bears a rosette of dirty-green leaves, 1–2 yellowish-green flowers with almost white lip with green veins. It is included in the Red Data Book of IUCN as an endangered plant. This orchid is protected in all the states where it is found.

FAMILY PALMAE

The only palmetto species, **blue palmetto** [*Rhapidophyllum hystrix* (Pursh) H. Wendl. et Drude], grows in the southwestern part of North America, the coastal plain of central and northern Florida, and the southern part of Georgia, Alabama, and Mississippi states. This is a relict plant, as indicated by the isolated area of distribution and primitive inflorescence. It has a short trunk, covered with up to 30 cm long needle-shaped remnants of leaves and tomentum, remnants of vascular tissues. The leaves are fan-shaped, with a diameter of 50–70 cm; inflorescence is short and thick, densely covered with unisexual flowers. The maturing fruits form a single mass which remains for a long time among the petioles. That is why seedlings of this palm are rarely found and mostly it is reproduced vegetatively. The palmetto cannot withstand competition and grows slowly. It is found on sandy or marshy soils along rivers and in shady places. The decline in its population is related to digging of the plants and commercial trade, as well as poor seed set. The natural plantations of this palm are endangered in the Torrey and Hilands-Hemmock parks (Florida). The blue palmetto is being cultured in Miami, Florida and in the National Arboretum of the United States. Under cultivation the palm grows well and fruits in open areas, in contrast to its natural conditions, with sufficient moisture. The seeds remain dormant for 2 years. In the USSR it is cultivated in Sukhumi and further south.

Tall roystonea [*Roystonea elata* (Bartam) F. Harper] is the royal palm of Florida. It has a straight trunk up to 30 m tall, numerous pinnate leaves, and large inflorescence. It grows in dense moist forests and on marshes together with other trees, shrubs, ferns, and epiphytes. Four populations of this roystonea are known, which consist of a few hundred trees (adult and young). All the populations are located on protected territories: Kolie-Seminol, the Everglades, and Facahatchi-Strand. There is one more location where only young palms have been found, in the Corkscrew Swamp reserve. In spite of conservation measures, these palms are becoming extinct as is the remaining vegetation of the protected areas. The reason is the lowering of the water table and frequent fires. The threat of population decline due to digging of young plants, which was quite serious in the past, has been practically reduced to the minimum, as the seedlings grow very fast under artificial cultivation (around 6 m within the first 6 years). The plant is protected in Miami (Florida) in the Fairchild tropical garden. It has been included in the Red Data Book of IUCN.

FAMILY PAPAVERACEAE

The prickly poppy genus (*Argemona*) of the family Papaveraceae distributed in North and Central America, is distinguished by a large diversity (about 50 species).

Arizona prickly poppy (*A. arizonica* G.B. Ownbey) is a herbaceous plant trailing on the ground; stems decumbent; sap is white, sepals in the shape of a horn, thin, 15 mm long; stamens are much smaller than in other prickly poppies. The plant grows on steep precipitous slopes of the northern wall of the Grand Canyon (Arizona state) and flowers from April to November.

The **pinnatisect prickly poppy** (*A. pleiacantha* Greene var. *pinnatisecta*) is an endangered endemic of New Mexico state. It is distinguished by its pale lemon-yellow sap and deeply cut leaves.

Both prickly poppy species have been included in the list of endangered plants of the United States. The list also includes **Walpol's poppy** (*Papaver walpolei* A. Pors), a plant with a discontinuous area of distribution: one habitat is located at the coast of the Bering Strait, a few habitats are in the extreme east of Chukchi Peninsula and in the northwest of Seward Peninsula (territory of the USSR), and another habitat occupies the narrow coastal belt of Alaska (USA). Conservation of this species requires the joint efforts of the USSR and the United States.

FAMILY POACEAE

The maize genus (*Zea*) of the family Poaceae comprises only the cultivated maize and two annual subspecies: the Mexican (from Mexico) and the luxuriant (from Guatemala). The genetic variation in maize is relatively small. Sometimes hybrids among annual subspecies are formed which are mostly found at the borders of fields. Unfortunately, these annuals disappear due to cattle grazing and change in agricultural technology.

In 1910, a plant was found in Mexico which was very similar to maize, **teosinte**, or **perennial euchlaena** (*Euchlaena perennis* A.S. Hitchc.). At one time, its only habitat was in Halisco state in an area of about 100 ha. By now, this plant has been preserved under cultivation. It reproduces vegetatively by cuttings of rhizomes. It has been established that teosinte hybridizes easily with cultivated maize as well as its annual subspecies. These hybrids are most resistant to pests and diseases, which made it possible a few years ago to replace the old maize crop by a new one. Teosinte has been included in the Red Data Book of IUCN.

Sharp-pointed orcuttia (*Orcuttia mucronata* Crampton) and all other orcuttia species as well as the monotypic genus *Neostapfia* are not related to the family Poaceae and constitute a separate group of relict origin.

In orcuttias, this is manifested in specific growth conditions and developmental cycles. They are found on highly moist areas after the lowering of the water level (it rises as a result of winter-spring rains) on soils with high alkali content. The inflorescence of these plants is very compact and sticky and has a strong smell. The fruits require a long dormant period of 6–7 months. If the winter rainfall is very low, the seeds of the previous year lose viability and do not sprout.

The sharp-pointed orcuttia was discovered in 1958 in the basin of a small lake in Solano county of California. It is a small, pubescent, yellow-green annual up to 12 cm tall. The leaves are needle-shaped, 1–4 cm long; inflorescence has 5–10 flowers, is racemose, with spirally twisted awns. The sharp-pointed orcuttia is disappearing as a result of agricultural and industrial use of the lands in California. At present, a project is being developed to create a network of protected areas to conserve the habitats in which orcuttia and other typical species grow. All the species of *Orcuttia* and *Neostapfia* genera are included in the category of endangered plants of the United States, and the sharp-pointed orcuttia is also included in the Red Data Book of IUCN.

Alexander's swallenia [*Swallenia alexandrae* (Swallen) Soderstrom et H.F. Decker], a perennial rhizomatous plant, is the only member of the genus. Its distribution is confined mainly to the Eureka valley in California. The largest population was found on the slopes of the Eureka sandy dunes in the southern part of the valley at an altitude of 900–1050 m, with a length of about 5 km, breadth 2 km, and height 250 m. As recently as 1950, this swallenia species was quite common on the northern slopes of large dunes. Isolated habitats were known in the Inio county and in the areas neighboring the Eureka valley. Now only a few specimens of this species have been conserved on the western slope. Under dormancy, Alexander's swallenia forms fairly dense growth with *Larrea* and is an excellent sand binder. It is a good fodder plant, and the endemic weevil *Miloderis nelsoni* lives on its outgrowths. The disappearance of swallenia is associated with destruction of its habitats: for several years the dunes are being exploited as amateur attractions; as a result, the vegetative cover is destroyed and the topography is distorted. At present, the plant is included in the category of endangered species. Entry of transport vehicles into the valley is prohibited since 1976. Alexander's swallenia is included in the Red Data Book of IUCN.

Texas Indian rice (*Zizania texana* A.S. Hitchc.) is distributed in the upper reaches of the San Marcos River in southern central Texas. It is a perennial cereal grass whose lower leaves are submerged in water and upper leaves are aerial. The stems terminate in panicles which are much above the water level. It was discovered in 1933, when its thickets were reported from the upper reaches of San Marcos as well as from the irrigation canals and below the dam on Soring Lake. At present it has been preserved only at the borders of San Marcos city. The population of this species has sharply declined for many reasons, including fluctuation in water levels, pollution and churning of water, destruction of the riverbed, swimming, boating, and other factors. As a result, the plant has changed from seed propagation to vegetative propagation during the last 20 years, which does not help in increasing its population. The Indian rice can grow in clean, constantly cold, and fast-flowing water. It forms large outgrowths on the stony bottom in the middle of the river as well as in smaller areas with a depth of 2–3 m. Sometimes it grows with pondweeds, wild celery, arrowhead, and hydrilla. Texas Indian rice is a valuable food and fodder plant. It is distinguished by a high nutrient content and can be cultivated for fodder. Its thickets are excellent shelters for aquatic birds. At present, this plant is cultivated in the botanical garden of the San Marcos University (Texas), where it is successfully producing fruits and is being multiplied through seeds. Under natural conditions, the population of *Zizania* is only 18,000, which undoubtedly calls for its inclusion in the list of endangered plants.

FAMILY PLUMBAGINACEAE

Coastal thrift or **sea pink** [*Armeria maritima* (Miller) Willd. ssp. *interior* (Raup) Porsild] is a rare endemic subspecies of Canada. It is a perennial herb with woody stocks and many basal leaf rosettes. The scapes terminate in a head surrounded by a sheath. This thrift grows on shifting sand dunes at the southern bank of Lake Athabasca (northwestern Saskatchewan province). Its distribution area extends to about 20 km in length. The plant forms 2 isolated populations. A few species and subspecies of plants growing in its association have not been found anywhere else, including *Achillea lanulosa* Nutt. ssp. *megacephala* (Raup) Argus, *Deschampsia mackenzieana* Raup, *Salix brachycarpa* Nutt. var. *psammophila* Raup, *S. turnorii* Raup, *S. tyrrelli* Raup, *Stellaria arenicola* Raup and *Tanacetum huronense* Nutt. var. *floccosum* Raup (Plate 33). The existence of a typical marine plant deep in the continent on an area 850 km away from the sea indicates relict origin of this habitat, possibly separated during the post-glacial period. The present condition of this thrift could be considered stable if the planned extension of mineral mining in the north of the province is not undertaken, which will definitely pose a threat to the existence of the entire community of endemic plants at the southern bank of Lake Athabasca.

FAMILY PORTULACACEAE

The western parts of North and South America are distinguished by a great variability in the members of the family Portulacaceae. The bitter-wort genus (*Lewisia*), endemic to the Rocky mountain region of North America, includes 20 species. These are perennial herbs with fleshy leaves and large white, pink, or reddish flowers. They are all ornamental. Most of the species (7 out of 10) included in the National List of Endangered Plants grow in California.

Cantelov's bitter-wort (*L. cantelowii* Howell) grows to a height of 15–40 cm. Its leaves are serrate, corolla is pink with darker veins. It grows on humid rocks in the northern part of the Sierra Nevada and flowers in May–June. The cotyledonous bitter-wort (*L. cotyledon* Robinson) is 10–30 cm tall and has numerous stems. The leaves are broad at the base, up to 10 cm long, slightly grooved and serrate. The inflorescence is a compact umbel or a cyme; corolla is white with pinkish or crimson tinge (Plate 39). The plant flowers during June–July. It is found on the rocks in northwestern California and southwestern Oregon. The **opposite-leaved bitter-wort** (*L. oppositifolia* Robinson) is 5–15 cm tall with simple stems. It has 1–2 pairs of cauline leaves, other leaves are basal. The sparse inflorescence bears 2–6 white or pink flowers. The plant flowers in April–May. It grows in humid areas of the rocks at the northwestern tip of California and in southwestern Oregon. **Tweedy's bitter-wort** (*L. tweedyi* Robinson) grows to a height of 10–20 cm. The basal leaves are 10–20 cm long, lanceolate, gradually narrowed into a broad petiole. The stems bear 8–9 pale pink or yellow flowers. Its distribution is restricted to the eastern part of the Cascade mountains (Washington state). It is a highly ornamental plant which has disappeared from nature due to extensive collection.

The spring beauty (*Claytonia*) species, which are more than 30, are succulent perennial herbs with a thick rhizome or tuber. Their leaves are basal, flowers are white, pink, or yellow, borne on one side of the stems.

Daisy-leaved spring beauty (*C. bellidifolia* Rydb.) has numerous basal leaves, broadened towards apex, 2.5–7.5 cm long, on petioles; the cauline leaves are narrow; corolla is white. It grows on rocks and gravel deposits at the latitude of the southern part of the Sierra Nevada in California, Oregon, and Wyoming states. The **golden lanceolate** variety of **spring beauty** (*C. lanceolata* Pursh var. *chrysantha*) has been found in Washington state. It is distinguished by bright yellow flowers (Plate 39). It flowers from April to July on humid forest glades and in meadows. Besides these plants, two more varieties and one species of spring beauty have been included in the National List of Plants of the United States.

FAMILY RANUNCULACEAE

About 100 species of the family Ranunculaceae are included among endangered plants in North America.

Noveboracense aconite (*Aconitum noveboracense* A. Gray) is a perennial herb with underground tuberous structures and dove blue flowers in racemes. It flowers from June to November and is distributed in a few states of the United States, including Wisconsin, New York, Ohio, and Iowa. Initially 27 habitats of this aconite were known, of which only 14 are still preserved; only 4 of these are on protected territories of federal importance. The changes in habitat and shrinking of habitats are responsible for the population decline of this aconite species. For instance, the rise in the water table in Wisconsin as a result of construction of the La Farge dam flooded 3 out of 9 habitats of aconite. The only habitat of the species in New York is near a highway and attracts the attention of plant collectors. The small population of aconite in Ohio is endangered because of the development of the city park, and in Iowa the cause of death of one of the populations is cattle grazing. This aconite grows on shaded and sufficiently moist areas of sandstone and rarely limestone rocks facing north and east in the forest communities rich in species. It is included in the National List of Endangered Plants of the United States as well as in the Red Data Book of IUCN.

Northern crowfoot (*Ranunculus hyperboreus* Rottb.) is a very rare species. Its habitat is limited to Thomson Bay at the southern banks of Lake Athabasca (Canada, Saskatchewan province). The crowfoot grows along the coasts of the bay in highly wet areas together with other rare species, including *Achillea lanulosa* Nutt. ssp. *megacephala* (Raup) Argus, *Deschampsia mackenzieana* Raup, *Salix brachycarpa* Nutt. var. *psammophila* Raup, *S. turnorii* Raup, *S. tyrrellii* Raup, *Stellaria arenicola* Raup, and *Tanacetum huronense* Nutt. var. *floccosum* Raup. All these plants are rare endemic species of Canada included in the List of Rare Plants of Saskatchewan.

FAMILY ROSACEAE

About 50 species of the family Rosaceae are included in the List of Endangered Plants of North America. **Graves' plum** (*Prunus gravesii* Small) is an endangered plant. It is a shrub up to 2 m tall with white flowers, in terminal clusters of 2–3 on the branches. It flowers in late May–early June, before unfolding of the leaves, and the fruits mature in early September. The only habitat is in Connecticut state in the northern coastal part of Long Island Sound (near Groton). The area of the shrubby outgrowths is about 100 m^2. This small territory is intersected with footpaths, as it is

PLATE 39. 1—Cotyledonous bitter-wort; 2—golden lanceolate; 3—California pitcher plant; 4—yellow-flowered side-saddle.

PLATE 40. 1—Alatamaha franklinia; 2—mountain side-saddle; 3—edible dioon;
4—Peruvian lily.

near a recreation area. According to several investigators, the entire area has only a single plant of Graves' plum of hybrid origin; one of its parents is **beach plum** (*P. maritima*). To protect the area, a fence enclosing the buffer zone was constructed in 1975. The plant is included in the National List of Endangered Plants of the United States. At present, it is successfully grown in several arboreta, where its ontogeny and morphology are being studied.

Robbins' cinquefoil (*Potentilla robbinsiana* Oakes) is a small, almost stem-less plant. Rosettes of its dry leaves of the previous year can be seen everywhere in early spring. New leaves appear in May, and if the spring is warm, the first flowers open at the end of May. In normal years, it flowers in early June, and mass blooming occurs in mid-June; isolated flowers are found till October. At present, this cinquefoil species has survived in a single habitat on Washington mountain in New Hampshire state. Its total population does not exceed 4000 plants, and although the plants flower and fruit, their population and area remain stable. It has been established that 50–70 percent of the plants bloom every year, each producing an average of 20 seeds, of which more than 90 per cent are viable. The seeds germinate in June–July after overwintering, but only 40 percent of the plants survive till autumn. The plants live up to 40 years. Robbins' cinquefoil never had a large area of distribution. Besides Washington mountain, it was also found on Mansfield mountain (Vermont state) and on the Franconia range (New Hampshire). It has been found that dense growth in the habitats of this cinquefoil destroys the substrate and disturbs the soil cover, causing washing and blowing away of seeds. After several unsuccessful attempts, the plant could ultimately be grown from seeds in artificial conditions and transplanted at one of the habitats on Washington mountain. A few plants have survived and are now producing flowers and fruits. Robbins' cinquefoil has been included in the category of endangered plants of the United States and is maintained in the Franconia Notch and Croford Notch paths of New Hampshire.

FAMILY RUBIACEAE

Storm's balmea (*Balmea stormae* Martinez) is an endangered species of the family Rubiaceae and is the only member of the genus. It is endemic to a small territory in southern Mexico (Michoacan state) and Guatemala (Wewetenango, Halapa, and Sacapa departments). This plant was first found in 1935 in Mexico but was described later, in 1941–1942, and named after the famous writer M. Storm, who not only collected herbarium specimens of this plant but described it poetically in her works, and also paid attention for many years to the catastrophic disappearance of balmea from the locations well known to her. Balmea is a small deciduous tree or shrub 4–10 m tall, with smooth thin bark and pale pinkish-cream wood. The leaves are 9–14 cm long, cordate at the base, with deciduous stipules. The inflorescence is an umbel, twice or thrice branched, each branch bearing one, barely two, deep red or dark wine-colored pentamerous flowers (giving a strong smell at night). The leaves fall at the time of fruit maturation. This balmea grows on lava deposits, forming a sparse plantation. At one time, this plant was widely used as a Christmas decoration due to its beautiful combination of large green leaves and bright flowers. For this, not only the branches were cut but entire plants were removed. The demand for balmea further increased when cutting of coniferous trees for Christmas festivals was legally banned in Mexico. The balmea population declined particularly sharply during the 1965–1971 period. Since 1974, this plant has been included in CITES. An attempt was also made to bring balmea under cultivation. It was found that it can successfully grow in the warm temperate zone as well as in dry subtropical regions.

FAMILY SALICACEAE

One of the members of the family Salicaceae, **silicole willow** (*Salix silicola* Raup), is distributed along the southern bank of Lake Athabasca, in northwestern Saskatchewan province, Canada on shifting dunes. It is a shrub, up to 3 m tall, with reddish chocolate-brown bark and whitish, densely pubescent branches. It is found sporadically on an area of about 20,000 ha in 2 populations 90 km apart. Different species of long-rhizome cereals, paper birch, crowberry, and a few endemic plants, whose distribution is limited to the southern banks of Lake Athabasca, grow together with it: *Achillea lanulosa* Nutt. ssp. *megacephala* (Raup) Argus, *Deschampsia mackenzieana* Raup, *Salix brachycarpa* Nutt. var. *psammophila* Raup, *S. turnorii* Raup, *S. tyrrellii* Raup, *Stellaria arenicola* Raup and *Tanacetum huronense* Nutt. var. *floccosum* Raup.

At present, this floristically interesting region is still maintaining its old character, as civilization has not affected it, due to isolation and inaccessibility. However, there is a real threat to the plant cover in connection with mining near Claffe Lake (northern Saskatchewan). It is proposed to declare the area of the southern bank of Lake Athabasca a protected park to conserve its flora and vegetation.

FAMILY SARRACENIACEAE

The family Sarraceniaceae has only 3 genera with 17 species, which are distributed in North and South America.

The only member of the pitcher plant genus (*Darlingtonia*) is the **California pitcher plant** (*D. californica* Torr.), a perennial up to 1 m tall. It was found in 1841 on a marsh surrounding a small tributary of the Upper Sacramento river (Shasta county) in California. Its pitcher-shaped leaves are bent in the upper part and resemble the shape of a cobra head; a dark red bifurcated leafy process forms at the leaf margin in the form of a fish tail. The inner surface of the leaf is covered with glands discharging nectar which attracts insects. The leaf surface is covered with hairs, facilitating the movement of insects into the pitcher. The flying insects sit on the leafy appendage leading into the mouth of pitcher. As soon as the prey reaches its margins, the hairs begin to bend and force the insect to move further inside. Bacteria living inside the pitcher decompose the insect (sometimes an unpleasant smell can be sensed near these plants). The plant reproduces through seeds. The pitcher plant has unusual colors. As was mentioned above, the appendage of its leaf is dark red. Isolated red flowers appear in spring, and the pitchers acquire similar colors towards the end of summer (Plate 39).

The California pitcher plant grows on marshes, near springs and water streams, and sometimes even floats on water. The distribution of this endemic plant covers the Pacific regions of Oregon and California. It forms about 250 small populations, of which 50 are in Oregon and 200 in California. The main areas of its growth are on lands under the control of the Forests Service. At present, this species is endangered, since its habitats are destroyed as a result of mineral mining, construction of roads, and other earth works. Collection of plants for sale poses a significant threat to the species: annually 10,000 plants are dug out. Considering that seed multiplication of this plant is difficult, and the demand in the world market is growing, very soon it may be on the verge of extinction. The California pitcher plant has been included in the National List of Endangered Plants of the United States and in Appendix 2 of CITES.

The generic name *Sarracenia* for the side-saddle flower was given by Tournefort in 1700 when he described the plant sent to him by M. Sarracin from Quebec province, Canada. The genus includes 9 species and is distinguished by its horizontal or erect pitcher-shaped leaves. All species of the genus are found exclusively in North America.

Among the rare endangered plants of the United States is the Alabama subspecies of the **Alabama pitcher plant** (*S. alabamensis* F.W. et R.B. Case ssp. *alabamensis*), which is distributed in central Alabama and on the Fall Line Sand Hills Plateau. At present, there are only about 500 specimens of this pitcher plant in nature, distributed in 6 areas. The Alabama subspecies is a perennial herb with a rhizome. It forms leaf rosettes every year, which broaden toward the upper side in the form of pitchers. Relatively short, bent, 17–50 cm long leaves develop in spring, narrow at the base and gradually broadened upward forming a 0.7–3 cm broad cavity. Usually they are light-colored, yellowish green with reddish veins. The summer leaves are erect, up to 72 cm long, pubescent, yellowish-green with a large hood; the pitcher margin is bright green. The insects fallen into the pitchers are digested with the help of enzymes and bacteria. Small leafy organs, phyllodes, form between the spring and summer leaves which perform the role of carbon assimilation; the flowers are borne on pedicels. The Alabama pitcher plant grows on marshes and marshy areas in sparse communities of Virginian magnolia, arundinaria, and alder. Among the species of the genus, this is the most tolerant to shade, but develops better in open places, particularly after fire and cutting. The population of this subspecies has declined significantly during recent years as a result of the introduction of honeysuckle in its habitat, with which it cannot compete, and also under the influence of herbicides and aerosols, agricultural operations, and drainage works. Commercial trade has caused maximum damage to the Alabama pitcher plant. The plant is dug out completely together with its rhizomes. Pieces of rhizomes, intact plants, and their seeds are sold. In 1973, the Alabama pitcher plant was included in the National List of Endangered Plants of the United States and in Appendix 1 of CITES. This plant also appears in the Red Data Book of IUCN.

Jones' side-saddle (*S. jonesii* Wherry) is an endemic of the United States. Its distribution is restricted to a small territory in the Blue Ridge mountains in North Carolina and adjoining parts of South Carolina (four habitats are known). It is a herbaceous perennial with erect pitcher-shaped leaves. The pitchers have a reddish tinge, similar to the red pitcher plant, but the plant is distinguished by the absence of sweet smell and several other characters. This plant lives on wet areas near mountain springs, rivulets, and on marshes. This side-saddle is endangered in North Carolina, and its total population in South Carolina comprises 1600 specimens. The main causes of the disappearance of its individual populations and decline in the population strength of the species are the developing tourism in its habitats and collection of plants for sale. The plant conservation laws operating in North Carolina do not ensure protection to this endangered plant. It is also not protected in South Carolina.

In view of the threatened extinction of Jones' side-saddle from its natural habitats, increased interest in insectivores the world over, and increased volume of trade in plants, this plant has been included in the National List of Endangered Plants of the United States and in Appendix 1 of CITES.

The **mountain side-saddle** [*S. oreophila* (Kearney) Wherry] was at one time found in 3 states: Alabama, Georgia, and Tennessee. At present, it has been conserved in 16 habitats in Alabama (about 700 plants) and one habitat in Georgia (not more than 100 plants). This side-saddle is a gravelly perennial with erect pitcher-shaped leaves, numerous evergreen phyllodes, green pitchers, and yellow flowers (Plate 40). The flowers appear after the pitchers are fully developed, in contrast to **yellow-flowered side-saddle** (*S. flava* L.), in which flowering occurs at the stage of trap formation (Plate 39).

The main areas of its growth are highly humid oak or pine forests on plains, sandy banks, flat sand deposits, and marshy areas near springs. The plant reproduces vegetatively and also by seeds, but its generation is difficult under competition in the natural communities. The poor natural regeneration, destruction of its habitats, and amateur and commercial collection of plants have reduced its area and population. It is included in the National List of Endangered Plants of the United States and in Appendix 1 of CITES.

FAMILY SCROPHULARIACEAE

Furbishia's lousewort (*Pedicularis furbishiae* S. Watson) belonging to the family Scrophulariaceae, is an endangered species. The distribution of lousewort covers the United States and Canada. All its habitats are confined to the St. John river, their total length in the United States being 200 km (Maine State), and in Canada 32 km (Brunswick province). In the United States, 18 populations with a total not exceeding 500 plants are known. As a result of construction of a hydroelectric power plant on the St. John river, 13 populations of 360 plants are threatened with flooding. Only 3 habitats (364 plants) have been reported in Canada, but they are 130 km downstream from the dam and information is not available on their present condition. Furbishia's lousewort is a perennial herb, 40–90 cm tall, with a terminal, dense inflorescence of greenish-yellow, bilabiate flowers. They are exclusively autotrophic plants, reproducing by seeds. They grow along the riverbank between a dyke ridge forming during spring floods and forest plantations at the base of the first terrace in the shrubby thickets of alder and willow together with other herbaceous plants. Since alder has dense foliage, it shades the lower tiers and prevents growth of herbaceous plants, including lousewort. Under such conditions, only a small number of louseworts flower and produce viable seeds. Favorable conditions for the plants of lower tiers in such areas are created only in the first year after frost damage to alder as a result of desiccating winter winds or ice formation on the plants in spring. Towards the end of the vegetative season, alder grows out because of fast development. Therefore, the condition of Furbishia's lousewort depends not only on anthropogenic but also on natural factors. In order to save this plant from complete extinction, it is proposed to transplant adult and young plants at places with similar habitat in areas not threatened with flooding. Furbishia's lousewort is included in the National List of Endangered Plants of the United States and in the Red Data Book of IUCN.

The list of endangered and vulnerable plants of the family Scrophulariaceae is the largest in the United States. It includes 143 names. Among them are a few species whose condition is unknown and a study of their natural habitats is essential.

The **narrow-leaved gerardia** [*Gerardia* (*Agalinis*) *stenophylla* (Pennell) Pennell] is found in Florida (Tampa and Hillsborough). It was seen for the last time in 1897. It is a slender, tender herb with narrow, almost filiform, mostly opposite leaves. The flowers are solitary, regular, on 1.2 cm long pedicels. The plant flowers at the end of summer or during autumn. The plant is included in the List of Endangered Plants of the United States.

Ludovician Indian paint-brush (*Castilleja ludoviciana* Pennell) is a member of a large genus which includes about 250 species. This is a herbaceous plant with tubular bilabiate flowers, enclosed by yellow or red bracts and upper leaves. The bracts are red-purple with 3 narrow lobes. Most probably, the plant has become extinct. It was last collected in Louisiana state in 1915.

Havard's seymeria [*Seymeria havardii* (Pennell) Standlay] is a herbaceous perennial, 40 cm tall; stem is covered with glandular hairs; leaves are pinnatisect. The flowers are on long pedicels, in loose cymes; corolla is tubular and yellow. The plant was collected in 1882 along the Rio Grande in southern Texas, and has not been found since then.

FAMILY TAXACEAE

The members of the family Taxaceae are distributed in the northern hemisphere, found in North America, Europe, and Asia. The ancient origin of Taxaceae is confirmed by paleobotanical data. The family comprises 5 genera and about 20 species, all of them evergreen trees or shrubs. The yew genus (*Taxus*) has 8 species, the rarest among them is the **Florida yew** (*T. floridana* Nutt.). It is included in the National List of Endangered Plants of the United States.

The genus stinking cedar (*Torreya*) comprises 6 species. Its area of distribution consists of two regions in North America: the western region is in California on the western slopes of the Sierra Nevada, and the eastern region occupies northwestern Florida and southwestern Georgia. Two more areas of stinking cedar are in East Asia: one in Japan and another in China and Burma.

Stinking cedar (*T. taxifolia* Arnott.) is a tree with pyramidal crown,

12–15 m tall and up to 60 cm in diameter. The bark is chocolate-brown with orange tinge. Its light green leaves are similar to those of yew, and a faint smell appears on rubbing. The stinking cedar grows in the eastern United States, forming the eastern part of its distribution area. It is confined to the limestone slopes of mountains. The area of distribution of this plant is a relict formation. Within its limits, the stinking cedar never existed in large numbers, but it was rejuvenated and maintained its position. Its balance has been disturbed by anthropogenic environmental changes. A few old trees were cut: a disease spread in the late 1950s and early 1960s which initially destroyed all the aerial parts of the trees. Even then it appeared that the stinking cedar would survive, as suckers appeared from the roots of adult trees. But with the death of roots, the young shoots also died. A few specimens of stinking cedar survived at some distance from each other. At present, this plant is included in the List of Endangered Plants of the United States. A few specimens of stinking cedar are maintained under cultivation. It was introduced into Europe may back in 1838, but is rare. One fruiting tree is growing in the Batumi botanical garden.

FAMILY THEACEAE

One of the most wonderful plants of the family Theaceae is the **Alatamaha franklinia** (*Franklinia alatamaha* Marshall). This was the only species of *Franklinia* discovered in 1765. Even then it was represented by a single population. Amateur gardeners almost completely destroyed it toward the end of the 18th century to the beginning of the 19th century. At present this franklinia is preserved only under cultivation. It is a deciduous shrub or a tree up to 10 m tall (Plate 40). The leaves are 6–20 cm long; flowers are large, about 12 cm in diameter, with gentle flavor, resembling the smell of blooming citrus, cream-white, with numerous golden stamens in the center. The plant flowers in April. The former habitats of this franklinia are the marshy areas along the Alatamaha river in Georgia state, where it grew on an area of 1–2 ha together with Virginian magnolia, liriodendron, and pond pine.

The Alatamaha franklinia is widely cultivated in the New as well as the Old World. It can be easily multiplied by seeds, requiring acidic soil for better growth and development. It is proposed to reintroduce this franklinia into its former habitats and organize a reserve there. At present, it is listed among the extinct plants of the United States and is included in the Red Data Book of IUCN.

FAMILY ZAMIACEAE

The family Zamiaceae belongs to the division Cycadophyta and includes 8 genera, one of which is the coontie genus (*Zamia*). About 30 species of this genus are distributed in the tropical and subtropical Americas over a large area from Chile and Brazil in the south to Mexico and Florida in the north.

Florida coontie (*Z. floridana* A. DC.) grows on the Florida peninsula and St. Simons Island (Georgia state) in subtropical pine forests on dry sandy soils, on coastal dunes among evergreen shrubs and palm outgrowths, and in broad-leaved tree plantations; in Georgia, it has been found in an oak-pine forest. Initially, coontie was a common species on the Florida peninsula and its underground tubers were used for food by the local population. The reduction in the habitats of the species is related to direct exploitation (digging of the tubers), large-scale development of tourism, and change in the habitat conditions. Coontie is protected by law in Florida, where damage or digging of the plant is banned. Sale of only cultivated specimens is permitted. It is proposed to establish a reserve (the area is being searched) to conserve this plant. This species is included in the Red Data Book of IUCN.

The genus *Dioon* comprises 4 species, 3 of which grow in Mexico and one in Honduras.

Spiny dioon (*D. spinulosum* Dyer) is a slender, unbranched, cylindrical tree with a crown of gracefully bent, spiny leaves, reaching a height of 15 m. It grows in hilly tropical forests under the canopy of taller trees. In vegetative condition it is similar to the tree fern. Large cones appear on the trees at the time of flowering and fruiting. The distinguishing feature of the genus from other Zamiaceae plants is the presence of outgrowths on the lower part of megasporophylls. The area of distribution of this dioon is confined to the territory of Veracruz, Oaxaca, and Yucatan states of Mexico, where tropical forests still exist. The plant is endangered. A few specimens are maintained under cultivation in the Acatlan botanical garden (Oaxaca).

Edible dioon (*D. edule* Dyer) is a stocky plant up to 1.5 m tall, with a branched trunk (Plate 40). It bears a cluster of hard, erect leaves at the apex, which spread from the center and form a funnel or nest-like structure. Since seeds sometimes germinate in the nest, the female plants look like 2- or even 3-tier structures. The edible dioon grows in dry open places in association with cacti and different xerophytes in Veracruz, Queretaro, and San Luis Potosi states (Mexico). The consumption of seeds by the local people is responsible for disappearance of this plant.

The genus *Ceratozamia* has 5 or 6 species, distributed in southeastern Mexico and Honduras.

Miquelian ceratozamia (*C. miqueliana* Wendl.) has a slender, 1–2 m tall stem, covered with a sheath of basal remnants of deciduous leaves. The crown is formed by a rosette of old gray and young light-green leaves. The cones have 2 appendages (2 on each scale). This ceratozamia grows in shady tropical forests on steep slopes. It is distributed in Veracruz state. The species is included in the List of Endangered and Vulnerable Plants of Mexico.

5
Central and South America

Central America and a large part of South America belong to the Neotropical kingdom. The flora has many families and genera common with the paleotropical genera, but the percentage of endemic genera is also fairly high. The non-tropical parts of South America belong to the Patagonian region of the Holantarctic floristic kingdom.

The vegetation of America has significantly changed as a result of human activity, and the flora of all the South American countries has become poor in species content and the population of many plants has sharply decreased. This holds true even for several earlier common plants, which were dominant at one time. Lists of plants requiring special conservation measures have been prepared in many countries.

FAMILY ACANTHACEAE

Dodson's balsam herb (*Dicliptera dodsonii* Wasshausen), is an endangered plant and has been included in the Red Data Book of IUCN. It is a twining herb with hexagonal stem and opposite leaves. The large orange flowers are surrounded with greenish-white bracts. A single specimen of this species was found in 1977 in a forest near the foothills of the Andes in central Equador (Los Rios province). Dodson's balsam can be cultivated in gardens as an ornamental plant.

FAMILY ADIANTACEAE

Lorentz's maidenhair fern (*Adiantum lorentzii* Hieron) is endangered in Argentina. This is a delicate terrestrial fern with rhizome and thin pinnate leaves. It is a highly ornamental plant and is, therefore, usually collected for decorating houses at the time of winter festivals.

FAMILY ALSTROEMERIACEAE

Earlier, **proud Peruvian lily** (*Alstroemeria pelegrina* L.), a plant with very beautiful flowers (Plate 40), was quite common in Brazil. Many of its habitats were destroyed due to collection of flowers and development of beaches; now only a few isolated habitats have survived.

FAMILY AMARANTHACEAE

The family Amaranthaceae includes **Haught's amaranth** (*Amaranthus haughtii* Standl.), an annual with unisexual flowers. At present, it is endangered in Peru.

FAMILY AMARYLLIDACEAE

The **bird-flowered lily** (*Amaryllis aviflora* Ravenna) is a rare plant. A very small population was found in Rosario de la Frontera (Salta province, Argentina). The search for plants at other places was unsuccessful.

The **Santa Catarina lily** (*A. santacatarina* Traub.) grows in marshes of Parana and Santa Catarina states (Brazil). Recently it was found in the northern part of Rio Grande du Sol state. The cattle eat its leaves and flowers, so the plant cannot produce fruits normally. The species has become very rare: only 3–5 plants have remained in each population.

Scopulate lily [*A. scopulorum* (Bak.) Traub. et Uphof] was found in a very limited area near Sorata in Bolivia. Most probably, its population is declining due to collection of flowers and digging of bulbs.

Many South American species of the blue amaryllis genus (*Griffinia*) have already become extinct as a result of habitat destruction. They cannot grow in open places and disappear on forest felling. The **hyacinth blue amaryllis** (*G. hyacinthina* Ker-Gawl.), **libonian blue amaryllis** (*G. liboniana* Lem.), **elegant blue amaryllis** [*G. concinna* (Mart. ex Roem et Schult.) Ravenna], and other species are on the verge of extinction. All species of the urn-flower genus (*Urceolina*) are endangered due to felling of rain forests, which were distributed in Bolivia and from Brazil to Panama.

FAMILY ARAUCARIACEAE

The plants of the family Araucariaceae are frequently called southern conifers. This ancient family has only 2 genera. The genus *Araucaria* comprises about 12 species, distributed in the southern hemispheres. These are large evergreen trees with regular whorled branching and skeletal crown. Usually the araucari as are dioecious plants, and the female plants in some species are much larger than male plants. The lower branches drop off with age and the trees carry a high, flattened, umbellate crown. Bark is thick, resinous, and grooved. The araucarias are long-lived plants; they fruit at the age of 40–50 years, reaching maturity at an age of 300 years; the maximum age recorded is 2000 years. The arancaria has discontinuous distribution (in South America and Australia), which indicates its ancient nature. The generic name was derived from Araucaro province in southern Chile.

Chile pine [*A. araucana* (Molino) C. Koch] is a very large, dioecious tree with a height reaching 60 m and diameter 1.5 m. Leaves are rough, prickly, dark green, spirally arranged, remaining on the tree up to 40 years. The cones are chocolate-brown, spherical, weighing up to 1.6 kg (Plate 41). It has discontinuous distribution, the area consisting of 2 parts. The smaller part is on the western coast of coastal Cordilera at an altitude of about 700 m, and the larger in the Andes at an altitude of 1600–1800 m. It forms forests, especially on volcanic soils. The seeds are edible; the beautiful wood is used for construction. The species is included in Appendix 1 of CITES.

FAMILY BOMBACACEAE

Among numerous members of the family Bombacaceae, the **West Indies balsa** (*Ochroma lagopus* Sw.), a more than 30 m tall tree with light bluish-gray bark and large palmate leaves, is endangered. On maturation, the fruits crack open and then resemble a hare's paw, as on the inner side they are covered with brown fuzz. 'Balsa', translated from the Spanish, means raft: its wood was used since ancient times for raft construction. The timber of West Indies balsa is one of the lightest in the world, very loose and soft, but acquires the hardness of oak on drying. At present, it has almost become extinct from the American forests, and is being maintained in small numbers only in the humid rainforests of Equador.

FAMILY BERBERIDACEAE

One of the members of the family Berberidaceae, **littoral barberry** (*Berberis littoralis* Phil.), a prickly shrub with flowers forming clusters in leaf axils, is endangered in Chile.

FAMILY BROMELIACEAE

The family Bromeliaceae has typical Neotropical area of distribution and comprises about 1000 species, predominantly herbaceous plants, mainly epiphytes. Their narrow, linear, frequently fleshy leaves are usually borne in dense rosettes from the center of which arises a long scape with bright-colored (pink, blue, violet) bracts enclosing small greenish or yellow axillary flowers. Leaf rosettes up to 1 m in diameter are found among epiphytic bromelias with their leaves broadened like a cup at the base, closed on the upper side, forming a pitcher-like structure. Water is accumulated in this pitcher, small plants and animals live in it, and the plant obtains nutrition from it with the help of scaly hairs covering the leaf bases. The members of Bromeliaceae are mainly distributed in the humid forests of tropical South America, especially in Brazil and Colombia in the Amazon basin. Pineapple originated from the terrestrial Bromeliaceae. Many members of this family are highly valued for their beauty. The collection of plants as well as forest felling caused a sharp decline in their population.

Dichlamydeous aechmea (*Aechmea dichlamydea* Baker) is a monocarpic epiphyte growing up to a height of 50 cm with lorate leaves, lanceolate bright red bracts, dove-blue sepals and white or bluish-lilac corollas. It is known from a very few habitats in Venezuela and on the Trinidad and Tobago islands. Its population is declining due to felling of trees on which it grows. It is an endangered plant and is included in the Red Data Book of IUCN.

The **upright-flower glomeropitcairnia** (*Glomeropitcairnia erectiflora* Mez) is included in the Red Data Book of IUCN as an endangered plant. It is a monocarpic plant with a cluster of numerous lorate leaves and bunches of yellow or white flowers surrounded with red bracts. It grows in Trinidad and Tobago and in Venezuela, growing on soil as an epiphyte in the forest of the spray (fog) belt.

Two rare vriesia species are also found here: *Vriesia broadwayi* L.B. Smith and *V. johnstonii* (Mez) L.B. Smith et Pittendr.

The **dependent-flower glomeropitcairnia** (*G. penduliflora* Mez) is a rare plant of the Dominican Republic and Martinique. The populations of all Bromeliaceae plants are very small.

FAMILY CACTACEAE

In Cuba, almost all cacti have local distribution and many are endangered, for example, *Melocactus matanzanus* Léon (*Mantanzas melocactus*, Plate 41), *M. guitart* Léon (*Guitart melocactus*, Plate 41) *M. acunai* Léon, and *Neolloydia cubensis*.

The **tree cactus** [*Cereus robini* (Lemaire) L. Benson]—a shrub or small tree 5–8 m tall, light bluish green—has been included in the Red Data Book of IUCN. The flowers are campanulate, chocolate-brown to green or purple. It grows in Cuba and Florida. Almost all its habitats in Cuba have been destroyed. In Florida, it is known from a few habitats. The species is endangered.

Horst's discocactus (*Discocactus horstii* Buining et Brederoo) grows only in the hilly steppes at an altitude of 100 m in Sierra de Barao (Brazil) together with small shrubs. The habitats of this rare plant are threatened due to mineral mining.

Bahian melocactus [*Melocactus bahiensis* (Br. et R.) Luetzelb.] is found only in the central part of Bahia in Brazil. It is endangered, as the local people burn grass and lower shrubs for pasture improvement in its habitats. Possibly, **pruinose melocactus** (*M. pruinosus* Werderm.) and **violet melocactus** (*M. amethystinus* Buining et Brederoo) have already become extinct. The central part of Bahia is very rich in melocacti, but all of them require protection. The **golden-yellow micranthocereus** (*Micranthocereus auri-azureus*) is also among the rare cacti (Plate 41).

FAMILY CAESALPINIACEAE

The **spiny brasiletto** [*Caesalpinia spinosa* (Mol.) Ktze.] is endangered in Chile, where its populations have declined as a result of commercial harvesting and export. Besides Chile, it is also found in the mountains of the Caribbean Sea at an altitude of 2400–3000 m. It has been introduced into cultivation.

The **prickly Brazil wood** (*C. echinata* Lam.) is a tree with valuable red wood and tripinnate leaves; the flowers are almost regular (Plate 42). Its major area of distribution is Brazil, where it has been almost completely felled for its valuable wood.

FAMILY CARICACEAE

The **Chilean papaw** [*Carica chilensis* (Planch. ex A. DC.) Solms], a dioecious tree, simple or sparingly branched, with lobed or divided leaves, is endangered in Chile.

FAMILY CUPRESSACEAE

The genus *Austrocedrus* of the family Cupressaceae is represented by only one species: **Chilean austrocedrus** [*A. chilensis* (D. Don) Florin et

Boutelje]. It is an evergreen photophilous tree with broad pyramidal crown, horizontal branches with numerous small twigs, covered with scaly dimorphic leaves and cones with woody scales. This species is endangered in Chile.

Among the disappearing plants of Bermuda is the **Bermuda cedar** (*Juniperus bermudiana* L.), an evergreen tree, 20 m tall and up to 1 m in diameter, with a broadly branched crown. Its bark is gray; leaves are scaly, closely adpressed to the branches; the wood has beautiful reddish chocolate-brown tinges. In the past, it formed forests along the hill slopes and borders of marshes, but was infested with accidentally introduced insects on the islands in the 1940s, and 90 percent of its plantations died. At present, it is being multiplied well at the protected places. The species is included in the Red Data Book of IUCN.

Bunch-fruited pilgerodendron (*Pilgerodendron uviferum* Florin) is a tree or shrub with a pyramidal crown; bark is dark red to chocolate-brown; leaves are small, scaly. The cones are ovoid. The plant grows in Chile on the western slopes of Andes, as well as on Chiloe Island. It is endangered.

Alerce (*Fitzroya cupressoides* J.M. Johnston) is a large, thick-stemmed tree up to 40 m tall; bark is reddish; branches are slender. The tree produces valuable wood and grows in Chile, along the Pacific coast from Valdivia to the Andes, on Chiloe Island, and in Argentina. Together with pilgerodendron, it is being conserved in the Los Alercales National Park (Chile) and Los Alerces National Park (Argentina).

FAMILY CYATHEACEAE

All members of the family Cyatheaceae are included in the list of species whose trade is regulated by CITES. The genus *Dicksonia* includes tree ferns growing up to a height of 4–6 m with numerous adventitious roots. The leaves are bipinnate or tripinnate, at the stem apex, hard. The ferns usually grow in mountain forests up to an altitude of 3000 m in wet ravines. The species endangered in Chile are *D. berteriana* (Colla) J. Sm. and *D. externa* Skottsb.; *D. esellowiana* Hook. In Argentina and *D. karsteniana*, *Cyathea flaccida*, and *Trichipteris tryonorum* in Venezuela appear to be protected.

FAMILY EQUISETACEAE

The horsetail genus (*Equisetum*) of the family Equisetaceae comprises rhizomatous perennial herbs; branches with distinct internodes and nodes with whorled leaves. They are all homosporous plants. The **giant horsetail** (*E. giganteum* L.) was quite common earlier in the forests of Buenos

Aires province (Argentina), but it has almost become extinct as a result of its collection for medicine.

FAMILY ERICACEAE

Gaultheria sphagnicola Rich. is an erect dwarf shrub with small, broadly ovate leaves and cream-red flowers in dense terminal panicles. It is found only on the Martinique and Guadeloupe Islands (the Lesser Antilles). In Guadeloupe, 2 populations with a total of less than 10 plants have survived on the slope of the Sastra volcano; 3 populations died during the volcanic eruption of 1977. In Martinique, it grows near the peak of Mt. Pelee volcano at an altitude of 1100–1200 m, but is being destroyed by priests collecting its beautiful flowers and edible fruits.

FAMILY ERYTHROXYLACEAE

The family Erythroxylaceae includes 4 genera and about 200 species, mainly spread in tropical America. One of the members of this family, **coca** (*Erythroxylon coca* Lam.), is a densely leafy shrub with alternate oblong leaves and small white flowers in leaf axils (cocaine is obtained from the leaves). The only area of its distribution was in Peru, Bolivia, and the high slopes of the Andes; now this plant is not found in nature, but is widely cultivated in Java, Sri Lanka, and Africa (Plate 42).

FAMILY FABACEAE

Balsam of Peru tree (*Myroxylon pereira* Klotsch.) is a tropical evergreen with imparipinnate leaves and whitish flowers (Plate 42). It has a very limited area of distribution in Salvador (the plant is being eradicated for obtaining the balsam of Peru). It is cultivated in Java and Sri Lanka.

FAMILY FAGACEAE

The southern beech genus (*Nothofagus*) is quite typical for the southern hemisphere, usually with large trees, growing up to 40–50 m, with angular trunk up to 2 m in diameter. The flowers are small, unisexual. The **sea-green southern beech** [*N. glauca* (Phil.) Krasser] and **Alexander's southern beech** (*N. alessandrii* Espinosa) are endangered in Chile.

PLATE 41. 1—Chile pine; 2—Matanzas melocactus; 3—Guitart melocactus;
4—golden-yellow micranthocereus.

PLATE 42. 1—Mahogany tree; 2—Brazil wood; 3—coca; 4—balsam of Peru tree.

FAMILY GOMORTEGACEAE

Keule (*Gomortega keule* H.M. Johnston) is a large evergreen tree with opposite entire aromatic leaves and small bisexual flowers in racemes. Keule is the only species of the single genus in the family Gomortegaceae. It grows only in the central regions of Chile.

FAMILY IRIDACEAE

Among the rare species of the South American continent is **Sabin's trimezia** [*Trimezia sabinii* (Lindl.) Ravenna], which has natural distribution in the forests of Salvador, Brazil. It is cultivated in some cities. **North's trimezia** [*T. northiana* (Schneev.) Ravenna] grows on rocky scraps above the sea near Rio de Janeiro (Brazil), on the neighboring islands, as well as in a few small areas further north. Because of economic development, the plant has already disappeared from some islands. The **woodland trimezia** [*T. silvestris* (Vell.) Ravenna] is a very elegant plant, found in small numbers in a few scattered habitats in Rio de Janeiro and Sao Paulo states (Brazil). The **variegated trimezia** [*T. variegata* (Mart. et Gal.) Ravenna] is found from Mexico to Costa Rica, but its area is shrinking (particularly in Mexico) due to forest felling. The **dwarf trimezia** (*T. humilis*) is still preserved in sufficient numbers on the rocks in the Rio de Janeiro region (Brazil).

Some forms of the wonderful species **blue trimezia** [*T. caerulea* (Ker-Gawl.) Ravenna] are distributed from Rio de Janeiro to Rio Grande du Sol in Brazil. This plant is endangered in nature, but is widely cultivated in several countries.

Other trimezias are also rare species: **sea-green trimezia** [*T. glauca* (Bak.) Ravenna], a polymorphic species found in Minas Gerais, Rio de Janeiro, and Sao Paulo states (Brazil); **glittering white trimezia** [*T. candida* (Hassl.) Ravenna]; **steyermark's trimezia** (*T. steyermarkii* Fost.); **yellow trimezia** [*T. lutea* (Klatt) Fost.], a herbaceous plant growing in fields in Brazil and central Colombia; and **minute trimezia** (*T. minuta* Ravenna), distributed in small areas in Apariman and Alcuco (Peru).

The **diminutive cypella** [*Cypella pusilla* (Link, Klotzsch et Otto) Benth. et Hook. f.] was at one time the most common species on the hills and fields near Porto Alegre and Rio Grande du Sol (Brazil), but it has now completely disappeared from this region. **Osten's cypella** (*C. osteniana* Beauv.) is also a rare species, a few specimens of which have been found in several places of the Minas Department of Uruguay.

The **extraordinary mastigostyla** (*Mastigostyla mirabilis* Ravenna) is a very beautiful plant. It has been found in a few places in the northern Tucuman province (Argentina). The species is suffering due to cattle grazing.

Libertia tricocca Phil. is found only in Chile, where a few small populations have survived. It is endangered due to trampling of grass and change in the environment due to human activity.

FAMILY JUGLANDACEAE

The **neotropical walnut** (*Juglans neotropica* Diels), a tree with thin, aromatic, compound, pinnate leaves, is found in the mountains of Equador and Peru at an altitude of 3000 m. At present, the species is endangered, as it has been destroyed almost completely for its valuable wood and tasty nuts.

The spurious walnut genus (*Oreomunnea*) comprises large trees, up to 47 m tall, with leathery leaves remaining on the tree for most of the year. The flowers are unisexual; fruit is small, roundish and winged. The **spurious walnut tree** [*O. pterocarpa* Oerst. (*Engelghardia pterocarpa*)] has been included in the Appendix to CITES.

FAMILY LAURACEAE

The species of the **Beilschmiedia** genus—[*B. berteroana* (Gay) Hosterm. and *B. miersii* (Gay) Hosterm.] are 10-15 m tall trees. They form humid laurel lowland forests in central Chile. Both species are endangered due to forest felling. The **ligulate avocado** (*Persea linguae* Nees), an evergreen tree of Chile, is also disappearing. **Glaziov's laurel wood** (*Ocotea glaziovii*) is very rare in Brazil.

The **chocolate-tree-leaved avocado pear** (*Persea theobromifolia* A. Gentry), found in the foothills forests of the Andes in central Equador, has been included in the Red Data Book of IUCN as an endangered species. It is a 30-40 m tall tree, with large crown. The leaves are opposite and elliptical. The flowers are grayish green; inflorescence in leaf axils. The fruit is large, reddish chocolate-brown, obovoid.

FAMILY LILIACEAE

Many species of the family Liliaceae are disappearing due to the destruction of their habitats.

The **Brazilian chlorophyte** [*Chlorophytum brasiliense* (Nees et Mart.) Ravenna] was at one time found among the limestone rocks near the city of Brasilia. Possibly, the distribution of this species was restricted only to this place. As a result of the construction of a cement factory here, the hillocks were completely leveled. The search for this plant at other places did not yield results.

Short-flowered paradise lily [*Paradisea stenantha* (Ravenna) Ravenna] is found on isolated low hills in the coastal formations in northern Peru. The roots of this plant are fibrous. It grows among mosses on rocks where erosion is common. The plant is eaten by goats.

Dwarf speea [*Speea humilis* (Phil.) Loes.] was found in Chile amid rocks. Trampling and goat-grazing may lead to the extinction of this plant. In contrast with most lily species, this speea has zygomorphic flowers imitating insects which attract the pollinators.

The genus *Gilliesia* comprises 4 species. All of them are endemic to Chile and, with the exception of **graminaceous gilliesia** (*G. graminea*), are endangered due to grass burning and cattle grazing.

FAMILY LOGANIACEAE

The butterfly bush genus (*Buddleia*) consists of about 100 species, which are widespread mainly in the tropics and subtropics of Asia, America, and South Africa. These are shrubs with simple opposite leaves, usually with white pubescence. Their flowers are borne in many-flowered spicate inflorescences. All species are ornamental. The **gray buddleia** (*B. incana* R. et P.) and **long-leaf buddleia** (*B. longifolia* H.B.K.) are endangered in Peru.

FAMILY MELIACEAE

The **mahogany tree** (*Swietenia mahagoni* Jasq.), an evergreen up to 15 m tall, belongs to the family Meliaceae (Plate 42). It is valued for its wood with narrow dirty-white sapwood and reddish-brown heartwood, of very fine texture. Its plantations were exploited for a long time, as the wood of this tree was widely used for internal fittings in ships, artistic works, etc. This beautiful tree was fairly widely distributed in the forests of South and Central America. At present, a few reserves of this species have survived in the inaccessible part of the Andes in Uruguay and Equador. The reserves of another valuable tree, the **large-leaved mahogany** (*S. macrophylla* King.), have been reduced significantly due to extensive exploitation.

FAMILY MELASTOMATACEAE

The **Brazilian spider flower** [*Tibouchina chamaecistus* (Naudin) Cogn.], a dwarf shrub with spirally twisted trunks and branches, is included in the Red Data Book of IUCN. Its leaves are small, acute, ovate, flowers are dark purple, axillary in groups of 1–5. This spider flower is found only in Martinique (the Lesser Antilles), in 3 habitats in the mountains. It is abundant in the upper parts of the slopes of Mt. Pelee volcano at an altitude of 900–1200 m. It is a highly ornamental plant with bright purple flowers (its local name is 'wild tulips'), which is becoming rare due to cutting and digging. Attempts to bring this species under cultivation have so far failed.

FAMILY MIMOSACEAE

A few South American leguminous species are on the verge of extinction, including **snow acacia** (*Mimosa lanuginosa* Glaziou ex Burkart), included in the Red Data Book of IUCN. This is a highly ornamental, sparingly branched shrub; leaves are pinnate with snow-white pubescence and globular fuchsin-red floral heads. It grows in savanna in a very limited area of Brazil.

The mesquite genus (*Prosopis*) includes prickly trees and shrubs, forming thickets. *Prosopis chilensis* Stuntz and *P. tamarugo* Phil. in Chile and *P. atacamensis* Phil. in Peru and Chile are endangered.

FAMILY ORCHIDACEAE

Central and South America are among the richest centers of orchid variability. Large-scale collections of orchids for commercial purposes have been made here since ancient times, and forest felling and other types of economic exploitation of the territory significantly changed or completely destroyed the habitats of these plants, many of which are on the verge of extinction.

The cradle orchid genus (*Anguloa*) comprises terrestrial plants which sometimes grow on bare rocks. The tubers bear 2–4 leaves, producing solitary flowers. They grow in the Andes from Venezuela and Colombia to Peru. The **Rucker's cradle orchid** (*A. ruckeri* Ldl.) with flowers olive-green outside and yellow inside with reddish chocolate-brown spots is found in Colombia; the lip is dark red to chocolate-brown (Plate 43). The other ornamental Colombian orchid is **Clowes' cradle orchid** (*A. clowesii* Ldl.) with 8–14 cm long tuber and 2–4 broadly elliptical, up to 50 cm long leaves. Flowers are lemon-yellow with a strong, pleasant smell (Plate 43).

The genus *Brassavola* comprises about 15 species, distributed in tropical America from Mexico and the Antilles to southern Brazil, Paraguay, and Bolivia. All of them are terrestrial orchids. The **Perrin brassavola**

(*B. perrinii* Ldl.) with 15–18 cm long tubers, terminating in a single narrow leaf, grows in Brazil. The inflorescence consists of 3–6 flowers; perianth is yellow, labellum white (Plate 43).

The **subulate-leaved brassavola** (*B. subulifolia* Lindl.) has been found only in Jamaica, where it is quite rare. Its inflorescence is a fairly dense raceme of 10–12 flowers.

The genus *Brassia* comprises epiphytic orchids and includes up to 38 species distributed from Mexico to southern Brazil.

The **warty brassia** (*B. verrucosa* Ldl.) is found in Guatemala and Mexico. It has oblong tubers and a pair of leaves. The perianth is bright green, with a few dark chocolate-brown spots at the base. Labellum is broadly oval, white, with dark green papillae.

The *Cattleya* genus is one of the most popular and well-known horticultural plants. All the wild cattleyas, about 40 species, are distributed in tropical America from Mexico to Brazil. They have cylindrical or fusiform flattened stems with 1–2 or more leathery leaves. The inflorescence is a terminal few-flowered raceme, sometimes a solitary flower. **Persival's cattleya** (*C. persivaliana* Rchb. f.), an endangered species, grows on rocks at an altitude of 1000–1300 m in Venezuela. **Trian's cattleya** (*C. trianae* Rchb.), with beautiful pink flowers having dark crimson labellum (Plate 44), is included in Appendix 1 of CITES. It grows in Colombia. **Bowring's cattleya** (*C. bowringiana* Veitch.), with many-flowered racemose inflorescence of minute lilac-colored flowers, is endangered in Honduras and Venezuela. **Aclandia's cattleya** (*C. aclandiae* Lindl.) is a rare endemic of Brazil. This is a small plant with relatively large olive-green flowers having fuchsin-red open labellum (Plate 43). **Skinner's cattleya** (*C. skinneri* Bat.) is the emblem of Costa Rica (Plate 44).

The genus *Epidendrum* includes about 800 species found in tropical America from the Florida peninsula to Bolivia and Paraguay. These are mostly epiphytic orchids. Usually they have a tuber with one or a few leaves or a creeping rhizome. The stems in some species are reduced, in others very long, with aerial roots. The flowers are aromatic, borne in inflorescences.

The species *E. mutelianum* Cogn. is included in the Red Data Book of IUCN as a vulnerable species. It is an erect herb, up to 40 cm tall, without corms; flowers are yellow and hard. It is found at a few places in Guadeloupe (the Lesser Antilles), where it grows at an altitude of 500–1000 m in small forests exposed to strong winds, usually as an epiphyte, rarely terrestrial. Because of the beautiful yellow flowers with violet-striped labellum and smell resembling that of lilies, this Guadeloupean endemic earned the name 'wealth of the island'. Six more endemic orchids are known from the island, and 14 endemic species are found on the Lesser Antilles. The epidendrum has very restricted distribution on the island and is endangered due to collection of plants.

The genus *Laelia* comprises 35 species distributed from Mexico to Brazil. All of them are epiphytes distinguished by a great variety of flower shapes and sizes. The plants are highly ornamental, the freshness of flowers remaining for a long time. In contrast with the cattleya species, the flowers of laelias are usually borne on long peduncles. Many species are widely known in cultivation as 'queen of orchids'.

The **purple laelia** (*L. purpurata* Lindl. et Paxt.) is found in a narrow coastal forest belt in eastern Brazil. This orchid is on the verge of extinction because of destruction of many habitats as well as indiscriminate collection (Plate 44).

Dayan's laelia (*L. dayana* Rchb. f.) is a beautiful small plant with flowers less than 12 cm long; the lip is erect with purple lines (Plate 44). It grows only in Rio de Janeiro state, Brazil. The **beautiful laelia** (*L. praestans* Rchb. f.) is found in Sierra de Moeda, Sierra Jerals de Minas, and the forest near Santa Teresita in Espiritu Santo. This is a small plant with dark violet-red flowers and almost black-violet tubular labellum. The forests are being cut at a very fast rate in the areas of distribution of these orchids, so the plants are endangered.

Jonghe's laelia (*L. jongheana* Rchb. f.) is a relatively small plant with large violet-pink flowers, having a fimbriate, yellow lip. Possibly, this species has already disappeared mainly due to commercial collection (even small specimens were collected and sold at a very high price). It is included in Appendix 1 of CITES.

Laelia sincorana Schltr. with subglobose, strong corms and fimbriate white flowers can also be included in the rare Brazilian laelia (Plate 45).

About 35 species of *Lycaste* are found in the mountains of Central America, on the Antilles and Bahamas, and in Brazil. These are orchids with tender deciduous leaves, originating from the apex of an ovoid corm. The flowers are solitary, on long pedicels, yellowish-pink, olive-green, or chocolate-brown to green.

The **fragrant lycaste** (*L. suaveolens* Summerh.) is included in the Red Data Book of IUCN as a vulnerable species. It is an epiphyte with large, compact, dark green, flat pseudobulbs. Each of them produces up to 10 solitary flowers. This plant is found in Salvador, possibly also in Honduras and Guatemala, but further information about its existence in these countries is not available. It grows in humid mixed forests of rough-leaved trees in the middle mountain belt on volcanic slopes. It is interesting for cultivation as well as developing hybrids because of its bright yellow or orange flowers.

The *Masdevallia* genus mainly comprises epiphytic orchids, but also orchids growing in the crevices of rocks or in the soil. They do not produce tubers, and the inflorescences bear 1–8 flowers each. The outer, perianth leaves terminate in a long cusp, and the lip is very small. About 150 species are found in Central and South America, in the Andes, where the climate is cooler.

The orchid *M. chimaera* Rchb. has dark chocolate-brown to red flowers with lighter spots; the lip has the shape of a slipper, light chocolate-brown in color (Plate 45). It grows in Colombia at an altitude of 1700–2000 m on rocks and trees. The **beautiful masdevallia** (*M. bella* Rchb.) produces a single yellow flower, covered with brown spots, and with very long spurs; lip is white, shaped like a fly. This epiphytic orchid grows in Colombia at an altitude of 1800–2000 m. Another species, **red masdevallia** (*M. coccinea* Lindl.), with large violet-red flowers, grows still higher up (2300–3000 m); the lip is light pink with white tip (Plate 45).

The **bicolored notylia** (*Notylia bicolor* Lindl.), a succulent epiphyte, 3.5–10 cm tall, is included in the Red Data Book of IUCN. Its inflorescence consists of lateral drooping panicles, each of which bears about 20 flowers. The plant is found in Costa Rica, Salvador (3 habitats), Guatemala, and Mexico, and possibly grows in Honduras and Nicaragua. This orchid has excellent adaptability to wild cypress (not found on other trees) and may disappear with felling of cypress.

The butterfly orchid genus (*Oncidium*) consists of about 530 species and is distributed in tropical America: from Mexico to Brazil and Paraguay, as well as Bahamas and Antilles. The flowers in this genus differ significantly in their size, but most species have yellow flowers with brown spots usually borne in many-flowered inflorescences on flexible branched peduncles. All plants are epiphytic.

The **golden butterfly orchid** (*O. varicosum* Ldl.) grows in the mountains of Brazil (Plate 46). The Kramer's butterfly orchid (*O. krameranum* Rchb. f.) grows at an altitude of 300–900 m in Equador, Costa Rica, and Colombia. It develops on old trees and woody creepers. One elliptical leaf originates from its bulb, which is covered with a blackish-violet marble pattern. The flowers very much resemble a butterfly: they are orange-golden with chocolate-brown spot; the lip is canary golden with a chocolate-brown spot (Plate 46). The male butterflies consider the flower another male and, protecting their territory, attack the flower and pollinate it in the process. This orchid is one of the most elegant and ornamental species in the genus. The butterfly orchid *O. papilio* Ldl. grows in Venezuela and Trinidad, and is very similar to *O. varicosum*. It is endangered due to commercial collection (Plate 45).

The **bird's-beak** (*O. ornithorhynchum* H.B. Kth.) with ovoid 2-leaved tubers is found in Guatemala and Mexico. The flowers are lilac-purple with wavy petals (Plate 46). **Heneken's butterfly orchid** (*O. henekenii*) is endangered in the Dominican Republic.

Up to 60 species of the genus *Sobralia* are found in the mountains of tropical America. These are terrestrial orchids with numerous shoots resembling a cane. A single flower remains open only for 3–4 days, but the flowers open alternately, which increases the blooming period.

The **yellow-white sobralia** (*S. xantholeuca* Hort. et Williams) has been included in the Red Data Book of IUCN as an endangered species. It is a large orchid with 150 cm tall, nonpubescent leafy stems, arising from rhizomes and covered with chocolate-brown-spotted leaf sheaths. The leaves are dark green, lanceolate; flowers are yellow. It is known from a very few localities in Guatemala (San Cristobal and Wewetenango) and Salvador; it may also be found in Honduras. It grows either as a terrestrial orchid or as an epiphyte in cool moist forests on rocks. It can easily disappear as a result of forest cleaning for coffee plantations.

The genus *Stanhopea* includes 50 epiphytic species, which are distributed in tropical America.

The **spider orchid** (*S. oculata*) is found from southern Mexico to Venezuela. It is a very beautiful epiphyte with waxy but rapidly wilting flowers emitting vanilla flavor. It is pollinated by bees.

While talking about American orchids, one cannot ignore the world famous genus *Vanilla*, which is widely used in industry. It includes more than 65 species, some of them widely cultivated. All the species of this genus are creepers with cylindrical stems, alternate, large and fleshy leaves, with aerial roots on each node. The flowers are large, yellow or yellowish-green, in inflorescences at the stem apex and in axils of upper leaves (Plate 46).

Common vanilla [*V. fragrans* (Salisb.) Ames] grows in tropical America from Mexico to South America and on the Antilles. It is widely cultivated. *Vanilla odorata* Presl. from Equador and **Mexican vanilla** (*V. pompona* Schiede) from tropical America, distributed in southeastern Mexico, Nicaragua, Panama, and Colombia (at an altitude of 800–1500 m), Trinidad and Guiana, have similar characteristics. The Mexican vanilla is a substitute for true vanilla (it has a flavor reminiscent of heliotrope).

FAMILY PALMAE

Two species of American palms are included in the Red Data Book of IUCN. One of them is **Ekman's pseudophoenix** (*Pseudophoenix ekmanii* Burret). This is a 4–6 m tall palm with simple stem. The main stem is about 20 cm in diameter, which thickens upwards, forming a bulge up to 80 cm in diameter, which is narrowed near the crown to 15 cm. The stem is very juicy; leaves are 1.5 m long; inflorescence is up to 80 cm long, drooping,

PLATE 43. 1—Perrin brassavola; 2—Clowes' cradle orchid; 3—Rucker's cradle orchid; 4—Aclandia's cattleya.

PLATE 44. 1—Dayan's laelia; 2—Skinner's cattleya; 3—purple laelia; 4—Trian's cattleya.

with numerous branches. It is found in the Dominican Republic, on the Barahoma peninsula in the extreme south, on dry Tertiary limestone rocks. It is believed that this is the wine palm described by the first investigators of the islands. A very good light-colored wine was prepared from the sap of its stems; sap extraction was responsible for the disappearance of this species. The **wine pseudophoenix** (*P. vinifera*) is a more common species, known from the Haiti islands; it suffers due to clearing of habitats.

Calyptronoma rivalis (O.F. Cook) L.H. Bailey is a medium-sized tree with the trunk up to 10 m tall and distinct leaf scars; the crown consists of 15–20 pinnate leaves. It grows in northwestern Puerto Rico, east of San Sebastian. This palm grows along rivers in humid forests on limestone rocks at an altitude of 300 m. Not more than 20 palms of all sizes were found in 1970. All of them are on the private land of a single owner who agrees to protect this plant. Possibly, this calyptronoma was more widely distributed in the past but has disappeared due to fire and felling.

Chilean palm (*Jubaea spectabilis* H.B. et K.) is a tree with more than 1 m thick and 15–18 m tall stem. The trunk is almost smooth, ash-gray, terminating in a large crown of pinnatisect leaves arcuately bent outwards, reaching 5 m in length. It grows in Chile, along the western coast, rising into the mountains up to 1200 m. It is widely used in the national economy. Its extensive exploitation (mainly for sap extraction) has lead to almost complete disappearance of this plant.

The mountain palm genus (*Chamaedorea*) has about 60 species distributed in the forests of Central and South America. These are shrubs or creepers with slender stems, pinnate leaves, and minute yellow or red flowers. **Tuerckheim's mountain palm** [*C. tuerckheimii* (Danm.) Burret.] is among the rare species of Guatemala.

The genus *Iriartea* is represented by trees with pinnate leaves. They grow along marshy river banks, which are periodically flooded, and develop adventitious roots. The **ventricose iriartea** (*I. ventricosa* Mort.) is endangered in Peru.

Other palm species of Central and South America also require immediate conservation measures. As a result of construction activity along the sea coasts, *Hippomane mancinella* and *Gaussia prinseps* H. Wendl. are endangered in Cuba; the endemic palm of Beata island (Dominican Republic), *Zombia antillarum* (Descourt) L.H. Bailey, known only from a few habitats, and *Gaussia attenuata* (O.F. Cook.) Beek in Puerto Rico are endangered.

The only population of *Itaya amicorum* H.E. Moore with less than 100 plants is known from Itaya river in Peru. *Chrysosallidosperma smithii* H.E. Moore, *Iriartella ferreyrae* H.E. Moore, and *Socratea sabazarii* H.E. Moore are also close to extinction in Peru.

The populations of wax-palm belonging to the genus *Ceroxylon*, growing at high altitudes in the Andes from Venezuela to Peru and Bolivia where the forests have been replaced by coffee plantations over large areas, have reduced sharply. The national tree of Colombia, *C. quinquiense* (Karst.) H. Wendl, and the **Andean wax-palm** *C. andicola* Humb. et Bonpl. (Plate 47) should also be included as endangered species.

Chunta macahuba (*Acrocomia chunta* Covas et Ragon), a palm with a small area of distribution, is endangered in Argentina, where its population has declined as a result of exploitation, like **assai palm** (*Euterpe edulis* Mart.) in which the apical buds are edible (Plate 47). In Guatemala and Panama, **Cook's colpothrinax** (*Colpothrinax cookii* Pead.), a plant found only at 2 distant localities, is on the verge of extinction.

FAMILY PINACEAE

Guatemala fir (*Abies guatemalensis* Rehder) is a tree up to 45 m tall. It has become a very rare plant in Guatemala as a result of intensive felling and destruction of the undergrowth by cattles.

Aycahuite pine (*Pinus aycahuite* Ehrenb.) is a tree 30–45 m high with a conical crown, gray bark, and very long, up to 30 cm, needles. It is distributed in the mountains of southern Mexico and Guatemala. It is being felled intensively for its valuable timber and bark, which is used for extracting tannins. This has severely damaged the reserves of this species.

FAMILY PIPERACEAE

The primarily tropical family of Piperaceae comprises about 10 genera and 3000 species.

One genus—the pepper elder (*Peperomia*)—is mostly represented by epiphytes or rock plants, rarely terrestrial semishrubs or herbs. The leaves are succulent; flowers are bisexual, small, and foul-smelling. The epiphytic species grow on trees in the forests of the fog belt, but some, for example, crystalline pepper elder, also grow in desert (in Chile). Seven species of pepper elder are endangered in Peru: *P. atocongona* Trelease; *P. crystallina* R. et P.; *P. limaensis* Trelease; *P. non-hispidula* Trelease; *P. seleri* C. DC.; *P. umbelliformis* C. DC.; and *P. pseudo-galapagensis* Trelease.

FAMILY PODOCARPACEAE

Only one of the 23 species of the genus *Dacrydium* is found in South America: **Fonk's dacrydium** (*D. fonkii* Ball.). This is a highly branched shrub growing on the ground, 30–80 cm tall, with spirally arranged leaves. It grows in southern Argentina, in the mountains of southern Chile, almost up to Terra del Feugo and Chonos island. It is endangered in Chile.

The genus (*Podocarpus*) is represented by dioecious coniferous shrubs or trees with narrow leaves. **Parlator's podocarp** (*P. parlatorei* Pilger)—the only coniferous tree in this part of the country—was found in the forests of northwestern Argentina. Now the forests in this area are almost completely cleared (the wood of podocarp is used to make paper, and the land used for plantation of fast-growing pine and eucalyptus). The **Chilean** or **plum-fruited yew** (*P. andina* Poepp.) growing in the Chilean Andes is a tree up to 15 m tall, with densely branched crown; needles are light bluish on the lower surface. This species as well as others—*P. salignus* D. Don and *P. nubigenus* Lindl.—are endangered in Chile.

The **Prince Albert's yew** (*Saxegothaea conspicua* Lindl.) is a beautiful monoecious tree up to 10 m tall with grayish-brown bark and whorled branches. It grows in dense humid mountain forests of the Chilean Andes and western Patagonia. It is endangered in Chile. It has been widely introduced into cultivation.

FAMILY POLYGALACEAE

The old specimens of the arboreal **Cowell's milkwort** (*Polygala cowellii*) have been preserved in the forests of Puerto Rico. However, all attempts for its restoration have proved unsuccessful.

FAMILY PORTULACACEAE

The purslane genus (*Portulaca*) of the family Portulacaceae comprises prostrate fleshy herbs with fairly large beautiful flowers. In Peru, the **hairy purslane** (*P. pilosissima* Hook.) is endangered. The rock purslane genus (*Calandrinia*) comprises about 150 species, distributed in the Andes and Australia. Four species of rock purslane are endangered in the Peruvian Andes: *C. alba* (R. et P.) DC.; *C. crenata* (R. et P.) Macride; *C. paniculata* (R. et P.) DC.; and *C. ruizii* Macride.

FAMILY ROSACEAE

Soap bark tree (*Quillaja saponaria* Molina) has evergreen leaves, and its bark contains saponin (Plate 47). It grows in the subtropical forests of the coastal Cordillera. Its population has sharply declined as a result of exploitation.

FAMILY RUBIACEAE

Red Peruvian bark (*Cinchona succirubra* Pav.) is an evergreen tree, with opposite, leathery, glossy, broadly elliptical leaves and light crimson red flowers, in terminal panicles on stem and branches (Plate 48). Its area of distribution is very restricted; it is found in Peru, Bolivia, Equador, and Colombia, on the slopes of the Andes, at an altitude of 1600–3200 m in wet forests. The red Peruvian bark trees were cut for obtaining quinine (an antimalarial drug). At present it is widely cultivated in southeast Asia and Africa.

Ipecac (*Caphaelis ipecacuanha* Willd.) is a small plant with long, slender rhizomes; stems are slender, up to 30–40 cm tall, with a few pairs of opposite, evergreen, broadly lanceolate leaves and a small head of minute white flowers (Plate 47). The plant grows on a wide territory on the upper reaches of the right tributaries of the Amazon (Brazil). Because of mass digging of roots as a medicinal raw material, the population of this species is declining. It is difficult to cultivate ipecac, although plantations have been laid in India, Indonesia, and Tanzania.

FAMILY SAPOTACEAE

The genus *Mimusops* comprises about 60 species of evergreen trees, distributed in the tropics of both the hemispheres. The **milk tree** (*M. balata* Miq.), up to 35 m tall, with simple entire leaves, is found in the forests of tropical America, mainly in Guiana and Venezuela. Its fruits are berries; wood is valuable, used for joinery. The bark and other parts contain a milky juice, close to Guttapercha and called 'balata'. Usually balata is used as an admixture in rubber to impart greater strength and viscosity; the fresh juice is used for preparing chocolate. To obtain balata, the trees are cut at the age of 30–40 years, as a result of which its natural population has sharply declined. This tree is cultivated in Brazil and the West Indies; experiments on its cultivation are in progress in western tropical Africa.

FAMILY SCHIZAEACEAE

The curly grass genus (*Schizaea*) consists of herbaceous, rarely creeping, ferns with erect rhizomes and repeatedly dichotomously branched leaves. Small plantations of **feathery curly grass** (*S. pennula*) have been preserved in Trinidad.

FAMILY SIMAROUBACEAE

The family Simaroubaceae includes **American bitter-wood** (*Simarouba glauca* DC.), having great importance as an oil and medicinal plant. It is a tree with small bisexual, regular flowers. It grows in tropical America, from southern Florida to Costa Rica and the Bahamas.

FAMILY THEACEAE

Foreror freziera (*Freziera forerorum* A. Gentry) is a dwarf tree with lanceolate, leathery leaves, having a symmetrical base. Flowers are axillary in clusters of 1–3. At present, only 3 partly dry trees have remained at the peak of Piko-Takarkun, the border range between Panama and Colombia, among the thickets of shrubs at an altitude of 1900 m. The plant may be dying due to climatic changes. It is not found under cultivation. It has been included in the Red Data Book of IUCN.

FAMILY ZAMIACEAE

Siliceous coontie (*Zamia silicea* Britton) is a stemless plant with underground, fusiform rhizome and 2–5 leathery leaves spreading on the soil, each having 8–16 leaflets. It grows in moist savanna and sparse, well-lit tropical pine forests. It is quite tolerant to long droughts and is fire resistant because of the stored material in the underground corm. It grows on Pinos Island (Cuba). *Zamia kickxii* Miq., *Z. multifoliata* A. DC., and *Z. ottonis* Miq. are also among the endemic Cuban species. *Zamia angustifolia* Jack. (the Bahamas), *Z. integrifolia* Ait. (Florida, Jamaica, Haiti), *Z. latifolialata* Prenleloup (Jamaica, Haiti, Puerto Rico), *Z. media* Jacq. (Haiti, Puerto Rico), and *Z. pumila* L. (Florida, the Bahamas) need to be protected.

Epiphytic coontie plants are found in the South American tropical forests. *Zamia poeppigiana* Mart. et Fichl. grows on soil, trees, and fallen tree trunks, in moist and shady mountainous tropical forests in eastern Peru and in Colombia. The **pseudoparasitic coontie** plant (*Z. pseudo-*

parasitica Jates) is found in Peru, Colombia, Equador, and Panama, as well as in dense forests on soil or tree trunks. The trunk of both the coonties is up to 1.5 m tall; leaves are up to 2 m long; cones are large, up to 40 cm. The names of the species are not correct, as both the species are facultative epiphytes.

Only one species of *Microcycas* is known: the **cork palm** [*M. calocoma* (Miq.) A. DC.], which is endangered. It is a beautiful plant with cylindrical stem, 6–8 m, rarely up to 10 m tall, with a diameter up to 30 cm. In contrast with all other sago palms, the trunk of cork palm is not covered by leafy bases, but is protected by cork. A large and long cone rises above the crown. With a relatively small weight, up to 9.5 kg, in the female plants it reaches a length of 94 cm. The cork palm grows in Cuba along the southern spurs of the Sierra de Los Organos, on the hills in tropical deciduous forests. Only small habitats are known; the total number of plants known is about 600. It is protected in all the locations and hence included in the Red Data Book of IUCN.

Mej's dioon (*Dioon mejaei* Standery et L.O. Williams) is endangered in Honduras. The plant is disappearing as a result of mass collection of its seeds by the local people.

6
Australia

The flora of Australia, the smallest and an isolated continent, is unique. It includes a whole series of endemic families. About 600 genera are typical of the flora of the Australian continent.

The natural conditions of Australia are also extremely variable. The northern part is in the tropics, and the southern part in the temperate zone. There are no high mountains, large rivers, or lakes in this country. A large part of the continent is covered with vast areas of monotonous plains, mainly dry. The highest places are near the eastern coast, where the mountain chains are continued from north to south, including the island of Tasmania.

At present, the flora of this continent includes about 20,000 flowering plants, of which 2206 or 10 per cent need protection. The Western Australian species constitute a large part of such plants (45 per cent), 12–15 per cent need protection in South Australia, Queensland, and New South Wales states, and 5–7 per cent need protection in the remaining states of Victoria and Tasmania and the Northern Territory. According to the data published by the Service of Australian National Parks and Wild Life, 78 plants have disappeared from the Australian flora, 221 species are endangered, and 842 species or 38.2 per cent of the total number of plants requiring protection are being protected in the national parks or other protected territories.

FAMILY ARAUCARIACEAE

The family Araucariaceae is represented by 2 genera in Australia: *Araucaria* and *Agathis*. One of the species from these genera is included in the list of endangered plants of Australia.

Bunya-bunya pine (*Araucaria bidwillii* Hook.) is the only member of the oldest section of Araucariaceae. Bunya-bunya is a tree growing in eastern Australia, in Queensland. It is a dioecious plant. The female plants reach a height of 40–50 m and a diameter up to 125 cm (Plate 48). The young trees have broadly pyramidal crown and lateral branches in whorls at the top. The leaves are ovate with a cusp, borne spirally in the upper part of the crown, and in 2 rows on the lateral branches. The cones are up to 35 cm long, weighing up to 3 kg (the largest cones among Araucariaceae). The bunya-bunya pine grows in the forests at the Pacific coast. It has a fairly large area of distribution, but the populations are small, and the wood of *Araucaria* is highly valued and used for furniture. The exploitation of plantations as well as change in the habitat conditions have reduced the total population of this relict plant. Several natural habitats of the bunya-bunya pine are protected in the National Parks of Queensland, particularly in the Bunya Mountains National Park.

In 1843, the English naturalist and traveller Bidwill, in whose honor this species has been named, transferred a few specimens to the Kew Botanical Garden (England). From here it has spread all over western Europe and into Russia. In the USSR, it is cultivated in the Batumi Botanical Garden.

The dammar pine genus (*Agathis*) includes 13 species. The area of its distribution covers the Malacca peninsula, Malayan Archipelago, Bismarck Archipelago, the Santa Cruz, New Hebrides, Fiji, and New Caledonia islands, and the southwestern islands of the Pacific Ocean. Members of this genus are also found in Queensland and New Zealand. The plants of this tropical genus are adapted to humid rainy tropical forests, sometimes entering the semi-evergreen rainforests with short dry season. They rise up to an altitude of 2000–2500 m along the hills, growing on rocks, as well as on podzol and calcareous soils. All the dammar species provide good timber. At present, most species are under cultivation.

The **big dammar pine** (*A. robusta* F. Muell.) is a tall tree with a height of 45–60 m and diameter up to 6 m. The trunk is cylindrical, with a scattered crown. The branches are also sometimes massive, like the main trunk; leaves are flat and broad, not needle-shaped as in other conifers, cones are globose ('Agathis' means 'ball of threads'). This species grows in the forests of the Pacific coast of Australia in Queensland, in the regions with a warm and humid summer and moderate winter: the annual rainfall here is 1000–1500 mm, most of it during the summer (December to March), the dry season being July to September. It is found in the valleys, on plains, and on coastal plains up to an altitude of 900 m. Its area of distribution is discontinuous, one part being in the northeastern corner of the continent, the other in the east, in the Bundaberg region and on Frazer Island (Queensland state). The present natural population strength of this species is unknown. However, its habitat conditions are deteriorating. Felling of the plantations with dammar pine leads to a decline in its population. It must be emphasized that this tree grows very slowly, and its young seedlings do not survive under other trees overgrowing them. It can

PLATE 45. 1—Sincoran laelia; 2—red masdevallia; 3—butterfly orchid; 4—chimera masdevallia.

PLATE 46. 1—Bird's-beak; 2—common vanilla; 3—golden butterfly orchid; 4—Kramer's butterfly orchid.

PLATE 47. 1—Assai palm; 2—ipecac; 3—soap bark tree; 4—Andean wax-palm.

PLATE 48. 1—Bunya-bunya pine; 2—giant byblis; 3—Australian pitcher plant; 4—red Peruvian bark.

be protected from extinction only by maintaining the natural plantations, as in the national parks of Queensland.

FAMILY ASTERACEAE

The family Asteraceae is represented by 88 genera and 500 species in Australia, of which 40 genera are endemic. The **Swan river daisy** (*Brachycome*) is an endemic of Australia, only 3 of its species have been found in New Zealand. These are perennial or annual herbs; flowers are white-light blue or purple in heads. Nine of the species of this genus are included in the List of Rare and Endangered Plants of Australia. Among them, **Muller's Swan river daisy** (*B. mulleri* Sonder) is an endangered plant of the South Australia state, found near Gawler. It is a perennial with simple or rarely branched stem and fairly large capitula. The leaves are long, narrow, pinnate. The plant is an ornamental, and collected for bouquets.

Most of the immortelle (*Helichrysum*) species are endemic to Australia.

Milligan's immortelle (*H. milliganii* Hook. f.) is a beautiful perennial with capitum, very prominent against the background of alpine meadows. It has an erect stem, up to 15 cm tall, and a rosette of basal leaves. Usually the plant forms a dense pad, consisting of numerous stems originating from a highly branched rhizome. The inner part of bracts is white, and they are twice as long as the white or cream-colored petals, which makes the inflorescence unusually attractive. The plant grows in the alpine belt of the mountain peaks of Tasmania.

Wasteland immortelle (*H. ericetum* W.M. Curtis) is a columnar shrub up to 3 m tall; the leaves are of cricoid type, narrow, rolled, short. The narrow floral heads carry 5–6 outer florets, surrounded by pale chocolate-brown, involucral bracts. The petals have white spreading ligules. The entire plant emits a pleasant flavor. The present condition of this plant remains fairly undefined. Its isolated habitats are known in Tasmania in the communities of mountainous heathlands on the Central Plateau around the Great Lake. Both immortelle species are included in the National List of Rare and Endangered Plants of Australia.

The daisy bush genus (*Olearia*) is fairly widespread in Australia and New Zealand. The **ericoid daisy bush** [*O. ericoides* (Steetz) N.A. Wakefield], a rare endemic species with a limited population, is a shrub up to 1 m tall, with tender tips of branches. The leaves are alternate, sessile, narrow linear, up to 5 mm long, hairy beneath, sticky, and with rolled margins. Often short branches with crowded leaves arise from the main branches, which very much resemble the branches of heath, with a strong smell. The capitula are about 20 mm in diameter, terminal, numerous, white or light bluish-purple. The shrub grows only in east and southeast Tasmania.

The genus *Podolepis* is endemic to Australia. The **mountain podolepis** (*P. monticola* R.J.F. Henderson) is a perennial herb with woody stems, rosettes of leaves and up to 20–50 cm long peduncles, bearing up to 10 capitula. Its distribution is limited to a few locations on the MacFerson range, at the border of two Australian states: Queensland and New South Wales. These areas are included in the Lamington National Park. However, the habitat conditions of podolepis are not uniform in the park. One of the habitats is at the upper slopes of the range at an altitude of 1100 m. Here podolepis grows on soil or on organic remains in the communities of closed forests of *Nothophagus*. Another habitat, where podolepis was found very recently, is characterized by a high degree of dryness and is located at an altitude of 1000 m in open eucalypt forests facing southwest. One section of the national park with the podolepis population is convenient for tourism, as the plant is very attractive.

FAMILY AUSTROBAILEYACEAE

A few ancient plants ('living fossils') with primitive characters have been preserved in the Australian flora. The **spotted austrobaileya** (*Austrobaileya maculata* C.T. White) is such a plant. This plant was discovered very recently in northern Queensland at the Aterton uplands in a tropical forest. It is a large climbing shrub with opposite, pinnately veined leaves. The flowers are solitary, bisexual, pale green; perianth is 12-petaloid, similar to leaves; anthers are flat, broad; carpels are like leaves folded into halves. This plant flowers quite rarely, and has juicy, berry-like fruits. It is still not known how it reproduces and maintains its life. However, austrobaileya has a great scientific value for studying the phylogeny of the plant kingdom and it is essential to determine its present condition and protect all its known habitats.

FAMILY BYBLIDACEAE

The members of genus *Byblis* are distributed in Australia and New Guinea. All of them are active insectivorous plants, as they catch their prey as soon as it is within reach.

There were rumors which still continue to spread about the Australian species **giant byblis** (*B. gigantea* Lindl.), that it is a man-eating plant. This is a small shrub slightly more than half a meter tall (Plate 48). Its narrow leaves are densely covered with sticky hairs and glands; it has been calculated that a single plant has up to 300,000 hairs and 2 million glands. The glands discharge a secretion which helps in digesting the prey. The branches and leaves of the plant form a dense, sticky barrier for insects. Sometimes snakes and frogs become the prey for byblis. This species is distributed in Western Australia, east of Perth, on a few low-lying areas as well as on the sandy plain in the interfluve of the Moor River and Eneabbe. It suffers due to collection by amateur florists. The annual demand for giant byblis in the world market is increasing. To protect it from complete extinction, the Government of Western Australia state covered this plant even in 1935 under the law for protection of Australian indigenous flora. At present, it is included in the List of Endangered Species of Australia. The habitats of the giant byblis are protected in the Moore River National Park.

Flax-flowered byblis (*B. liniflora* Salisb.) is very similar to the *B. gigantea*, and can be distinguished only in the flowering condition. It grows in the tropical parts of Australia and in New Guinea. At present, it is not being exported due to difficult conditions of collection, however, both the byblis species are included in CITES to ensure control on its export.

FAMILY CALYCANTHACEAE

The family Calycanthaceae includes 3 genera and 8 species. The taxonomic position of the members of this family is of great interest, although there is no doubt that they are closest to Magnoliaceae. Their geographical isolation is even more surprising: they are found in North America, China, and Australia.

Australian idiospermum [*Idiospermum australiense* (Diels) S.T. Blake] differs from the other members of the family by the structure of its wood. This is an evergreen tree up to 40 m tall and 1 m in diameter (at chest height). The bark is light gray; leaves are dark green; flowers are red to purple; fruits are olive-green to chocolate-brown with 1–2 seeds, each having 3–4 large cotyledonous leaves. The leaves and stem have secretory cells with essential oils, emitting a pleasant smell. Idiospermum is known from a few locations in the northern part of Queensland: Russel River, Noah Creek, and Deintry, between Cooktown and Kerns. It grows in the lowland rainforests on alluvial soils, developed from metamorphic rocks. The annual rainfall in its habitat is 3000 mm, and the dry period continues from July to October. Under such conditions, the plant was discovered by Diels in 1902; it was established even then that its seeds are poisonous for cattle. Felling of the rainforests, further agricultural use of the territory, construction, and other activities caused idiospermum to disappear from the field of vision of botanists. A report about poisoning of cattle by its seeds in 1971 led to its rediscovery, but this time 160 km north of the first habitat. The present status of the species is still undefined, but its scientific importance for studying the botanical-geographic and historical associations of the flora of different continents is beyond doubt. It is being maintained in cultivation in Brisbane. The plant is included in the List of Endangered Species of Australia, and in the Red Data Book of IUCN.

FAMILY CAMPANULACEAE

The family Campanulaceae is distributed mainly in the northern hemisphere, but a few endemic members can be found in Australia and even in Tasmania. One of them is the **New Zealand blue bell** [*Wahlenbergia saxicola* (R. Br.) A. DC.]. It is a small perennial herb with delicate rhizome and spreading vegetative shoots; leaves are in rosettes; flowering shoots are erect with a few pale light-blue campanulate flowers. It grows on hilly meadows and rocks. The area of distribution of New Zealand bluebell includes the hilly regions of Tasmania. The plant is quite easy to cultivate (usually the seeds are sown and germinate fast), but does not live long. It is cultivated in the Launceston Botanical Garden in Tasmania and in Edinburgh (Great Britain).

FAMILY CASUARINACEAE

The family Casuarinaceae includes more than 60 species of trees and shrubs, belonging to the only forest-oak genus (*Casuarina*). They are distributed in the subtropics and tropics of the southern hemisphere. A large number of the species grow in Australia, Tasmania, and New Caledonia.

The rarest species of the genus—**fibrous forest-oak** (*C. fibrosa* C.A. Gardner)—is endangered. This is a 50–150 cm tall shrub with numerous stems, arising from an underground, woody tuber. The branches have a large number of needle-like twigs, as a result of which the plant resembles pine; the leaves are scaly, and the 'cones' are covered with numerous intertwined hairs. At present, not more than 100 specimens of this species have survived. It was first discovered in 1926 by C.A. Gardner in Western Australia, not far from Perth, in an area of about 600 ha of open wasteland along with members of Proteaceae (*Banksia, Dryandra*), Myrtaceae (*Eucalyptus*), and other plants. Subsequently, it was found in 1928 and 1949; till 1969 it was considered to be extinct. In 1969, during a study of the vegetation of the reserve organized for conservation of the flora of this region, the fibrous forest-oak was found again, and the floristic reserve was named in honor of the first discoverer of Casuarinaceae. Unfortunately, this small area is the only habitat of this forest-oak, now surrounded by farms and agricultural fields. Very little is known about the biology of this species. It has been established that the plant is resistant to fire: it is regenerated through the shoots developing from the underground tuber. Since it is pollinated by wind, the plant normally produces fruits but an important condition for seed germination is fire. The plant count at the area showed that at present there are very few young specimens of this forest-oak. The frequency of fires necessary for maintaining the stability and variability of the vegetational cover in the reserve and conserving as also increasing the forest-oak population is still unknown. This plant is included in the category of endangered species and in the Red Data Book of IUCN, as well as in the List of Rare and Endangered Plants of Australia together with three other species: **cypress-shaped forest-oak** (*C. chamaecyparis* Poiss.) growing in Queensland; **multibranched forest-oak** (*C. ramosissima* C.A. Gardner), in Western Australia; and **bicuspid forest-oak** (*C. bicuspidata* Benth.) in South Australia. The only habitat of the last species is known only from the type collections, i.e., from the place it was first collected, the Flinders Islands. Therefore, the specimen was described as the species. Its present status has not been determined. The bicuspid forest-oak is a shrub or small tree with ascending, grayish, grooved twigs with 10 teeth on the upper part of each segment of the twig. The male inflorescence is terminal on long shoots, and female inflorescence on dwarf shoots. The 'cones' are oval-roundish, 2–4 cm long.

FAMILY CEPHALOTACEAE

The members of the family Cephalotaceae are insectivorous plants, in which the leaves in the form of pitchers play the role of traps. The only plant of this family—**Australian pitcher plant** (*Cephalotus follicularis* Labill.)—grows in Western Australia. This is a small herbaceous plant with basal leaves, some of which are simple, assimilating, and others modified into pitchers (Plate 48). The pitcher-shaped leaves have short petioles, with crestate processes on the outer side, a ridged ring along the margin, and a lid at the top which is usually open. If the prey falls into the pitcher, the lid immediately closes. There are multicellular glands embedded in the tissue on the outer side of the lid and inner side of the pitcher below the ring. There are no glands in the lower part of the inner surface of pitcher. An insect fallen into such a pitcher cannot escape. It is digested by microorganisms. It is interesting that the plant secretes a substance inhibiting the growth of putrefying bacteria. The floral stalk or scape of Australian pitcher plant terminates in a compound raceme bearing up to 20 flowers. The number of maturing seeds is small and germination not very low. Australian pitcher plant is distributed in Western Australia sporadically. It is found among the dense growth of shrubs and sedges around marshes and along water streams where the water always filters through the soil. Its main habitats are concentrated in the coastal region (from Manypeaks mountain to the Donnelly and Yallingap rivers). The present condition of Australian pitcher plant raises doubts about its fate. This rare endemic of Australia with extremely limited distribution is in great demand in the world market for cultivation. The entire plant and its rhizomes are used in trade. The exploitation of Australian pitcher plant was not regulated till now, although the species was covered by the law on Conservation of Natural Flora of the Western Australia State even in 1935. The major export was to the United States. The demand for it increased steadily because of low survival of the Australian pitcher plant. To prevent its extinction, the Australian pitcher plant was included in Appendix 2 of CITES in 1979.

Ample experience on the cultivation of this plant has been gained in several botanical gardens of England, where it was introduced in the last century. There are special clubs of amateurs and collectors of insectivorous plants, including Australian pitcher plant, who exchange information on cultivation and plantation material. In Australia, this species is included in the List of Endangered Plants.

FAMILY CHLOANTHACEAE

The independent status of the family Chloanthaceae has been recognized by almost all taxonomists; some call it Dicrastylidaceae. We support the Australian taxonomists and separate Chloanthaceae as an independent family. The main genus of this family is *Chloanthes*. Ten more genera, with almost 50 species, belong to this family. All the Chloanthaceae plants are typical only of the Australian flora.

The *Chloanthes* genus has 4 species, of which **red chloanthes** (*C. coccinea* Bartl.) is included in Appendix 2 of CITES because it is exported. This is a branched, 25–60 cm tall shrub. The leaves are narrow linear with rolled margins, 1–2.5 cm long and 2–3 mm broad; they are leathery and hairy beneath; flowers are borne in a short, leafy, terminal cyme, glandular, densely hairy on the lower side, purple or scarlet. Chloanthes is endemic to southwestern part of Western Australia. It is a fairly common plant on small areas between Lake Grace and Kukerin, as well as between Corrigin and Kulin; a few habitats are known in the environs of Bridgetown and Koyonup. The species is geographically isolated, located 2500 km west of the main area of the genus. Such isolation of the genus is scientifically interesting.

The genus *Dicrastylis* is the largest and includes about 20 species growing predominantly in the dry regions of Western Australia. Only a few dicrastylis species enter South Australia. Almost all species of the genus are commercially exploited. These are densely woolly or tomentose semishrubs and shrubs with opposite leaves and paniculate inflorescence.

Beveridge's dicrastylis (*D. beveridgei* F. Muell.) is a small shrub with whitish pubescence on the branches and grayish pubescence on leaves. The leaves are sessile, 10–30 mm long and 2.5 mm broad, the youngest leaves are golden. The racemose inflorescence bears a few flowers, initially dense, subsequently gradually elongated and converted into a loose spike, 5–10 cm long, with strong golden pubescence. Calyx is short, golden-pubescent; corolla is scarlet, pubescent on the outer side. The plant flowers in summer. It grows in South Australia, in the environs of Ooldea, and in Western Australia.

Costell's dicrastylis (*D. costelloi* Baylei) is an erect shrub, about 50 cm tall with dense grayish pubescence. Leaves are linear-lanceolate, up to 18 mm long and 2 mm broad, with rolled margins. The flowers are borne in spicate inflorescence, arranged along a common axis. The calyx and corolla are as in the previous species, but with whitish pubescence. The shrub grows in Western and South Australia (near Lake Eyre).

Whorled dicrastylis (*D. verticillata* J.M. Black.) is a small shrub with sparse whitish pubescence. Leaves are whorled, in groups of 3, linear, up to 10 mm long, with curved margins. Flowers are borne in a few terminal verticils on branches, surrounded by pubescent bracts. Corolla is pubescent on the inner and outer sides. The plant is found only in South Australia north of Muret Bay.

Doran's dicrastylis (*D. doranii* F. Muell.) is a common South and Western Australian species. It is a shrub up to 1 m tall with sparse, white-grayish pubescence. It differs from *D. verticillata* by narrow paniculate inflorescence. In South Australia, the plant is protected in the Simpson Desert Park.

Three more dicrastylis are rare endemics of Western Australia and are included in the List of Endangered Species of the continent.

The endemic Western Australian genus *Lachnostachys* includes 9 species. These are shrubs with white tomentum of branched hairs; stems, leaves and spicate inflorescence are pubescent. Each flower appears to be submerged in a cotton pad. Most species are exploited for commercial purposes.

Mullein-leaved lachnostachys (*L. verbascifolia* F. Muell.) is a woolly, pale gray plant with a pale violet corolla visible only in the spicate inflorescence (Plate 49). It grows in sandy plains between the Clack Line railway and the granite mountains Tarin Rock (Western Australia). Sometimes it forms thickets. The annual plant collection for sale is reducing its population; therefore, the entire genus has been included in Appendix 2 of CITES.

The genus *Newcastelia* includes about 10 species of woolly or tomentose shrubs with opposite, entire leaves and dense, spicate inflorescence. All of them are adapted to dry habitats. The West Australian species are included in Appendix 2 of CITES.

Dixon's newcastelia (*N. dixonii* F. Muell. et Tate) is an erect, densely pubescent shrub. The leaves are alternate or whorled, sessile, 7–15 mm long; flowers are few, calyx is up to 3 mm long, corolla is 3 times longer, with lanceolate lobes. In Western Australia, it grows in the Great Victoria Desert, and in South Australia, near Renmark and on the sandy islands near Crystal Brook.

Golden-hair newcastelia (*N. chrysotricha* F. Muell.) has been so named because of the golden color of spicate inflorescence and dense yellowish pubescence of calyx. It grows in the Great Victoria Desert (Western Australia) and near the Birksgate range (South Australia).

The **golden-leaf newcastelia** (*N. chrysophylla* C.A. Gardner) is a rare species with limited distribution, adapted to the territory of Shark Bay (Western Australia). This plant is densely pubescent, especially its leaves. Under strong light, they appear golden. The **branched hair newcastelia** (*N. cladotricha* F. Muell.), living in the southeastern regions of Queensland state, is a species whose population has significantly declined. Its hairs are highly branched and intertwined, forming a woolly layer covering all the plant parts. Golden-leaf newcastelia and branched hair newcastelia are included in the List of Endangered Plants of Australia.

FAMILY CUPRESSACEAE

The cypress pine genus (*Callitris*) of the family Cupressaceae is distributed exclusively in the southern hemisphere: in Australia and New Caledonia. It includes about 20 species. Among the rare species is the native oblong cone **cypress pine** (*C. oblonga* Rich.), a small tree with a columnar, glaucous crown and roundish black cones on branches and trunk. It originated from Tasmania and its distribution is limited to the northeastern regions. It is represented by small populations, the present size of which needs to be determined. This species is included in the National List of Endangered Plants of Australia. In the USSR, Tasmanian cypress pine is cultivated in the Batumi Botanical Garden.

FAMILY CYATHEACEAE

Three species of the family Cyatheaceae are included under the rare plants of Australia. **Celebes tree fern** (*Cyathea celebica* Bl.), **ponja** (*C. cunninghamii* Hook. f.), and **feline tree fern** [*C. felina* (Roxt.) Morton]. They are all up to 20 m tall plants, with pinnate leaves. As the tropical regions in Australia are in Queensland state, these tree ferns are distributed only there, and ponja is also found in Victoria state.

The existence of tree fern is threatened, on the one hand, by the fairly rapid reduction in the areas of tropical forests and their subsequent use for road and industrial development or agriculture and, on the other, by felling of tree ferns, digging of their rhizomes, and their use as a substrate for orchid cultivation. This has become the main reason for including all species of tree fern in Appendix 2 of CITES.

FAMILY DICKSONIACEAE

In Australia, **Young's dicksonia** (*Dicksonia youngiae* Hook. et Bak.) has been included among the protected plants of the family Dicksoniaceae. Its trunk reaches 4–6 m in height and a large number of adventitious roots twine around the bases of fallen leaves. The trunks are mostly covered with a thick mass of epiphytes, and the leaves are bipinnate. It grows in moist ravines of coastal mountainous regions of Queensland and New South Wales. It is being protected in several national parks.

Hairy culcita (*Culcita villosa* C. Chr.) is a member of another genus of Dicksoniaceae. This tree fern has been found in the tropical forests of southeastern Queensland.

These species are not the only ones endangered. The roots of almost all Dicksoniaceae plants are used as a substrate for cultivating orchids and bromeliads and, therefore, collected and exported to many countries of the world for botanical gardens and private collections. With a view to regulate the harvests of the Dicksoniaceae plants and controlling their export, all species of the family have been included in Appendix 2 of CITES, since 1974.

FAMILY DILLENIACEAE

The tropical family Dilleniaceae is represented by 5 genera in Australia, the largest among which is the garland flower genus (*Hibbertia*), comprising about 80 species. This is an almost entirely Australian plant; only 2 species have been found in Madagascar and one in New Caledonia. The garland flower is a highly branched shrub or semishrub; sometimes creepers and herbs are found among them. Their flowers are yellow or white, solitary. Twenty-six of its species have been included in the List of Endangered Plants of Australia, a majority of which are from Western Australia and the Northern Territory.

The rarest species, with only one habitat in the valley of Swan River (southwestern Australia) is **slender garland flower** (*H. leptopus* Benth.), a very elegant shrub with slender, naked branches. The bark is whitish-gray; leaves are narrow linear. Flowers in large numbers are on thin peduncles. The present condition of the species is unknown.

The **vilious garland flower** (*H. lasiopus* Benth.) from Western Australia, found east of Perth, is endangered. It has short, pubescent shoots; in addition, there are long, erect hairs. The leaves are up to 5 cm long, sometimes serrate and hairy. The flowers are borne on strongly hairy peduncles surrounded with broad, chocolate-brown bracts. A few species of garland flower are highly ornamental and their branches are used for making bouquets.

FAMILY EPACRIDACEAE

Australia is the center of origin of the family Epacridaceae; only a few members are found in Oceania, on the Antarctic Islands, and in South America. The family includes more than 20 genera. These are mostly shrubs, rarely trees with beautiful spicate or paniculate inflorescence of white, red, rarely dove-blue, green or yellowish flowers. Many Epacridaceae species are morphologically very similar to the European or South African species of Ericaceae or included among the Australian 'heathlands'. Because of their ornamental value, some species of Epacridaceae are cultivated.

Andersonia is an endemic relict genus of Western Australia. Members of this genus have several primitive characters, such as sheathing leaf bases, and lack of difference between the assimilating and floral leaves. Almost all andersonias are shrubs with solitary pink or dove-blue flowers arranged in capitate inflorescence at the end of branches. Seven of its species are included in the list of plants identified for protection.

Large-flowered andersonia (*A. grandiflora* Stschegl.) is a highly branched shrub up to 15 cm tall; a few scarlet flowers together with leaves form dense heads. It grows in the extreme southwest of Australia.

The Australian heath genus (*Epacris*) is distributed only in Australia and New Zealand. Almost all its species are endemic except 2–3, which are common to both countries. The Australian heaths are small shrubs with few flowers. Ten of its species have been included in the List of Endangered Plants of Australia.

Bearded Australian heath (*E. barbata* Melville) is an endemic of Tasmania. It grows on sandy coasts in the southeastern part of the island, sometimes forming thickets, for example, near Oyster Bay. It is an evergreen 60–120 cm tall shrub. The leaves during its growth season differ in shape and size. The most typical leaves are oval-elliptical, 5–10 mm long, narrowed at the base into a petiole, and at the apex into a cusp. The flowers are arranged in a cone-shape or capitate inflorescence at the end of branches. The floral leaves (bracts) gradually pass into a purple calyx, covered with hairs; corolla is white; anthers are dark-colored.

Most of the rare species of this genus are concentrated in Tasmania. The endemic Tasmanian genus *Prionotes* is represented by only a single species—**waxy prionotes** (*P. cerinthoides* R. Br.). This is a shrub with tender branches, prostrate or twining around the stems of dead trees up to a height of 6–9 m. The branches are intertwined and form a dense green mass bearing bright crimson red flowers (Plate 49). The shrub grows in moist forests of Research Bay, on the slopes of Wellington and La Perouse mountains, and near the Macquaire Harbor. One more rare species—**paniculate richea** (*Richea scoparia* Hook. f.)—a tree or up to 1.5 m tall shrub, also grows on Wellington mountain. Its leaves are clustered at the tips of branches; flowers are whitish-pink or orange, in dense racemes or panicle (Plate 49). In Tasmania, this plant was also found on the Great Western Tiers at an altitude of 900–1200 m. Now it has survived only on inaccessible slopes.

The Australian currant genus (*Leucopogon*) is the most widespread member of Epacridaceae. Its species are mainly typical to the flora of Western Australia. A few Australian currants are found in New Zealand. These are shrubs with small, usually white flowers. Out of almost 150 species of the genus, 38 are included in the list of endangered plants.

Only one location of the **disconnected Australian currant** (*L. interruptus* R. Br.) is known in Western Australia at the southern coasts, in the King George Sound region. This small shrub bears leaves only on the current year branches. The flowers are small but numerous, arranged in thin racemes.

Covered Australian currant (*L. obtectus* Benth.) is a rare plant of Western Australia. Its small population has been found on sandy and calcareous soils between the Moore and Murchison Rivers. It is a shrub up to 1.5 m tall with cream-yellow flowers, which are arranged in groups of 2–3 on a short peduncle and covered with cordate-ovate leaves.

FAMILY EUPOMATIACEAE

The family Eupomatiaceae comprises a single genus with 2 species, which are distributed in the coastal zone of eastern Australia from the Cape York peninsula in the south to eastern Victoria and in the eastern part of New Guinea. The rose-bushes are trees or shrubs, sometimes growing up to a height of 15 m. The anatomical and morphological structure of wood, flowers, and fruits indicate their primitive nature and ancient origin. Taxonomically, they are closest to Magnoliaceae.

Bennett's rose-bush (*Eupomatia bennettii* F. Muell.) is an endangered species of Australia. It was discovered in the second half of the 19th century. Most probably, it was never widespread and remained a relict plant in the form of admixture in tropical forests of the coastal regions in northern and eastern of Queensland (Cape York peninsula and Brisbane region) and in northeastern New South Wales. It is a small tree or shrub with a unique, tuber-like main root, from which a few trunks originate. The leaves are simple, pinnatinerved, leathery. The flowers are borne on long, leafy shoots and are bisexual; stamens are flattened on the lower side, inner stamens modified into brightly colored aromatic staminodes; carpels are many (Plate 49). The disappearance of this ancient relict species is related to change in environmental conditions, mainly forest felling. At present, this rose-bush is protected in the Queensland National Park.

FAMILY FABACEAE

In Australia, the leguminous plants are widely distributed in different communities and are distinguished by great generic and specific variability. The family includes about 100 genera, among which few are endemic, typical to the Australian flora. Reduction in the population of most plant species is due to sheep grazing and change in the habitat conditions. A few species have commercial value.

Aote genus (*Aotus*) is endemic to Australia. It is represented by shrubs with twig-like (virgate) branches and simple exstipulate leaves. The flowers are borne in racemes or umbels. Five West Australian species are included in the List of Rare and Endangered Plants.

The **keeled aote** (*A. carinata* Meisn.) is included in the category of endangered plants. The branches and leaves of this shrub are densely covered with long, soft, silky hairs. The leaves are whorled, in groups of 3, folded all along their length and forming a keel on the lower side. Flowers are in groups of 3–4, pedicellate. The plant has been found in extreme southwestern Australia.

The **poison bush** genus (*Gastrolobium*) is mainly distributed in Western Australia. It is represented by shrubs, in which the leaves with short petioles are arranged in whorls. Flowers are yellow, sometimes with purple-red keel; inflorescence is umbellate. Fourteen poison bush species are included under rare plants.

Pyramidal poison bush (*G. pyramidale* S. Moore) is a tall and beautiful shrub. Its young branches are softly pubescent. Leaves are arranged in whorls of 3, leathery, glabrous, with thick margins. Bright yellow flowers with a red keel are arranged in dense, short umbels or heads. The bush grows in the coastal region of the Great Australian Bight (Heinis Beach, South Coast, Western Australia).

Hooked poison bush (*G. hamulosum* Meisn.) has grayish, fluffy branches and whorled leaves; 2–3 pairs of flowers form a bunch. A species endemic to Western Australia, it is found on clayey areas in southwestern floristic province.

The endemic dogwood genus of Australia (*Jacksonia*) comprises leafless shrubs or semishrubs with hard cylindrical or winged branches; the scales perform the functions of leaves. Flowers are yellow with a tinge of purple, inflorescence is umbel or a spike.

Leafy dogwood (*J. foliosa* Turcz.) is a plant up to 30 cm tall with woody stems arising from a thick rhizome. Lower part of stems is covered with simple, dentate leaves, upper part is leafless with many hard, flat branches, arranged in whorls in the form of panicles. The entire plant is silky-pubescent. Flowers are terminal on branches. This rare endemic of Western Australia grows in the inner regions: Kalgoorlie, Menzies, Leonora. Its populations are very small. Six more dogwood species are included among the rare plants of Australia; all of them have been included in the List of Endangered Plants.

The Australian bean-flower genus (*Kennedia*) is represented by herbaceous, often shining plants. Leaves are simple or ternate, broadly elliptical, stipulate. These are highly ornamental plants. Two Western Australian species of bean-flower are endangered and 3 other rare species are included in the National List of Endangered Plants.

Northcliff bean-flower [*K. glabrata* (Benth.) Lindl.] is a hairy biennial with creeping stems; erect peduncles are up to 15 cm long, with leafy stipules at the base, and 4–8 red flowers (Plate 50). It grows in southwestern Australia in the crevices of granite rocks north of Pt. D'Entrecasteaux. **Large-leaved bean-flower** [*K. macrophylla* (Meisn.) Benth.] is a tall, biennial, and densely hairy plant. Its leaflets are ternate and very broad; stipules are often fused, up to 2.5 cm [sic] in diameter; flowers are red, arranged in panicles (Plate 50). The plant grows along the borders of southwestern Australia (Warren region). Both species are endangered plants of Australia.

The largest endemic genus of family Fabaceae is the **Victorian wallflower** (*Pultenea*), which includes about 100 species, distributed all over Australia. These are shrubs bearing yellow-orange flowers with a tinge of purple, rarely pink. The National List of Plants includes 32 of its species, 7 of which are in the category of endangered plants.

Sticky Victorian wallflower (*P. viscidula* Tate) is a shrub up to 1 m tall. Its branches are pubescent, viscid; leaves are 8–12 mm long, furrowed, with short petioles, and pubescent on the lower surface. Flowers are yellow, with chocolate-brown bracts, borne on the peduncle in groups of 2–5. The plant flowers in October–November. It is distributed on Kangaroo Island (South Australia). The **three-lobed Victorian wallflower** (*P. trifida* J.M. Black) also grows with it and is another South Australian species included in the category of endangered plants. This is a prostrate shrub with dark green, 4–5 mm long leaves. Yellow flowers with cream-colored keel are arranged in capitate inflorescence. The entire plant is strongly pubescent. It flowers in October–November.

Cup-shaped Victorian wallflower [*P. calycina* (Turcz.) Benth.] has short, grayish, pubescent branches and glabrous or silky, oblong-linear, acute leaves. Its flowers are aggregated at the end of branches. This is an endemic of Western Australia.

The member of the monotypic genus *Ptychosema*—**small ptychosema** [*P. pusillum*] Benth.—is included among the endangered plants. It is known only in the Gingin region, where it grows on sands in the communities with bordered eucalyptus and banksia. This is a small, 5–10 cm tall, slender, mildly hairy plant; leaves are up to 1 cm long, imparipinnate, consisting of 7–11 leaflets; flowers are dark red, chocolate-brown, or yellow, solitary, terminal.

The indigenous Australian Darling river pea genus (*Swainsona*) includes about 60 species, one of which has been found in New Zealand. These are mostly pubescent herbs. Their leaves are imparipinnate with numerous leaflets; flowers are violet-purple, red, white, or yellowish, borne in racemes. Seven of its species have been included in the National List of Rare Plants.

Green Darling river pea (*S. viridis* J.M. Black) is a spreading, subglabrous plant. Leaves have 7–11 leaflets, are 5–10 mm long, glabrous or with short hairs along the margins and midrib; stipules are large. Flowers are purple, arranged in racemes, in groups of 5–8. The plant flowers in August–September. This is an endemic of South Australia (Flinders Range adjoining part of the southeastern region). It is an endangered species.

FAMILY FAGACEAE

The southern beech genus (*Nothofagus*) includes about 14 species, half of which grow in Guinea and New Caledonia, and others in South America, Australia, and New Zealand. **Moore's Southern beech**

(*N. moorei* Krasser), an evergreen tree, has been included among the endangered plants of Australia, whose trunks are usually densely covered with epiphytes. Its distribution is limited to small areas in the environs of Brisbane (Queensland), where it forms a part of forest plantations. It is being conserved in the Lamington National Park.

FAMILY FRANKENIACEAE

The family Frankeniaceae includes about 4-5 genera, of which only one, the sea heath genus (*Frankenia*), is represented in the flora of Australia. The Australian species of sea heath are endemic. Most of them are distributed in the coastal regions of South or Western Australia. Fifteen species are included in the List of Rare and Endangered Plants.

The **diminutive sea heath** (*F. parvula* Turcz.) is known from only one location in Western Australia, on the Eyre peninsula. It has a short, creeping, multibranched, herbaceous stem; the flowers are arranged in capitate inflorescence. The **bracteate sea heath** (*F. bracteata* Turcz.) is a Western Australian endemic. Its stems are woody at the base, flowers are in dense heads, with a rosette of leaves below, similar to bracts. A few of its habitats are east of Perth. The present status of the species is not determined.

FAMILY GOODENIACEAE

The family Goodeniaceae includes about 15 genera, most of them endemic to Australia. A few species of the family are known in New Zealand, along the coast of tropical and subtropical Africa, Asia, and America. These are herbs, semishrubs, and rarely shrubs. The plant tissues contain juice. The flowers are yellow, dove-blue, white, rarely red or purple, arranged in spicate, umbellate, or paniculate inflorescences. The plants are ornamental.

The endemic genus *Dampiera* includes almost 50 species of herbaceous or shrubby plants with purple, dove-blue, or white, rarely yellow flowers, forming paniculate or spicate inflorescence. The List of Rare and Endangered Plants of Australia includes 10 species of this genus. They are all endemic with limited distribution.

Varied-leaf dampiera (*D. diversifolia* De Vriese) is a prostrate herbaceous perennial; each plant often occupies a large area. Basal leaves are oblong-lanceolate, 4 cm long; other leaves are linear, 1.2 cm long, leathery, finely serrate. Flowers are dove-blue, glabrous, on short pedicels. This Western Australia endemic is found along the coast of the Great Australian Bight.

The genus *Goodenia* of the Australian flora is represented by herbs or semishrubs. It produces a single or a few flowers, borne in umbels or panicles. Nineteen species of the genus are included in the List of Rare and Endangered Plants.

The **four-chambered goodenia** (*G. quadrilocularis* R. Br.) has a glabrous, poorly branched, up to 0.5 m tall stem, leafy in the lower half. Its leaves are petiolate, ovate, and serrate. Flowers have bracts. Calyx is tubular; corolla is finely pubescent, upper lobes of corolla are recurved. The plant grows in the southern regions of Western Australia: Lucky Bay, King George Sound.

Chambers' goodenia (*G. chambersii* F. Muell.) is a shrub with fine pubescence. Its leaves are minute, short-petiolate, and serrate; flowers are yellow, with purple streaks. A few populations of this plant have been found in the desert regions in the northern part of South Australia.

The Australian genus *Leschenaultia* includes 20 species, which are mainly distributed in Western Australia. These are herbs or shrubs with narrow linear leaves of ericaceous type. The flowers are few. Seven Western Australian species are rare and endangered plants; their extinction is related to mass collection and commercial sale.

Green-flowered Leschenaultia (*L. chlorantha* F. Muell.) is a short, spreading, multibranched shrub with narrow leaves. Flowers are pale green, rarely conspicuous. This rare plant has been found in the Murchison river valley. **Superb Leschenaultia** (*L. superba* F. Muell.) is a shrub with a height of 0.7 m, twig-like branches, and soft, narrow, up to 2 cm long leaves, crowded at the end of branches, terminating in large, solitary, yellow to red flowers with winged petals. It grows in the ravines of Mt. Barren in shrubby communities.

The most widespread genus of the Goodeniaceae family is the fan flower genus (*Scaevola*). Its members have been found not only in Australia but also in other continents and islands of the Pacific Ocean. All the Australian species of this genus are indigenous to this continent. They are shrubs or herbs with alternate leaves and a few tubular flowers. Seven species of the rare indigenous plants with limited distribution have been included in the National List of Endangered Plants.

Small-leaved fan flower (*S. parvifolia* Krause) is a herbaceous, multibranched plant up to 30 cm tall. The leaves are lanceolate, mostly reduced to small bracts. It grows in Western Australia along the southern coasts of the Eyre region.

FAMILY HAEMODORACEAE

The family Haemodoraceae occupies an intermediate position between

PLATE 49. 1—Bennett's rose-bush; 2—waxy prionotes; 3—mullein-leaved lachnostachys; 4—paniculate richea.

PLATE 50. 1—Large-leaved bean-flower; 2—Northcliff bean-flower; 3—ground sword lily; 4—kangaroo's foot plant.

PLATE 51. 1—Beautiful sword lily; 2—pink-flowered eucalyptus; 3—almond-leaved eucalyptus; 4—sooty macropidia.

PLATE 52. 1—Gardner's rhizanthella; 2—large juniper myrtle; 3—rock lily; 4—spicate juniper myrtle.

families Orchidaceae and Liliaceae and includes 6 genera. The sword lily genus (*Anigozanthos*) includes up to 1 m tall, evergreen perennial herbs. They have narrow, sword-like leaves and tubular flowers borne in panicles. Flowers have an amazing variety of color: green, buff-red, yellow and black, red and yellow. The genus comprises 11 species, and all of them are cultivated as garden plants since ancient times for their colorful flowers, highly valued for their ornamental qualities, and an object of commercial trade. The last factor is the main cause for a sharp reduction in their populations under natural conditions. About half of the species have a limited distribution and have been included in the List of Endangered Plants.

Anigozanthos pulcherrimus Hook. is a herbaceous plant with fleshy leaves. The numerous hairs covering the leaves on both sides impart to them a glaucous appearance. The peduncle and paniculate inflorescences are yellow, hairy, with bright yellow flowers (beautiful sword lily, Plate 51). Its area of distribution is restricted to southwestern regions of Western Australia. The extremely attractive inflorescence of this plant were collected for bouquets and the seeds sold, which was the main cause of a decline in its population strength. The extent of exploitation of the plants of this genus is evident from the sale of **kangaroo's foot plant** (*A. manglesii* D. Don), a plant included in the emblem of Western Australia state: 200,000 live bouquets were prepared for sale annually. The flowers have a fantastic resemblance to the kangaroo's paws. The contrasting colors of different plant parts also attract attention: green with metallic hue of perianth, reddish stems and peduncles, gray or pale green limbs of flowers (Plate 50). The kangaroo's foot plant grows in the sandy plain through which the Murchison River flows.

Anigozanthos humilis Lindl. with yellow, dark orange, or red flowers and dense woolly pubescence (ground sword lily, Plate 50) is also endangered. It grows in the extreme southwest of the state on sandy soils. **Gabriel's lily** (*A. gabrielae* Domin.), **Kalbarii lily** (*A. kalbariensis* S.D. Hopper), **Preiss' lily** (*A. preissii* Endl.), and others have also been included among the endangered plants. To conserve the natural flora of Western Australia, all plants of this genus were declared protected plants way back in 1935. However, their exploitation continues, as demand in the world market increases every year. In 1979, all species of this endemic genus were included in Appendix 2 of CITES, and the most rare and endangered (5 species) in the National List of Endangered Plants.

The only member of the monotypic genus *Macropidia* is the sooty **macropidia** [*M. fuliginosa* (Hook.) Druce], also called 'black kangaroo's foot'. Actually, it has nothing black except the branched hairs that cover the pale green flowers, leaves, and stems (Plate 51). In contrast with the *Anigozanthos*, macropidia have deeply divided perianth lobes and

3-seeded fruits. A few of its habitats are confined to the wastelands and sparse forests along the west Australian coasts between the cities of Perth and Geraldton. Macropidia has low reproductive ability, which decreases further because of the collection of its shoots with flowers for sale. The species is included in the List of Rare and Endangered Plants of Australia, and also in Appendix 2 of CITES.

FAMILY HERNANDIACEAE

The family Hernandiaceae includes only 4 genera and up to 70 species of arboreal plants. The genus *Hernandia* with 25 species is the main genus on the basis of which this family has been named. **Cudgerie** (*H. bivalis* Benth.)—an evergreen tree up to 35 m tall—belongs to the category of endangered plants of Australia. Its distribution is limited to the regions north of Brisbane up to Bundaberg (Queensland), where it is found in the coastal tropical forests. The population of this hernandia is small, which puts a question mark on the possibility of its survival.

FAMILY IRIDACEAE

The family Iridaceae does not have a great variety in Australia, although even here it has its endemic species and even genera.

Among the rare endemic plants of Tasmania is the **Tasmanian isophysis** [*Isophysis tasmanica* (Hook.) T. Moore], a herbaceous perennial with short-branched and woody rhizome. A number of overlapping leaves develop from the bud on the rhizomes. As in most members of Iridaceae, they are folded lengthwise, hard, slightly pointed, and lustrous. The peduncle bears pinkish to chocolate-brown bracts and a terminal flower bud with one more pair of bracts. The flower is quite large and dark purple, almost blackish. In contrast with other species of Iridaceae, the ovary of Tasmanian isophysis is superior. It grows in mountains, facing the western coasts, and on marshy areas along sea coasts in the west and extreme southwest of the island. It is included in the List of Rare and Endangered Plants of Australia.

FAMILY LAMIACEAE

The family Lamiaceae includes 20 genera in Australia, a few of them endemic. The genus *Hemiandra* with 25 species is among the endemic genera. They are shrubs or semishrubs with woody branches; leaves are opposite, entire, narrow; flowers are few, with a pair of floral leaves (bracts) at the base. Three of its species are included in the List of Rare and Endangered Plants of Australia.

Gardner's hemiandra (*H. gardneri* O.H. Sargent) is an endangered plant with a few populations at the southwestern tip of the continent. Its disappearance is associated with commercial exploitation (the flowering shoots are collected for sale).

The endemic Australian genus *Hemigenia* includes more than 20 species of shrubs with whorled (in threes) or opposite leaves. The flowers are borne in leaf axils terminating the branches. Eight species of this genus are included among rare or endangered plants.

Smooth hemigenia (*H. glabrescens* Benth.) is an endemic plant from the southwestern part of Western Australia. It is a shrub with slender, smooth branches; leaves are opposite, serrate, glabrous. The flowers are small, on short peduncles, bilabiate, and pubescent on the outer side.

The mint bush genus (*Prostanthera*) is endemic to Australia. It comprises small shrubs or semishrubs with oil glands and very strong smell, producing few flowers. Sixteen of its species have been included in the National List of Endangered Plants of Australia.

The **cryptandroid mint bush** (*P. cryptandroides* Benth.) has been included in the category of endangered plants. The outgrowth of this shrubby plant is called heathlands in Australia. It has slender, smooth, or slightly glandular branches; leaves are on short petioles, linear and serrate. The flowers are on peduncles with bracts fused with the calyx. A few of its populations are known in New South Wales on the Sandstown Hills plateau and in the northwest of the state in the Hunters river valley.

FAMILY LAURACEAE

Dodder-laurel genus (*Cassytha*) is particularly distinguished in the family Lauraceae, which is represented by parasitic herbs. Morphologically they resemble the dodder and are often confused with it. The stems of dodder-laurel are creeping, thread-like, yellowish or pale green, resembling a wire, and highly branched; leaves are scaly. The stems and branches have numerous haustoria, with the help of which they suck water and nutrients from the host plants. They are perennial plants, parasitizing on trees, shrubs, and perennial herbs, sometimes even on their own type. Besides parasitization, they also maintain the ability to photosynthesize, which makes them distinctly different from dodder. The genus comprises up to 20 species, growing in tropics and subtropics, mainly in Australia, 3 of which are included in the National List of Endangered Plants of Australia: *C. nodiflora* Meisn., *C. racemosa* Nees, and *C. tepperana* Tepper. The former 2 species are endemic to Western Australia and are found in its southwestern corner, and the third species is endemic to the Kangaroo Islands (South Australia state).

FAMILY MIMOSACEAE

Almost three-fourths of the species of the large *Acacia* genus, which includes about 800 species all over the world, are found in Australia. Most of them have no leaves, their function being performed by phyllodes, which are modified flattened petioles of variable shape, often thick, with or without spines. Many acacias have a limited distribution, growing in specific habitats, but almost all are affected by commercial activity. The List of Rare and Endangered Plants of Australia includes 105 acacia species, 3 of which are on the verge of extinction, whereas *A. anomala* A.B. Court is already included in the category of extinct plants.

The **leafless acacia** (*A. aphylla* Maslin) is a poorly studied species of Australia. Its relationship has not yet been established and its evolution and geography are unknown; also, there is practically no information on its biology and physiology. It has a very limited distribution, and the total strength of its 3 known populations does not exceed a few dozen plants. The species, discovered in the 1960s, is a shrub up to a height of 2 m, without leaves, and even its phyllodes are reduced to scales (hence the species named). The plant was found 80 km away from Perth in the southwest of Western Australia state on granite slopes of Darling Range. This acacia grew in the surroundings of sparse (open) eucalyptus forests, occupying rock crevices, which is not typical for acacias. It is in a critical condition: the largest population (about 100 plants over an area of a few ha) is on private land used for agriculture, and is not protected. Two other populations (each with 30 plants) are included in the camping reserve forests. One of them suffered heavily at the time of road construction in 1975, and has been reduced to 12 plants. Observations on its regeneration showed that it produces seeds every year, but not a single new plant was found during the last 4 years. This plant is also maintained under cultivation, as it is of particular interest for ornamental gardening because of its unusual appearance (without leaves and covered with numerous bright golden flower heads). The leafless acacia flowers in August–September, and the fruits mature from December to March. It has been included in the Red Data Book of IUCN.

Acacia peuce F. Muell. is a tree up to 15 m tall with dense green crown, short horizontal branches, and needle-like, 8–12 cm long phyllodes. Its distribution is limited to the central regions of Australia, mainly found within the limits of Simpson deserts. It grows in arid conditions, where the annual precipitation does not exceed 200 mm, mean maximum temperature in the hottest months is 37°C, and mean maximum temperature of the coldest months 5°C. In the localities where it grows, there are no other trees which attain such a size. Different cereal species and shrubs of the family Chenopodiaceae are found in the acacia community. Because of the use of *Acacia peuce* for wood, and its green parts for cattle feeding, the natural population of this species has been reduced. Only a few isolated populations have remained. This acacia is cultivated from seeds in the Adelaide Botanical Garden. Its cultivation has a great importance for studying its biology and ecology, origin, and geographic isolation.

Menzel's acacia (*A. menzelii* J.M. Black) is a small, smooth shrub with slender branches. The phyllodes are cylindrical, 1.5–3.5 cm long, sharp along the margins; the floral heads have up to 25 flowers on 1 or 2 peduncles. The fruits are linear. It flowers at the end of winter, in August–September. The area of its distribution is in South Australia. A few habitats are known near the cities of Murray Bridge and Monarto South. It has been included in the category of endangered plants. This category also includes **hairy acacia** [*A. pubescens* (Vent.) R. Br.], an endemic of New South Wales, growing in the Blue mountains. Its branches, trunks, petioles, and peduncles are hairy; leaves are bipinnate with 6 to 20 pairs of leaflets; floral heads are small. It is being protected in the Blue Mountains National Park.

FAMILY MYRTACEAE

The endemic Australian genus *Calytrix* includes about 40 species, most of which are distributed in Western Australia. The calytrices are found among 'heathland'. Their leaves are small, hard, entire; flowers are on short peduncles, few in number. Nine of its species have been included in the List of Rare and Endangered Plants of Australia.

Larch-shaped calytrix (*C. laricina* R. Br.) is a multibranched shrub with spreading branches. Its height does not exceed 50 cm in dry open places, and reaches 180–120 cm in highly moist conditions. Its small flowers are clustered at the ends of branches; leaves are thin, linear, arranged in whorls of threes. This west Australian endemic species is distributed along the gulf of Carpentaria and in the Arnhem Land area.

The genus *Darwinia*, including about 30 species, is endemic to Australia. These are shrubs, but morphologically resemble heaths. Their leaves are small and entire; flowers are in capitate inflorescence, floral leaves (bracts) large and brightly colored. Eighteen species of this genus are included in the List of Rare and Endangered Plants. The decline in the population of some of the species is because of collection of their flowering shoots for commercial trade within the continent as well as abroad.

Darwinia macrostegia (Turcz.) Benth., growing in Western Australia in the Stirling and Barren mountains (southwestern part of the continent), belongs to the category of rare plants. These are shrubs up to 2 m tall, with oblong-elliptical, slightly pointed leaves with incurved margins. The spathes are campanulate; inner bracts are pale yellow with red streaks, outer bracts are more reddish; flowers are white, numerous. The attractive color combinations in darwinia attract collectors; the cut flowering shoots are sold in large quantities.

Thomas' darwinia (*D. thomasii* Benth.) is a rare plant of Queensland. It grows on sandy-stony areas in the upper reaches of Cape River. This is a shrub with slender, smooth branches; leaves are opposite, ovate, crescent-shaped; flowers are large, pink, and with pedicels.

The eucalyptus or gum tree genus (*Eucalyptus*) with about 600 species occupies a prominent position in the family Myrtaceae, out of which only 5–6 species grow within Australia, including 2 in New Guinea and the Philippines. All eucalyptus are trees or shrubs from 2–3 m to 100 m tall. Their leaves are glaucous because of waxy coating, oblong, evergreen, somewhat vertical; therefore, the eucalyptus forests appear to be very light-colored. The plants emit a strong smell because of the presence of glands releasing essential oil in the tissues. The eucalyptus flower has a unique structure: it does not have sepals and petals; numerous stamens emerge from the calyx tube and are located around a disc which functions as a nectary. The anthers give a beautiful appearance to the flower. A capsule with a large number of seeds develops within the calyx tube after pollination. The eucalyptus make a unique landscape and grow in different conditions: from semi-deserts to wet rainy forests. Because of its fast growth, strong suction force, and resistance to droughts and excessive moisture, they have long been used in forest tree and ornamental gardens in almost all the continents. The heterogenity of eucalyptus is also manifested in their distribution and present condition: some species are found in certain regions over large areas, while others have a limited number of habitats and are more frequently exposed to the dangers of extinction. At present, 120 species are facing such a danger, 7 of which are included in the Red Data Book of IUCN.

Argophloia eucalyptus (*E. argophloia* Blakely) is a tree up to 35 m tall with a highly raised crown, white bark, green and hard leaves, and 6–9 flowers in umbellate inflorescences. It is widely cultivated for shade, protection from winds, and as a source of hard wood for construction. The area of distribution of the species is located in Queensland and forms a 30–50 km long and 10–15 km wide belt. This eucalyptus grows in areas whose altitude does not exceed 350 m. Together with bastard myall and crestate casuarina, it forms open plantations. Its habitats are characterized by warm temperate climate: average maximum temperature of the hottest month being 32°C, mean minimum night temperature of the coldest month 5°C, and annual precipitation 600–700 mm. The cause of disappearance of the species in natu e is its limited distribution, felling of the plantations, and use of the clearings for agriculture. At present, a project is under preparation for development of a reserve, covering the area with natural plantations of this eucalyptus species.

Carnaby eucalyptus (*E. carnabyi* Blakely et Steedman) is a typical plant of shrubby thickets in marshes. It is 3 m tall and span of the spreading branches is 6–9 m. Its filaments are cream-yellow to white. At present, only one plant of this species is surviving (the species was discovered in 1937). This is an adult plant, standing alone on an area near the city of Piavaning, 150 km northeast of Perth (Western Australia). Sandy soils, 470–500 mm rainfall in a year, mostly during winter and summer temperature reaching 37°C are the conditions of its habitat. The only surviving plant is growing on a private land, and if it dies, the species will fall into the category of extinct plants.

Crenulate eucalyptus (*E. crenulata* Blakely et Beuzev.) is a tree from 6 to 12 m tall with leaves bright green above and silvery glaucous beneath, crenulate. Its distribution is limited to a small area in the state of Victoria. It grows in plains or slightly sloped areas near rivers, sometimes in depressions with higher moisture content or even with free water. It forms plantations together with ovate eucalyptus, remaining all the time under its canopy. It is planned to organize protection to the habitat of crenulate eucalyptus, which is the remnant of natural plantations that suffered heavily in 1972.

Curtis' eucalyptus (*E. curtisii* Blakely et C.T. White) is a plant forming shrubby thickets in marshes, 2–12 m tall. A few populations of this species have survived in the environs of Brisbane (Queensland), where it is being seriously threatened with construction activity. A few specimens have been found on sandy hillocks near Plankett. This eucalyptus mostly grows on well-drained soils together with other eucalyptus species. At present, a small population of Curtis' eucalyptus is included in the territory of a park under the administrative control of Queensland. It is proposed to create a reserve forest for protecting the shrubby thickets including this species.

Froggat's eucalyptus (*E. froggattii* Blakely) is a 6–9 m tall tree with paniculate inflorescence bearing 7–11 flowers. It is found sporadically in a small region in the central part of Victoria state (the annual rainfall here is 385–525 mm, mostly during winter, when frost is very frequent; the temperature rises to 37°C in summer). It grows together with other eucalyptus species. The measures to protect this endangered plant have not yet been evolved.

Pink-flowered eucalyptus (*E. rhodantha* Blakely et Steedman) is a shrub by growing in marshes, 2.5–3 m tall. Its campanulate flowers have numerous reddish stamens with yellow anthers (Plate 51). It is found on a small territory in the southwest part of Western Australia, 200–300 km north of Perth (the annual rainfall in this region is 380–500 mm, mainly during winter when frost also occurs; summers are dry and hot). The shrub grows on sand, forming small but almost pure communities. It is proposed to organize a reserve for conserving this species in nature. It is widely cultivated in Australia and California as a beautiful flowering and drought resistant plant.

Steedman's eucalyptus (*E. steedmanii* C.A. Gardner) is a 8–12 m tall tree with yellow, sometimes pinkish or red stamens and green, tetrahedral, winged calyx, which is prolonged toward the base. It has almost completely disappeared from the natural habitats, but is being maintained under cultivation. It is endemic to Western Australia. The area of its distribution is limited to a small territory southwest of Farrestian. This region has a low rainfall (300 mm in a year), with winter frosts and hot summer. Shrubby thickets of marsh or semi-arid short-stem eucalyptus plantations are predominant in this area. Cattle grazing was common in these areas during the 1920s and 1930s, which most probably was responsible for the sharp reduction in its populations.

The **Risdon gum tree** (*E. risdonii* Hook. f.), growing in the extreme south of Tasmania near Hobart on the hills and in the valley of the Derwent River, on Risdon, is a rare plant of Australia. The communities of this endemic Tasmanian eucalyptus are quite distinct because of the light bluish-silvery leaves and branches. They occupy sunny and warm slopes where the temperature never falls below 15°C. Usually it grows up to a height of 8 m, but sometimes up to 15 m. The young and adult specimens are practically indistinguishable. The leaves are opposite, and each subsequent pair is turned at 90°C in relation to the previous pair of leaves. Young leaves on the tips of branches are pink and silvery on the lower side with pinkish margins, their major part being light blue-silvery. The leaves emit a faint astringent smell. The flowers are small, up to 2 cm in diameter; their bunches are very attractive in leaf axils. This eucalyptus is being cultivated in the Canberra Botanical Garden (Australia).

Cordate eucalyptus (*E. cordata* Labill.) has a few local populations in southeastern Tasmania. It is found in mixed eucalyptus forests or farms, small pure plantations in regions with a high rainfall in plains as well as on mountains (up to an altitude of 750 m). Under favorable conditions, it can grow up to a height of 35 m, but it is better known as a small tree, 6–15 m tall with the crown at a lower level or as a dense shrub. The leaves on young as well as old branches of the young and adult plants are cordate, silvery, opposite. The flowers are borne in threes; fruits globose or hemispherical. A few specimens of this eucalyptus, raised from seeds, are being cultivated in private collections.

The **cneorum-leaved eucalyptus** (*E. cneorifolia* DC.), a shrub or tree up to 12 m tall, with a red bark, is found in the shrubby thickets of South Australia (Kangaroo Island). The leaves are linear-lanceolate, 5–9 cm long and 6–9 mm broad, acute. The umbels comprise 3–12 axillary flowers, on 2–4 mm long pedicels. It is the main source of eucalyptus oil on the island.

The **sugar gum** (*E. cladocalyx* F. Muell.) grows in the Flinders Range mountains, on the Eyre peninsula and Kangaroo Island (South Australia). Its sweetish young leaves are relished by sheep and cattle. It is therefore, often cultivated. Under natural conditions, it grows to a height of 25–35 m. It has a dense crown because of dark green foliage; young bark is smooth, and whitish, and the old bark is grayish. The leaves are broadly lanceolate, paler on the lower side than on the upper side; umbels consist of 4–14 flowers.

All these eucalyptus species are included in the List of Endangered Plants of Australia.

The family Myrtaceae includes the tea tree genus (*Leptospermum*). This common name has been derived because, in contrast with the eucalyptus flowers, it has fully developed perianth. The tea tree species are mainly distributed in Australia and New Zealand. Eight species have been included in the List of Rare and Endangered Plants of Australia: 3 of them are found in Western Australia, 3 in Queensland, and 2 in Tasmania.

Florid tea tree (*L. floridum* Benth.) is a Western Australian endemic species from the Swan River valley. It is an erect shrub, up to 3 m tall. Such large specimens flower very rarely; a few flowers develop on younger and short plants, with 3–4 bracts.

The genus *Melaleuca* is entirely Australian. It includes more than 100 species of trees and shrubs with leathery leaves and capitate or spicate inflorescences of red, white, or yellow flowers. The National List of Endangered Plants includes 23 species of this genus, mainly, the Western Australian endemics.

Melaleuca elachophylla F. Muell. is a short shrub with alternate leaves, pink or purple flowers, and small capitate inflorescences, found in the valley of the Fitzgerald River (Western Australia). *Melaleuca polycephala* Benth. is covered with numerous small, dense heads of pink flowers. It grows on sandy soils in southwestern Australia, in the Coolgardie region.

The Western Australian genus *Scholtzia* includes shrubs with umbellate inflorescence of white or pale pink flowers. The rare species *S. uberiflora* F. Muell., growing in the Murchison River valley, appears to be covered with a transparent pale pink layer at the time of flowering.

The endemic Australian juniper myrtle genus (*Verticordia*) comprises almost 60 species. The shrubby growths of juniper myrtle are called 'heathlands'. The plants of this genus are almost smooth, except for fine hairs at the ends of small entire leaves. The flowers are arranged in racemes, umbellate or spicate inflorescences of most variable colors. The freshly cut branches are used for making bouquets. A large number of juniper myrtle species have limited distribution and are confined to the southwestern tip of Western Australia. The number of such plants is not very large and, because of exploitation, the population is declining fast owing to the lack of seed setting. The collection of shoots for commercial trade for the internal as well as world market has increased so much during recent years, and the condition of some species has reached such a critical state, that on the suggestion of Australia, the entire genus of juniper myrtle has been included in Appendix 2 of CITES; 11 species of juniper myrtle have been included in the List of Rare and Endangered Plants of Australia.

Large juniper myrtle (*V. grandis* Turcz.) grows on the sandy plains along the western Australian coast. This shrubby plant flowers over a large part of the year, but is particularly beautiful during summer, when the red inflorescence with large flowers sharply contrasts against the gray background of other plants growing in this region (Plate 52). The population of the species is decreasing regularly as a result of exploitation.

Spicate juniper myrtle (*V. spicata* F. Muell.) is a Western Australian endemic growing on sandy soils in the Murchison River valley. It is a highly ornamental plant, especially in late spring, because of dense spicate inflorescences (purple to yellow), which are subtended by leaves crowded below the inflorescence (Plate 52). It is sold commercially.

Bright-colored juniper myrtle [*V. nitens* (Lindl.) Schau.] is a shrub up to 1.5 m tall, multibranched, with a large number of bright orange flowers, which contrast with the bright green leaves in the beginning of summer. It grows on marshy lands. The distribution of this species is limited to southwestern parts of Western Australia. Its shoots are collected for sale.

Golden juniper myrtle (*V. chrysantha* Endl.) is a short, 30–60 cm tall, shrub with linear, keel-shaped leaves, crowded on short branches, the flowers are bright yellow. It has been found in the inner sandy regions of Western Australia.

FAMILY ORCHIDACEAE

There are not many groups of plants in the plant kingdom in which adaptation to the mode of life and pollination processes would have got as large a variability of forms as in the members of Orchidaceae. In Australia, orchids are represented by a large number of endemic genera and species. Two saprophytic, monotypic genera (each with a single species) are particularly interesting: *Cryptanthemis* and *Rhizanthella*. The plants of these genera are entirely without chlorophyll, therefore, their existence depends on symbiotic activity of the fungi associated with them. Both these orchids are endangered and have been included in the National List of Endangered Plants of Australia.

Slater's cryptanthemis (*C. slateri* Rupr.) has not only underground rhizomes, but also flowers that develop under the soil. The pollination method of this orchid has not yet been studied but, most probably, it is very effective, as almost all mature capsules contain seeds. It is easy to detect it, as a fruiting stem covered with numerous colorless scales appears above the soil surface for seed dispersal. Cryptanthemis was first found in New South Wales in the regions adjoining the Pacific coast, and later it was also found in Queensland. The population strength of this species is small in both places.

Rhizanthella gardneri R.S. Rogers is another saprophytic orchid, leading a subterranean life. A few habitats of this plant are known in the southwestern end of the continent, exclusively confined to the melalenca communities. Rhizanthella is a succulent herb with short, thick, horizontal, branched, colorless rhizome. A very strong smell of formalin is released on injury to the plant. This orchid flowers in May–June. At this time, reddish-purple capitate inflorescences up to 60 cm long, surrounded by 6–12 large floral bracts, appear above the ground. Each inflorescence, with a diameter reaching 70 cm, bears 50–100 tubular purple flowers, which are pollinated by insects (Gardner's rhizanthella, Plate 52). These orchids can be conserved only by organizing reserves in their habitats.

Norton's adenochilus (*Adenochilus nortonii* W.V. Fitzg.), one of the 2 less known species of the genus, is distributed in New South Wales. It grows in moist ravines on sphagnum moss in the Blue Mountains and Barrington Top Mountain above 900 m. This is a delicate herbaceous plant up to 20 cm tall. Its stem with a single leaf arises from a thin rhizome. The flower is white, 1–2 cm in diameter, with short hairs at the tips of floral segments; lip is trilobate, and the middle lobe is shorter than the lateral ones; lip and column have reddish spots. It flowers from November to December (almost nothing is known about its pollination). The species is included in the List of Endangered Plants of Australia.

The **globular bulbophyllum** (*Bulbophyllum globuriforme* Nicholls) is one of the smallest orchids of Australia. Its rhizomes are attached to the bark of trees. The tubers are globose and green; the only leaf is linear-subulate. A few white flowers have yellow or crimson red streaks on the spur and calyx; lip is crimson red or pale yellow, oblong, and grooved.

The only known habitat of this orchid is in northern New South Wales and southern Queensland. It is very similar to the **minute bulbophyllum** (*B. minutissimum* F. Muell.), from which it can, however, be easily distinguished by the spherical tubers, larger size, and color and shape of the flower. The minute bulbophyllum has a flat tuber, which forms a dense cluster of irregular shape, up to 6–10 cm^2 in size, which may have up to 200 tubers. The tubers contain assimilating tissue and perform the role of green leaves. They are flat, and a small lanceolate leaf is located in the center. Sometimes a short glandular peduncle with the only flower, an orchid in miniature, appears at its base. The flowers are pink with 8 reddish veins on the sepals and 1 vein on the petals; the lip is red, ovate-oblong. This bulbophyllum usually grows on the trunks of *Ficus* trees. Its area occupies the territory from the northern coast of New South Wales in the south to the cities of Milton and Kiama. The plant is cultivated, but it is difficult to cultivate, and the orchid survives for a short period.

Bottle-shaped bulbophyllum (*B. lageriforme* F.M. Bail.) does not form large clusters of tubers. A few leaves, up to 4 mm long, and a peduncle bearing 1–4 campanulate pale-green flowers with chocolate-brown streaks arise. The plant grows at an altitude of 750 m at the northeastern border of Queensland. All these bulbophyllums are rare and included in the List of Endangered Plants of Australia.

Wedge-shaped burnettia (*Burnettia cuneata* Lindl.) is a rare plant of several states: Tasmania, Victoria, New South Wales. In Tasmania and Victoria, it grows in swamps and marshes in the communities with cajaput tree (*Melaleuca*), and in New South Wales in the coastal marshes, in the alpine zone and in plains. Sometimes it forms easily-noticeable thickets on the areas in shrubby communities developing after fire. Most probably, the destruction of the dense grass cover facilitates growth and development of burnettia, whereas construction of drainage channels and drying of marshes and swamps with further cleaning of habitats for agriculture reduce its area and population. This burnettia is included in the National List of Endangered Plants. It is a herbaceous terrestrial plant up to 10 cm tall. There are no leaves at the time of its flowering, the flowers are white with reddish or purple chocolate-brown spots on the perianth; lip is bilobate with a cuneate incision (hence the species name). The plant flowers in early spring: from September to the end of November.

The genus *Calochilus* includes 11 species distributed in Australia, New Zealand, New Caledonia, and New Guinea.

Rich's calochilus (*C. richae* Nicholls) is endangered in Australia. This orchid was first found in 1928, and thereafter it was not seen by anybody for 40 years. It was rediscovered in 1968 at the same place: in central Victoria. This location so far remains the only habitat for this plant. It has a solitary greenish flower with purple veins and spots on the perianth; the upper tepal is ovate, slightly bent, forming a sort of hood over the flower, and the lateral tepals are broadly ovate, with a purple lip.

Campynema lineare Labill is an endemic plant of Tasmania. It grows in the mountains of the Central Plateau at an altitude up to 1200 m at moist places as well as along the sea coast from the northwest to the south. In spite of its wide distribution, it is found quite rarely at these places. At the same time, it is an uncommonly small plant. Its most distinct parts are the stamens: bright orange or pink-orange, masked by the green of the perianth; sometimes the perianth is yellow or chocolate-brown to purple. Flowers are solitary, rarely in groups of 3-4 on an erect stem, not exceeding 20 cm. The stem base is surrounded by scaly leaves and a dense layer of filamentous remnants of those scales. Like other rare plants of Australia, this species requires protection.

Among the endangered species of the genus *Corybas* in Australia is *C. despectans* Jones et Nash, which also grows in the coastal region of South Africa (Indian Ocean).

A few corybases grow in Tasmania. *Corybas aconitoflorus* Salisb. has been found in the forests of *Eucalyptus obliqua*. Its solitary leaves appear in early autumn, in April-May. At this time, a heavy rainfall is received and western winds from the Indian Ocean are predominant. The first flowers open at the end of May. They have a large, reddish tepal, which almost completely covers the lip; lateral tepals are quite conspicuous, and are bent downward along the sides of the ovary. This plant not only flowers profusely but also produces fruits every year. **Fimbriate corybas** [*C. fimbriatus* (R. Br.) Rupp.] is the most beautiful and delicate of all the corybases in Tasmania. It grows in sparse forests of the almond-leaved gum *Eucalyptus amygdalina* (Plate 51). Its flowers appear fire-opal in bright sunlight, the solar rays being reflected by the fimbriate lip. The **clawed corybas** (*C. unguiculatus* R. Br.) has been found along the sea coast in the communities of eucalyptus and casuarinas. The cordate leaf is the characteristic feature of this species. Besides, this plant has a well-developed bract. The flowers are deep red, drooping. The plant flowers from May to July.

The genus *Cryptostylos* includes about 20 species, five of which grow in Australia, others in Taiwan, the Philippines, Malaysia, New Guinea, New Caledonia, and Sri Lanka. Four Australian species are endemic to Western Australia, and the fifth and the rarest species, **Hunter's cryptostylos** (*C. hunteriana* Nicholls) is found sporadically in Victoria and New South Wales. This is a small, leafless, saprophytic plant. The perianths are very narrow and inconspicuous, but the lip is well developed. The pollination method of this plant has been described repeatedly in the literature. The males of *Lissopimpla semipunctata* are the main pollinators. The plant flowers in summer. It has been included in the category of endangered plants and has been included in the National List of Endangered Plants of Australia.

The spider orchid genus (*Caladenia*) is considered to be fairly large among the orchid family. It includes 60 species, of which 3–4 have been found in New Zealand, 1 in Java, and others in Australia; most species (32) are endemic to Western Australia. The spider orchids are extremely variable in color and shape of flowers.

Multiclavate spider orchid (*C. multiclavia* Reichenb. f.), a small herbaceous plant covered with glandular hairs, is a rare species of Western Australia. The flowers are solitary, yellow to chocolate-brown, with a horizontal lip, unusual for orchids, almost at a right angle to the ovary. The plant has been found east of Perth.

Gladiolate spider orchid (*C. gladiolata* R.S. Rogers) grows in South Australia at two places. This is a perennial, densely hairy plant with a height of 8–18 cm. The leaves are covered with hairs, which differ in shape and size; leaves are oblong-lanceolate to ovate. The flowers are solitary, rarely in twos, yellowish-green. The plant flowers in September.

Bicalliate spider orchid (*C. bicalliata* R.S. Rogers) is a small, delicate, hairy plant, up to 10 cm tall. Its leaves are linear-lanceolate, 6 cm long, trinervate; flowers are solitary, cream-colored with reddish veins. The plant flowers in September. It grows in South Australia, Kangaroo Island, and York Peninsula.

Besides these spider orchids, 22 more species, 15 of which are in Western Australia, are included in the List of Endangered Plants of Australia.

Rock lily (*Dendrobium biggibum* Lindl.) is the emblem of Queensland. It is a terrestrial plant with large (15–45 cm long) tubers and a solitary stem, bearing 3–4 lanceolate leaves and a floral cluster of 10 purple flowers in the upper part (Plate 52). It is fairly widely distributed in Queensland (north of Cooktown), but its population has significantly reduced because of commercial trade. The plant is included in the National List of Endangered Species.

Diplocaulobium masonii (Rupp) Dockr.—a small epiphytic orchid with pseudobulbs forming clusters—is included among the very rare plants with a single habitat. Each pseudobulb is up to 5 cm long, fusiform, slightly angular, and sometimes even winged. The single leaf is broadly lanceolate, up to 2 cm long. The flowers are solitary, outer tepals narrow, about 2 cm long, inner slightly shorter; inner side of perianth is white and outer side is yellow with dark brown markings on the lip. The plant was first found in 1950 in the northern part of Queensland at Cape Tribulation.

The dragon's head genus (*Diuris*) is among the typical Australian genera. It comprises a few species of beautiful terrestrial orchids, 2 of which are endangered. One of the species—the **fastidious dragon's** head (*D. fastidiosa* R.S. Rogers)—has already disappeared. This plant was found in 1923 as a single specimen near Melbourne (Victoria). In 1925, 5 plants were found at the same place in a patch of about 10 m², only 2 plants in 1926, and 5 plants again in 1927. Subsequently, this area was covered with gravel (a railway line was constructed). Most probably, this orchid was at one time much more widely distributed, over the entire plain west of Melbourne. But the natural vegetation here was replaced by introduced species with the beginning of establishment of farms. Initially, these were cereal communities on the basalt rocks. The years of 1925 and 1927 were favorable for the growth and development of plants, and under such conditions the dragon's head not only formed aerial branches but also produced flowers. At the same time, 2 other species of the same genus—*D. pedunculata* R. Br. and *D. palustris* Lindl.—also flowered.

The fastidious dragon's head is a terrestrial plant with the leaves resembling those of cereals (therefore it is difficult to detect it if not flowering). The stem, 5–20 cm tall, terminates in a raceme inflorescence, bearing 1–3 small yellow flowers. It is included in the Red Data Book of IUCN.

Large-flowered epiblema (*Epiblema grandiflorum* R. Br.), a member of a monotypic genus, has been found only in Western Australia, where it grows on marshy areas and humid plains in communities with the bottle brush tree and Australian honeysuckle, and even in the thickets of reed at the southwestern tip of the continent. This is a delicate herbaceous plant up to 70 cm tall with leaves up to 30 cm long. Flowers numerous (up to 10), light bluish to violet, with dark violet veins; lip is broadly ovate, violet. Its pollination mechanism is unknown. The plant is included in the List of Endangered Species in Australia.

Galeola cassythoides (A. Cunn.) Reichb. f. is a creeping terrestrial orchid, climbing along the trunks of living or dead trees, up to 5 m long. It is entirely without leaves, but the flowers are borne all along the stem on short branches in numerous multiflorate panicles. The flowers are bright yellow, small. The distribution of galeola is stretched from central New South Wales to Queensland. The plant is included in the National List of Rare and Endangered Species of Australia.

The small genus of flower of sadness (*Lyperanthus*) includes 3 species with limited distribution. One of them, **Australian flower of sadness** (*L. nigricans* R. Br.), has a single green roundish leaf at the stem base. The stem is thick, covered with scaly pale purple leaves terminating in a sparse inflorescence. The flowers are purple with white, sometimes pink. This orchid is among the early-blooming and long-flowering plants. It grows in the shady forests in the southwest of Western Australia. Its flowers turn black on drying (hence the species name).

The **Forrest's flower of sadness** (*L. forrestii* F. Muell.) was found on Barker Mountain (Coolgardie region, Western Australia). Most probably, this only habitat of this species is now destroyed.

Almost all species of the tway-blade genus (*Liparis*) are very rare, now on the verge of extinction. The **Swenssen's tway-blade** (*L. swenssonii* F.M. Bail.), which was collected only once in the Emu valley in west Queensland, is no exception. This is an epiphytic orchid, having ovoid, green, and smooth tubers, slightly compressed laterally, and covered with the remnants of leaves. The only leaf is linear-lanceolate, sword-shaped, slightly bent, and 7–15 cm long. The racemose inflorescence has 12 or more pale yellow flowers, only the lip being deep orange at the base.

Among the protected plants of Queensland is **Tancarvill's limodorum** or **marsh lily** (*Phaius tancarvillae* Bl.), growing here at the southern border of its area of distribution.

The generic name *Prasophyllum* originates from the single leaf morphologically resembling the leaf of leek. This is a characteristic property of all members of the genus, of which 2 species are already believed to be extinct. One of them—*P. concinnum* Nicholls—grew in the south of Tasmania, and the other—*P. diversifolium* Nicholls—at the extreme southwestern tip of Victoria. Besides these species, the List of Rare and Endangered Plants of Australia includes more than 20 other species, including *P. validum* R.S. Rogers. Its racemose inflorescence consists of 28 greenish flowers. The plant flowers in October–November. Its only habitat is known in South Australia in the environs of Melreus, on the Flinders Chase range. It is being conserved in the National Park of the same name.

Fitzgerald's sarcochile (*Sarcochilus fitzgeraldii* F. Muell.) is a plant well known to amateur floriculturists and plant collectors. It is multiplied in glasshouses and is highly valued for its white flowers with red centers, which are borne in erect racemes (Plate 53). It is exported quite widely from Australia, which was one of the reasons for including this species in the list of rare plants of the continent. Under natural conditions, it grows in dark ravines on moss cover in the hilly eastern region in the northeast of New South Wales and in the south of Queensland.

Hartmann's sarcochile (*S. hartmannii* F. Muell.) is an object of trade, whose distribution is similar to that of Fitzgerald's sarcochile (Plate 53).

The genus *Thelymitra* has a characteristic property that its flowers do not have a differentiable lip. Almost all species of the genus grow on clay or wet sandy soils in the coastal plains. The **epipectoid thelymitra** (*T. epipactoides* F. Muell.), one of the 8 species of this genus, is included in the List of Rare and Endangered Plants of Australia. This plant has a strong, upto 50 cm tall stem. The leaves are linear, inrolled into a tube at the base. The inflorescence consists of 6–18 large, grayish-green, almost regular flowers with pink streaks. The plant flowers in September and early October. It is found in several coastal regions of the southeastern part of South Australia and southern Victoria.

PLATE 53. 1—Fitzgerald's sarcochile; 2—Hartmann's sarcochile; 3—red Australian honeysuckle; 4—tricolored vanda.

PLATE 54. 1—Broad-leaved isopogon; 2—mongaensis warratah; 3—bilobate petrophile; 4—Menzies Australian honeysuckle.

The tropical, mainly asiatic genus *Vanda* is represented only by a few species in Australia, among which the **tricolored vanda** (*V. tricolor* Lindl.), found in the Northern Territory (Plate 53), and **Whitaen's vanda** (*V. whitaena* Herbert et Blake) from the northern regions of Queensland, are included among rare species with limited distribution. These are epiphytic orchids with numerous linear, fleshy, inrolled leaves, which are clustered. The flowers on a long multibranched peduncle are large. The tricolored vanda has white flowers with yellow spots, and Whitaen's vanda produces yellow to chocolate-brown flowers.

FAMILY PALMAE

In Australia, the family Palmae includes the **marie fan palm** (*Livistona mariae* F. Muell.), a relict of the tropical flora in Ehremey. It has been preserved almost right in the center of Australia, in dry regions where drought prevails for a larger part of the year. Its location is cut off from the closest point of the main area of distribution of the genus by more than 1000 km toward south. The habitat of fan palm has many mountain ranges (including McDonnell Mountain) of pre-Cambrian quartzite and sandstone; the height of the mountain above the surrounding area is 1200–1400 m. The region is intersected by the beds of seasonal rivers of the Finke River basin. The fan palm was found in 1872 in the Palmer River valley near Hermansberg. In 1894, a special expedition surveyed the valley. Since then, the conditions under which the fan palm grows have been studied along with its state, source of regeneration, effect of fire, frequently recorded in the Palmer River valley, and accumulation of litter-fall. The flora of the valley has been studied in detail. It includes 333 species of higher plants (about one-fourth of the entire flora of Central Australia). About 10 percent of these species are considered rare or have limited distribution in the Palmer River valley. The growth of the ancient member of the family Cycadaceae, McDonnell's Queensland nut (*Macrozamia*), together with fan palm is most interesting.

The marie fan palm is the main palm, with a trunk up to 20 m tall and 30 cm in diameter. The bisexual flowers are arranged in long panicles. The fruits are globose, 13–15 mm in diameter. The fan palm has a superficial, poorly developed root system, which is sensitive to long flooding and various soil erosion processes. Periodical rise in the level of ground water in the arid zone and short surface flooding of the land after rain are essential conditions for the survival of this plant and seedling growth every year. The unique water regime in combination with the unusual stratification of the underlying mountain rock, not exposed to erosion, has been recorded over the territory of about 60,000 ha in the Palmer River valley.

There are about 1500 adult fruiting palm trees at the age of 100–300 years on this territory; their total population together with young specimens is about 3000. The counting of adult and young trees showed that their total number has increased over the last 50 years and there is no immediate danger of its extinction. The greatest danger for this species is the lowering of ground water level, destruction of rocks, or pollution of the territory. At present, there is no deterioration in the environment. The river valley is included in the Fink Gorge National Park, which was established in 1967.

The marie fan palm is being successfully cultivated in Miami (Florida, USA) in the Adelaide Botanical Garden (Australia), and in Bogor and Kebumen (Java, Indonesia). It is included in the Red Data Book of IUCN.

FAMILY PODOCARPACEAE

The genus *Microstrobos* of this family comprises only 2 species, which are ancient in origin and rare because of their limited distribution. Both species are endemic to Australia.

Fitzgerald's microstrobus [*M. fitzgeraldii* (F. Muell.) Garden et Johnson] is a dioecious, highly branched shrub, up to 3 m tall. It has slender, long, drooping branches and subulate, 3–4 mm long spirally arranged leaves. It grows in the crevices of rocks, near large waterfalls, where the air is saturated with fine droplets of water. Abundant growth of these densely branched shrubs are suspended from the stone wall along both sides of streams. The area of distribution of the species is located in New South Wales, in hilly regions, in the outskirts of Sydney and slightly to the west, up to an altitude of 1000 m. The specific conditions needed for the growth of Fitzgerald's microstrobus, negligible seed productivity (4–5 seeds in a cone), and absence of its basic adaptation for seed dispersal partly explain the low level of its natural population. Attempts to introduce this plant under cultivation were unsuccessful. Fitzgerald's microstrobus was included in the National List of Endangered Plants of Australia. Presently, this plant is not protected.

Hooker's microstrobus (*M. niphophilus* Gardn. et Johns.) differs from the previous species by short branches and scaly leaves up to 1 mm long. It grows in Tasmania, on wet marshy plateaus, rarely in dense forests, at an altitude of 1000–1400 m. This plant also has low seed productivity and there are no cultivated specimens. However, it is not yet included in the List of Endangered Plants.

FAMILY PROTEACEAE

The family Proteaceae includes 38 genera, most of which are endemic to

Australia. The largest, Australian honeysuckle genus (*Banksia*) includes about 100 species. These are small trees or 6–8 m tall shrubs with stiff leaves with sharp teeth along the margin. The shape of bright red and yellow inflorescences of this tree resembles a bottle brush of various shapes; oval, cylindrical, candle-like. Australian honeysuckle grows in eucalyptus forests, forming undergrowth, on poor sandy soils, or in shrubby communities together with the bottle brush tree and cajuput tree of the family Myrtaceae. Their ornamental inflorescence not only attracts nature lovers but also serves as an item of commercial trade within the country as well as for export. Dry inflorescences are used on a large scale for bouquets and for decoration. Their collection and sale prevent fruit formation, reduce fertility of plants, and ultimately reduce their population strength. In order to save Australian honeysuckle from extinction, the genus is included in Appendix 2 of the CITES. Twenty-two species of this genus have been included in the National List of Endangered Plants of Australia.

Tricuspid Australian honeysuckle (*B. tricuspis* Meisn.) is a shrub up to 4 m tall with slender, glabrous, somewhat grayish branches. The leaves are needle-like, narrow-linear, 5–10 cm long, with inrolled margins. The inflorescence is oblong-cylindrical, bright yellow, 12.5–15 cm long and 10 cm in diameter. The ovary is up to 4 cm long, curved, with a small oval disk at the tip. It grows on laterite soils along the slopes and foothills of the Gardner range (Western Australia).

Red Australian honeysuckle (*B. coccinea* R. Br.) is a shrub up to 1 m tall. Its inflorescences are cylindrical, at the ends of branches; flowers are in uniform vertical rows, grayish, but this color is not at all distinct at full bloom due to numerous bright red pistils, which are strongly protruding (Plate 53). The plant flowers in late spring. It grows on sands and stony areas between King George Sound and the Olfield River, as well as near Barren Mountain (Western Australia).

Menzies Australian honeysuckle (*B. menziesii* R. Br.) is a tree 9–12 m tall with thick and tomentose branches. Leaves are short-petiolate, serrate, and pubescent along veins on the lower side. The inflorescence is 10–12.5 cm long, bordered with bracts. Perianth is tubular; pistil is red, up to 4 cm long (Plate 54). The plant flowers in the beginning of autumn. It is distributed in Western Australia between the Swan River and Cape Rich.

Good's Australian honeysuckle (*B. goodii* R. Br.) has short, woody stems, terminating in cone-shaped, up to 20 cm long inflorescence with yellow pistils. A few leaves on slightly hairy petioles up to 45 cm long arise from the trunk. The lamina is half the length of the entire leaf, with irregularly serrate margins and yellow midrib. The distribution of this Australian honeysuckle is limited to a small territory between Olbani and Porongerap range, where it grows on sand in sparse forests of eucalyptus and other Australian honeysuckles.

The genus *Conospermum* is endemic to Australia. It includes about 50 species, which are most widely distributed in Western Australia and have limited range. These are shrubs or semishrubs with entire leaves. The flowers are dove-blue, lilac-pink, or white, arranged in short dense spicate inflorescences in the form of compact heads. Most species are in demand in the world market. About 50,000 bouquets of **incurvate conospermum** (*C. incurvum* Lindl.) are sold annually. Because of mass-scale exploitation of the conospermums, 37 species of this genus are included in Appendix 2 of CITES, and 7 rare species that have extremely limited distribution and are endangered are included in the National List of Australian Plants.

The **weak conospermum** [*C. debile* (Kipp.) Meisn.] has slender, glabrous, up to 30 cm tall stems. The leaves in the lower part are long-petiolate, linear-lanceolate; those in the upper part are sessile, narrow linear, up to 5 cm long. The inflorescence is paniculate, sparsely branched, with short spike; flowers are dove-blue. It grows in the Coolgardie regions (Western Australia).

Scaped conospermum (*C. scaposum* Benth.) is a herbaceous plant, in which almost all parts except the older leaves are covered with hairs. The flowering shoots are branched, with a bract; perianth is hairy, bilobate. A few habitats of this endangered plant are known in the outskirts of Perth (Western Australia).

The endemic genus *Dryandra* with about 60 species is distinguished for great variability. Dryandras are very similar to the Australian honeysuckle and differ from them only in the shape of inflorescence (the dryandras have globose inflorescence). Some of the species are export items. Two species are in maximum demand in the world market.

Beautiful dryandra (*D. formosa* R. Br.) is a 2.5–4.5 m tall shrub. Its branches are pubescent and often long-haired; leaves are tripartite or with broad sword-shaped segments. The capitate golden inflorescences are surrounded with floral leaves (bracts), and pistils are scarlet. The shrub grows in the elevations in the Olbani administrative regions and in the Stirling Ranges (Western Australia). The **multiheaded dryandra** (*D. polycephala* Benth.) has slender and glabrous branches and narrow leaves, which are divided into a few segments, stiff, and with inrolled margins. The floral heads are small, but numerous; the pistils are longer than perianth. A few of its habitats are confined to the poor gravelled soils in the southern province (west of Perth) in Western Australia. Both species are included in Appendix 2 of CITES, since 1979, and in the list of Rare and Endangered Plants of Australia. This list includes 22 other dryandra species from the West Australian flora, including *D. pulchella* Meisn., a dwarf shrub with densely pubescent branches. Its leaves are

7.5–15 cm long, pinnate, with numerous pointed segments and incurred margins. The heads are surrounded with floral leaves (bracts); perianth tube is silky, up to 2.5 cm long; pistil about 4 cm long. The plant has limited distribution in Western Australia and is included in the category of endangered species.

The endemic genus *Franklandia* of Western Australia includes only 2 species: *F. fucifolia* R. Br. and *F. triaristata* Benth. These plants have retained several primitive characters. Both species require protection, as they have limited distribution in the extreme southwest of the continent (Warren, Stirling, Eyre), especially the triaristate franklandia, a disappearing plant with only one habitat, between Bunbury and Nannup. It is included in the List of Rare and Endangered Plants of Australia. This is a shrub 60 cm to 1.5 m tall. Its leaves are about 30 cm long, adpressed to the stem, incised in the upper part into numerous narrow segments. The flowers are 5–6 cm long, cream-white inside and red-pink outside, tubular in the lower part and with 4 equal-sized divergent lobes in the upper part, functioning as nectaries. It flowers from August to October. The fruits are awned, up to 15 cm long, trifurcate in the upper part, with spirally arranged hairs.

The genus *Grevillea* is large (including about 250 species) and widely distributed on the territory of Australia. The National List of Rare and Endangered Plants includes 60 species of this genus. Most of them grow in Western Australia and have one or a few habitats each. The members of this genus are shrubs or trees with alternate leaves, highly variable in shape. Their flowers are borne in pairs on peduncles, forming racemose inflorescences that are umbellate or oblong.

Grevillea flexuosa (Lindl.) Meisn. is on the verge of extinction. Its branches are oblong, glaucous; leaves are pinnate, in 4–6 pairs; axis is flexuous. The racemose inflorescence is cylindrical, dense, up to 4 cm long, and the pistil is filamentous. It has been found in the Swan River valley (Western Australia).

Grevillea scabra Meisn. from the outskirts of Perth (extreme southwest of Western Australia) is among the endangered plants. This is a shrub with sessile leaves having inrolled margins. Inflorescence is umbellate; perianth is hairy, recurved; pistils are long and projecting.

Grevillea dryandroides C.A. Gardner is a low, highly spreading shrub. Leaves up to 12 cm long, deeply divided into numerous narrow segments with recurved margins. The cymes spread on the ground; flowers are red with long styles, quite conspicuous on yellow sandy soils among other shrubs in the area of its distribution: Bollidu Kadu (Southwestern Australia). It flowers from September to October and February to March.

Grevillea thyrsoides Meisn. has prostrate stems. Its leaves are pinnate (6–14 pairs), narrow, and hard. The flowering shoots are ascending, leafless, up to 30 cm long, bearing 1–3 terminal racemes. It is distributed in the southwestern part of Australia between Dundagaran and the Smith River.

Grevillea rogersii Maiden is a rare species. It is a branched shrub, 50–80 cm tall. Its leaves are 3–12 mm long and 1–2 mm broad, hard and prickly. One to three flowers are located at the end of short lateral shoots; calyx is scarlet or pink, pubescent outside; tube is 10–14 mm long, swollen at the end, and glandular. The pistil is up to 2 cm long; ovary is hairy; fruits are glabrous and ovoid. The plant flowers from October to December. It has been found on Kangaroo Island in the Birchmoor Lagoon, between Kingscot and Vivon Bay and near Cape Board (South Australia).

Grevillea treueriana F. Muell. is a plant of dry habitat. It grows in the Great Victoria desert region of Ooldea (South Australia). This is a shrub with 3–6 cm long, hard pinnate leaves, twice grooved on the lower side and acute, bilobate or trilobate at the apex. The multiflowered racemes are 5–8 cm long; calyx is silvery, calyx tube is swollen, 10 mm long, glandular; ovary is pubescent. The plant flowers in September–October.

The wooden cherry tree or hakea genus (*Hakea*) is endemic to Australia and is characterized by a large species variability (120 species), 80 percent of all the species growing in Western Australia. Out of the 17 hakea species included in the List of Rare and Endangered Plants of Australia, 13 are endemic to Western Australia. The hakeas are sparsely branched shrub or small trees with very hard, woody, 2-seeded fruit capsules, which remain on the branches till death or fire.

Round-winged hakea (*H. cycloptera* R. Br.) is a shrub with oblong, pubescent, 5–15 cm long leaves. Flowers are borne in axillary inflorescences; perianth is tubular, scarlet, 4 mm long. Fruits are 35–40 mm long and up to 30 mm across, with a horny tip. The shrub flowers during summer from November to February. It grows in the southern part of Eyre peninsula (South Australia).

Crestate hakea (*H. cristata* R. Br.) is a shrub up to 2.5 m tall, glabrous. The leaves are broadly ovate, serrate along the margins, hard. Flowers are small, borne in short racemes; fruits are ovoid with a broad, slightly curved beak; fruit wall has a hard dented wing or crest along the sutures. It is distributed in the Swan River valley in southwestern Australia.

Obtuse hakea (*H. obtusa* Meisn.), is a spreading shrub up to 1 m tall; young branches are silky-pubescent, adult branches glabrous; leaves are on very short petioles. Red flowers are arranged in dense racemes. Fruits have a short, straight beak. The shrub grows at the southern coast of Southwestern Australia.

The genus *Isopogon* comprises 32 species, 26 of which are endemic to Western Australia, 2 to Victoria, 2 to the Northern Territory, and 2 to South Australia. Fifteen rare and endangered species need to be protected, 12 of which are from Western Australia. These are shrubs up to 1 m tall, with hard leaves and dense capitate inflorescence; their flowers are yellow, pink, or lilac-colored.

Broad-leaved isopogon (*I. latifolius* R. Br.) is a shrub with thick reddish chocolate-brown branches. Leaves are leathery, sessile, long and broad. Inflorescence is terminal, capitate, at the base surrounded with bracts as if forming a disk; perianth is pink, tubular (Plate 54). The shrub flowers from September to November. A few populations of this species are known in the Coolgardie region (Western Australia). Its population is declining because of collection of inflorescences for commercial sale.

Adenanthoid isopogon (*I. adenanthoides* Meisn.), a rare plant of Southwestern Australia, grows west of the Murray River. It is a short shrub, all parts of which are covered with long and erect hairs. The leaves are clustered, thin, up to 2 cm long. Cone like inflorescence is compactly enclosed by bracts. Another rare species, *T. tridens* F. Muell., also grows in this region, in the Murray River valley and in the sandy plain near Diamond spring. It received its specific name because of the shape of the leaves: narrow-cuneate, with 3 teeth, on long petioles. The inflorescences are cone-like, with numerous bracts.

The honey flower genus (*Lambertia*) is endemic to Australia. All the members of this genus are found in Western Australia. Six Western Australian species are included in the National List of Rare and Endangered Plants. The **spiny honey flower** (*L. echinata* R. Br.) is included in the category of endangered plants. It is a shrub 1 m tall, with woody, pubescent branches; leaves are cuneate, 4 cm long, dentate, and forming a rosette; flowers are rose-pink, in groups of 7, with a common sheath of bracts. The pollen sacs are spiny with a short and thick cusp. The spiny honey flower grows on the slopes of coastal rocks in Lucky Bay and in the King George Sound region.

The geebung genus (*Persoonia*) includes 72 endemic Australian species, of which one-third are distributed in the southwestern botanical-geographic province of Western Australia, one species in the tropical regions of Northern Australia, and others in southeastern Queensland, along the coast and plateau of New South Wales, in the east and south of Victoria, and in moderately warm regions of South Australia and Tasmania. The plants of this genus retain several primitive characters: e.g., a single axillary flower, numerous ovules, and partly open carpel. Ten species of this genus are included in the National List of Endangered Plants.

The **short-style geebong** (*P. brachystylis* F. Muell.) grows in a natural habitat in the Murchison River valley (Western Australia). This is a short shrub in which young branches and leaves are pubescent, whereas the adult branches and leaves are glabrous. The leaf is linear, acute,

2.5–5 cm long, on short petiole. The flowers are white or yellow, with silky pubescence. **Fringed geebong** (*P. marginata* R. Br.) is a rare plant of New South Wales. It is distributed west of Sydney and the rocky Barren Mountains and on the rocks of the Kugegong River. It is a shrub with pubescent branches, and oval to nearly ovate, leathery, shining leaves. A characteristic feature of this species is the limb modified in the form of a horn.

The small endemic Australian genus *Petrophile* includes 36 species, distributed mainly at the southwestern tip of Western Australia. The List of Rare and Endangered Plants of the continent includes 5 Western Australian and 1 South Australian species. The **multifid petrophile** (*P. multisecta* F. Muell.)—a branched shrub with 4–6 cm long tripartite leaves—is included in the category of endangered plants. The flowers are yellow, arranged in dense spicate and capitate inflorescences, inside linear silvery bracts or scales; the calyx is silvery, about 15 mm long. A few small habitats of this species are known on Kangaroo Island in the Eleanor River valley (South Australia).

Bilobate petrophile (*P. biloba* R. Br.) is one of the most attractive plants of southwestern Australia. It is distributed in the Darling Range, mainly around the city of Darlington. This is a shrub with erect branches up to 1.5 m tall and lavender-pink racemes; flowers are densely hairy with protruding yellow or orange pistils (Plate 54). It is one of the few plants found on rocks; most species of the genus grow on sand in the regions where winter rains are common.

The most wonderful plant of the Australian continent after eucalyptus and acacia is considered to be **mongaensis warratah** (*Telopea mongaensis* Cheel), a plant reaching up to 4 m in height, with straight, leafy, succulent stem, terminating in capitate, crimson-red inflorescence with a diameter up to 7.5 cm (Plate 54). It grows in sparse eucalyptus forests in the north of New South Wales in the Blue Mountains and is found very rarely. Not many Australians have seen warratah under natural conditions; poachers have almost completely destroyed this wonderful plant. The species is included in the List of Endangered Plants. A few habitats of warratah are included in the Blue Mountains National Park.

The woody pear genus (*Xylomelum*), comprising only 5 species, 2 found in the western part of the country and the others in Queensland and New South Wales, is also endemic to Australia. The woody pears are trees or tall shrubs with dense spicate inflorescence. The **narrow-leaved woody pear** (*X. angustifolium* Kippist.) and the **western woody pear** (*X. occidentale* R. Br.) from Western Australia (Plate 55) are endangered. The former is a small tree up to 2.5 m tall. Its branches with leaves and pubescent fruits are used for dry flower bouquets. The small area of distribution of this species covers the Ironstown Mountains and Murchison

Basin. Branches with fruits of western woody pear are exported, but they are not as attractive as those of narrow-leaved woody pear. Its distribution is more restricted (King George Sound, Geographer Bay, Swan River).

Native pear or **woody pear** (*X. pyriforme* Sm.) is called so because of the pyriform shape of its fruits. This is a common plant of Queensland and New South Wales, but its population has noticeably decreased because of irrational exploitation. Besides the collection of branches with fruits, its wood is also used for fencing. Since the distinction between the species is not always very clear, it was decided to include the entire woody pear genus in CITES.

FAMILY RHAMNACEAE

In Australia, the family Rhamnaceae is represented by 12 genera, 7 of which are endemic. The endemic genus *Cryptandra* is one of the most abundant in the family. The thickets of cryptandra resemble heathland. They have narrow, cylindrical, leaves which are whitish tomentose on the lower side and have recurved margins. The flowers are borne in spikes or capitate inflorescences, surrounded by chocolate-brown bracts. Sixteen species are included in the List of Rare Plants of Australia.

Alpine cryptandra (*C. alpina* Hook. f.) is known only from the Central Plateau near the Great Lake and the Great Western Tiers Range (Tasmania) at an altitude of 1500 m. This is a small, branched perennial, with spreading branches which supposedly form a pad of about 30 cm diameter. The leaves are triangular-oblong, borne very closely on young branches. The white flowers are subtended by chocolate-brown bracts. The fruits are capsules.

The **smooth cryptandra** (*C. nudiflora* F. Muell.) is a Western Australian endemic, growing in the Murchison River valley. Its distinguishing feature is the presence of a tubular calyx and absence of pubescence on the disc and ovary.

The native hazel genus (*Pomaderis*) is mainly distributed in Australia. All its species are endemic, a few growing in New Zealand. The native hazels are shrubs with leathery leaves. The flowers are borne in umbellate or paniculate inflorescences. Bracts are leathery, chocolate-brown, dropping off by the time of full bloom. The **large native hazel** (*P. grandis* F. Muell.)—a Western Australian plant from the Coolgardie region—belongs to the category of rare species. This is a tall shrub or a small tree. Its leaves are ovate-lanceolate with wavy margins, silky-white on the lower side. The paniculate inflorescences are multiflorate. The large native hazels and 12 other species of the genus are included in the National List of Endangered Plants.

The endemic Australian genus *Stenanthemum* is represented by a single species in the List of Rare and Endangered Plants. This is the **pimeloid stenanthemum** [*S. pimeloides* (Hook. f.) Benth.], on a small tree-like plant with numerous spreading branches, up to 30 cm long. The leaves are obcordate, pubescent, and grooved; flowers are borne in capitate inflorescence, on the lower side covered with white bracts, sharply contrasting with the green assimilatory leaves. The plant is very attractive in bloom. It grows along the eastern coast of Tasmania in Cowles Bay and Spring Bay.

FAMILY RUTACEAE

The members of the family Rutaceae are fairly widely distributed in Australia and play a significant role in the composition of its flora. The abundant endemic genus is that of Australian native rose (*Boronia*): it includes about 70 species. Some of them are found all over Australia, others are extremely rare or represented by a few populations. Independent taxonomic status of some species is doubtful. All species of native rose are attractive, especially in bloom. Flowering shoots are sold not only within the continent but also outside, which is causing serious damage to the populations of Australian native rose. Besides, all its parts are used in the perfume industry as a source of essential oils. As a result of this, the population and area of almost all the species of this genus are decreasing; the earlier common species are becoming rare, and the rare species extinct. The depletion of population of some species leads to increase in the demands for others and, therefore, the extinction of native rose species cannot be prevented without adopting strict control measures on the state and exploitation of their populations. Already the native roses are protected by law in some states of Australia, and this genus is included in Appendix 2 of CITES so as to regulate its trade. The native roses are shrubs or semishrubs with opposite leaves, bright flowers, borne in racemes or umbellate inflorescences. Like other members of the family Rutaceae, the vegetative organs of native roses have glands producing essential oils.

Keys' native rose (*B. keysii* Domin) is among the endangered plants. It was first found by James Keys in 1909 in wet eucalyptus forest. It could be collected again from the same place only in 1971. Its habitat is located in the environs of Lake Kutaraba in southeastern Queensland (150 km north of Brisbane) and is occupied by a low eucalyptus forest with dense undergrowth. The soil is sandy, rich in humus, and quite wet; the annual rainfall is 1600 mm.

Key's native rose is a shrub up to 2 m tall with branches wide apart, covered with pinnate leaves having 3–9 leaflets. The bright pink flowers are borne in umbels and their petals have a fringe of long white silky hairs. The seeds develop after pollination. Although the population of this species is small, it has been pointed out by the investigators that this native rose multiplies normally in its habitat. Numerous attempts were made to artificially spread the area of its distribution by sowing the seeds in similar conditions in other regions of Australia, but they have all failed, and the reason for such a biotopic specificity is not yet known. Most probably, the seoret lies in some ecological peculiarities of the area, which has been included in the Kulula National Park for its conservation. Considering the present status of Keys' native rose, and its scientific and economic value, the Commission on Endangered Plants of IUCN has included it in the International Red Data Book. The **kindred native rose** (*B. affinis* Benth.), whose distribution is limited to the Arnhempland peninsula in the Northern Territory, has also been included among the endangered plants.

Riverside native rose (*B. rivularis* C.T. White) is a delicate, fragrant plant of open habitats. Under favorable conditions, it branches profusely and grows up to a height of 4.5 m. The leaves are pinnate, dark green. Flowers are purple, in racemes of 9 flowers. The flowering continues from spring to late summer. It grows along the banks of rivers and streams, near marshes, in sparse communities of trees and shrubs. A few of its habitats are known in Wide Bay on Fraser Island and near Tin Can Bay (Queensland).

Edward's native rose (*B. edwardsii* Benth.) is a small, erect shrub with pubescent branches. The leaves are trilobate, almost sessile, 4–6 mm long. The flowers are pink, in groups of 1–3 terminating the branches. The shrub flowers from October to December. The area of distribution of this species is in South Australia (Barker Mountain, Lofty Range, Kangaroo Island, Encounter Bay).

The **large-pistil native rose** (*B. megastigma* Nees) is in great demand in the world market because of its large flowers, dark purple outside and yellowish inside with long purple pistils (Plate 55). It flowers from June to October. It gives a tender pleasant smell. A total of 19 species of native roses are included among the Rare and Endangered Plants and the entire genus is included in Appendix 2 of CITES.

The endemic Australian genus *Acradenia* is represented by only one species—Franklin's acradenia (*A. franklinii* Kipp.). Its area of distribution is limited to Tasmania, along its western coast. It is a fairly tall shrub, from 2.5 to 4 m, with opposite ternate leaves on short petioles. The umbellate inflorescences are terminal on branches. All parts of the flower have barely noticeable or dense pubescence. The flowers are white.

The endemic Australian genus fuchsia (*Correa*) includes about 10 species. The flower of Australian fuchsia is the emblem of the Society of

Naturalists of Victoria. It is a 1-2 m tall shrub with broad leaves and large, drooping, campanulate flowers of bright colors, combining red, yellow, white, and green.

Decumbent Australian fuchsia (*C. decumbens* F. Muell.) is a low shrub with prostrate branches and densely pubescent twigs. The leaves are narrow ovate, 1.5-3.5 cm long and 3-6 mm broad, with the upper surface glabrous and grooved, and lower surface tomentose, somewhat curved. The flowers are conical, solitary, on short pedicels. The calyx is 4-fid; corolla is fused, red, 25 mm long; stamens are exserted. The plant is distributed in South Australia on Kangaroo Island and in the Lofty Range mountains and flowers almost round the year. Besides this species, 4 more species of Australian fuchsia are included in the National List of Endangered Plants.

The endemic genus *Crowea* includes only 3 species, the rarest among which is the **narrow-leaved crowea** (*C. angustifolia* Sm.), found in Western Australia. This is a shrub up to 0.5 m tall, very attractive at the time of flowering because of bright red or white flowers. It suffers heavily due to mass collection of the flowering shoots for bouquets. Uncontrolled cutting of the plants is endangering the existence of the species in nature, and results in greater exploitation of other less widely distributed crowea species: the **wingless crowea** (*C. exalata* F. Muell.) and **willow-leaved crowea** (*C. saligna* Andrews) (Plate 55).

The wingless crowea grows in Victoria on rocky areas of the eastern slopes of Mount Cober and the Pine Mountains, and in New South Wales. The leaves of this short shrub (up to 1 m tall) are flat, linear, 2-3 cm long. The flowers are large, bright pink, up to 2 cm in diameter.

The willow-leaved crowea is still a fairly common plant of the coast and adjoining areas of New South Wales, where it grows together with wingless crowea; rarely it is found in Queenslands. This is a shrub with willow-like leaves. The flowers are light purple or pink, terminal on axillary branches. In order to conserve the ancient flora of Western Australia, the narrow-leaved crowea was included in the List of Endangered Plants of the state even in 1935. The wingless crowea is also included among protected species. In view of the difficulty in differentiating the species, increasing exploitation, and need for strict control on the trade of this endemic Australian genus, all the crowea species have been included in Appendix 2 of CITES since 1979.

The genus *Eriostemon* includes about 20 species. Almost all of them are Australian, except a few found in New Zealand and New Caledonia. These are shrubby plants with simple, entire, glandular leaves; flowers are solitary, white-pink, rarely dove-blue or in umbels of a few flowers. Six eriostemon species need protection.

Eriostemon obovalis A. Cunn. is a rare species of New South Wales. It grows in the undergrowth of mountain forests west of Sydney. It is a shrub up to 1 m tall with obcordate, thick, flat petiolate leaves. The flowers are small; filaments are flattened and ciliate.

The endemic Australian genus *Geleznowia* with only one species— *G. verrucosa* Turcz.—is entirely confined to the sandy plains in the northern part of Western Australia, where it has a limited distribution. This is a shrub with appressed leaves and tuberiferous glands in the lower part. The flowers are borne in groups of 1–3 at the tips of branches and are quite conspicuous because of their petaloid bracts. The flowering shoots not only attract nature lovers but also are an object of trade. At present, the species is included in Appendix 2 of the CITES.

The **Tasmanian mountain myrtle** (*Pheballum montanum* Hook.), a member of a large genus, is mainly distributed in Australia. Nineteen of its species are included in the List of Rare and Endangered Plants of Australia.

The Tasmanian mountain myrtle is an endemic Tasmanian species, growing among boulders and stones in the mountains at an altitude of 750–1200 m in northeastern Tasmania and in the Western Tiers Mountains. This is a spreading, highly branched shrub, 15–50 cm tall. The leaves are crowded, densely covered with almost cylindrical glands, and with a slight groove on the upper surface; scars remain after the leaves fall off. The flowers have short pedicels, are in small groups terminating axillary branches; they are white or pink, pentamerous. The plant is cultivated by rooting or cuttings, mainly in private collections.

Ralston's Tasmanian myrtle [*P. ralstonii* (F. Muell.) Benth.] is a tall shrub with linear leaves. The flowers are green or reddish, on short pedicels, in umbels. It grows in the extreme southeast of New South Wales, near Twofold Bay, as well as in the foothills of Castle Rock Mountain. The species is included in the category of endangered plants of Australia.

FAMILY SAPINDANACEAE

About 20 genera, 5–6 of which endemic, belonging to the family Sapindanaceae, are found in Australia.

The hop bush genus (*Dodonea*) comprises shrubs, sometimes very tall and much branched. The young branches, and sometimes the entire plant, are sticky. The leaves are pinnate, rarely simple. The male and female flowers are borne in short panicles or racemes on different plants. Eleven of its species have been included in the List of Rare and Endangered Plants.

Microzyga hop bush (*D. microzyga* Endl.) is a 50–150 cm tall shrub with pinnate leaves; inflorescences have few flowers; fruits are winged. The shrub grows in southeastern Queensland along the eastern coast. The **ericoid hop bush** (*D. ericoides* Miq.) is a rare endemic of Western Australia (environs of Perth). It is a short shrub with slender branches and dentate linear leaves; fruits are angular, without wing-like processes.

FAMILY SCROPHULARIACEAE

The family Scrophulariaceae is represented by herbaceous plants, rarely shrubs. Among the rare species is *Ourisia integrifolia* R. Br., an endemic plant of Tasmania. It is found in mountains at wet places, mainly along the banks of rivers and streams, near waterfalls, among stones and boulders. The vegetative shoots of this perennial herb are spreading and branched and produce roots. The leaves are crowded at the end of branches, opposite, broadly lamellate, thick, and glabrous. The solitary flowering shoot bears 1–2 flowers and a few sessile bracts. The flowers are white with bluish-purple tinges and dark purple stamens. The species is included in the List of Rare and Endangered Plants of the continent. It is being cultivated in the Kew Botanical Garden (England).

FAMILY STACKHOUSIACEAE

Most of the species of the family Stackhousiaceae are distributed in Australia, one each found in New Zealand and Malaysia. The family includes 3 genera. The genus *Stackhousia* consists of 4 species, distributed in South Australia. The **annual stackhousia** (*S. annua* W. Barker), an annual herb, 12 cm tall, is included among the category of endangered species. Its stem is poorly branched and covered with leaves; inflorescence is dense, cylindrical. It is found in the coastal regions in the southwest of the York Peninsula and in the southeast of Eyre Peninsula, possibly also on Kangaroo Island. It grows on moist sandy soils in open herbaceous-shrubby communities. Only a few populations of this endangered plant are known. Its present condition is related to commercial activity and cattle grazing.

FAMILY STERCULIACEAE

The genus *Lasiopetalum* is endemic to Australia. The lasiopetalums are shrubs with distinct pubescence; leaves are leathery and glabrous; flowers borne in drooping heads. Most species of the genus are distributed in

Western Australia. Thirteen species are included in the National List of Rare and Endangered Plants.

Bracteate lasiopetalum [*L. bracteatum* (Endl.) Benth.] is a shrub up to 4 m tall; its stems, lower leaf surface, bases of bracts, and styles are densely hairy. The pink flowers are borne in much branched inflorescences. The shrub grows on granite outcrops near water streams in the Darling Range (environs of Perth). It is included in the List of Endangered Plants of Western Australia.

This family also includes another endemic genus, *Thomasia*, which differs from *Lasiopetalum* by the structure of its calyx and general growth habits. The areas of distribution of both genera and their ecology are very similar. The thomasia genus includes about 30 species, 12 of which are included in the List of Rare and Endangered Plants of Australia. The category of endangered plants includes **mountainous thomasia** (*T. montana* Steud.), whose habitats are confined to the foothills of the rocky Mount Bakewell and the Swan River valley.

FAMILY TAXODIACEAE

The genus *Athrotaxis*, with only one species, *A. laxifolia* Hook., is distributed in the southern hemispheres. (The fossil remains of the members of this genus are found in the Tertiary deposits of Tasmania and Australia.) The present area of distribution of this species includes several hilly regions of western and central Tasmania, where small populations are adapted to the mountain slopes at an altitude of 900–1300 m. *Athrotaxis* is a small tree with beautiful pyramidal crown. Its light wood is used in the transport industry and for decorative work. At present, individual habitats of this plant are being protected in the Tasmanian reserves. The species is included in the National List of Endangered Plants of Australia, requiring constant control of its condition.

FAMILY THYMELAEACEAE

The family Thymelaeaceae includes the beautiful flowering endemic shrub, **inflated rice flower** (*Pimelia physodes* Hook.). Its greenish-yellow flowers are borne in capitate inflorescences, surrounded by bracts of various colors from purple to apple-green (Plate 55). The plant appears particularly beautiful in autumn. It grows on sandy, stony, or rocky elevations, where shrubby communities are predominant. The area of distribution of rice flower is limited to Australia between the Gardner River and Ravenshorpe. The limited area of distribution and strictly adapted habitats indicate its low population. Considering these circumstances, the

State of Western Australia included it among the protected plants of the State by a special law in 1935. At present, the rice flower has become even more rare because of collection of fresh shoots for bouquets and commercial trade, as well as digging of plants for plantation into private gardens and nurseries. In order to control its export, it has been included in Appendix 2 of CITES. It has also been included in the List of Endangered Plants of Australia along with 10 other species of the same genus.

FAMILY TREMANDRACEAE

The family Tremandraceae includes 3 genera, of which the genus *Tetratheca* is represented by a few dozen species. These are short shrubs with narrow glandular leaves of heath type and pink or purple flowers. They grow in arid regions.

Halmanturin tetratheca (*T. halmanturina* J.M. Black) is included among the rare endemic plants of South Australia. It is a short and rough shrub with sessile glandular hairs. Its leaves are reduced to scales. The flowers are borne on 5–6 mm long pedicels with scaly bracts, which are red or white; flowers are up to 13 cm long; ovary is glandular-hairy. The plant flowers throughout the spring and summer. It was found on Kangaroo Island. The habitat of **Gunn's tetratheca** (*T. gunnii* Hook. f.) is located in northern Tasmania, and that of **remote tetratheca** (*T. remota* J. Thompson) in the extreme west of Western Australia. Both species are very rare.

Glandular tetratheca (*T. glandulosa* Sm.) is a rare plant of New South Wales (it grows in the Sydney region). It is a branched, densely pubescent shrub up to 60 cm tall, with glandular hairs. The leaves are pointed, glabrous, ciliate and serrate along the margins; petals are large, ovate; seeds are hairy. The Western Australian endemic **deltoid tetratheca** (*T. deltoidea* J. Thompson), which was at one time growing in the environs of Coolgardie, has been included among the extinct plants.

A total of 18 tetratheca species are included in the List of Rare and Endangered Plants.

FAMILY WINTERACEAE

The family Winteraceae includes the endemic genus *Tasmannia*, which was named after the island on which its members were first found. It was later found that the distribution of *Tasmannia* is much wider. The genus includes 6 species, distributed up to northeastern Queensland all along the coast and on the Great Dividing Range. Two species of the genus—*T. stipitata* (Vickery) A.C. Sm. and *T. insipida* R. Br. ex

PLATE 55.   1—Large-pistil native rose; 2—willow-leaved crowea; 3—western woody pear; 4—inflated rice flower.

PLATE 56. 1—Chatham island forget-me-not; 2—tripistillate cheirodendron; 3—Brock's bluebell; 4—parrot's bill.

DC.—have retained the most primitive characters of flowering plants. Two tasmannian species are included in the List of Endangered Species.

Purple tasmannia [*T. purpurascens* (J. Vickery), A.C. Sm.] is an aromatic tree. Its leaves have transparent dots; flowers on peduncle originate from the base of a deciduous scale and are pale purple. The **stipitate tasmannia** [*T. stipitata* (J. Vickery) A.C. Sm.] is a tall shrub; its leaves are oblong-lanceolate, pointed, up to 12.5 cm long, on a short petiole. The flowers are large, with numerous stamens and one or a few free carpels. Both species grow in New South Wales in the Northern Plateau. They are being conserved in the Barrington Tops National Park.

FAMILY ZAMIACEAE

The family Zamiaceae belongs to class Cycadophyta. In Australia, the family is represented by 3 endemic genera.

The fern palm genus (*Bowenia*) includes 2 species. The underground organs of these plants resemble the taproot system of flowering plants. The stem is short and branched; leaves are long-petiolate, bipinnate, arising from the branches together constituting the entire aerial plant part. The male and female 'cones' or strobiles are right at the soil surface under the leaves.

Beautiful fern palm (*B. spectabilis* Hook.) was found and described in the mid-19th century from Queensland. It has large, glossy, dark green leaves, repeatedly divided in the form of feathers. It grows at the northeastern coast of Australia along the margins of tropical rainforests, in isolation or in groups. The natural population of this species is declining as a result of digging of the plants by amateur gardeners for transfer to private collections.

Serrulate fern palm [*B. serrulata* (Bull) Chamberlain] has an underground, almost globose stem, whose lower part is prolonged into a taproot, and 5–20 short branches growing upward. Leaves develop whose segments have minutely serrate margins. The plant grows in the lower parts of slopes or on flat streaks of coastal elevations. Its largest concentrations were found in the lower tier of eucalyptus forests, and isolated specimens are found in the areas of highly destroyed rainforests. Till recently, serrulate fern palm formed dense thickets, and its isolated populations had up to 1000 plants. At present, its population has been greatly reduced; isolated habitats are found from the central parts of northern Queensland to the environs of Port Kerns, as well as on the coastal islands. The Hinchinbrook Island National Park and the Bifield State Reserved Forests have been established to conserve its natural habitats. The latter forest is proposed to be converted into a national park. The serrulate fern palm is being maintained in 10 botanical gardens of the world.

The genus *Lepidozamia* is endemic to the tropical and subtropical parts of eastern Australia. Two species of this genus were discovered and described in the middle of the 19th century by E.L. Regel, Director of the Petersburg Botanical Garden, on the basis of glasshouse specimens.

Perouskll's lepidozamia (*L. peroffskiana* Rgl.) is a beautiful palm-like plant with long shining leaves and large, up to 1 m long strobiles. It is found scattered in a narrow belt from north to south in the subtropical parts along the eastern coast of Australia in humid sclerophyllous forest. It is being conserved in the Franklin National Park (Queensland).

Hope's lepldozamia (*L. hopei* Rgl.) is the tallest plant among the living Cycadophyta: its height reaches 18–20 m. The crown consists of numerous leaves. Its distribution is limited to wet tropical forests in the extreme northeast of Australia.

The macrozamia genus *Macrozamia* includes 14 species, distributed in Australia from the moderately warm regions to the subtropics in sclerophyllous communities. The **common macrozamia** (*M. communis* L.A.S. Johnson) is spread the farthest south. It is a common plant in the narrow coastal belt north and south of Sydney.

Moore's macrozamia (*M. moorei* F. Muell. ex C. Moore) and Riedl's macrozamia (*M. riedlei* C.A. Gardner) grow in southwestern Australia. The former species has a large crown, consisting of up to 150 leaves, which are located on a massive trunk up to 7 m tall. Almost all macrozamias grow in the coastal regions, and only **McDonnell's macrozamia** [*M. macdonnellii* (Miq.) A. DC.] is found in the dry regions of central Australia in the communities of short eucalyptus and xerophytic grasses, and sometimes together with mariae fan palm. The habitats of the first 2 ancient species as well as almost 300 other rare or poorly distributed plants are included in the Finke Gorge reserve (Northern Territory).

The present state of all the Cycas in Australia is not similar. Some species are widespread and there is no immediate threat of their extinction, whereas the habitats of others are very localized and their population strength not high. The serrulate fern palm, Moore's macrozamia, and a few other Australian Zamiaceae members are particularly endangered. Most of the macrozamia fern palms are endangered due to burning of grass, clearing of soil, cattle grazing, and their elimination as poisonous plants. Almost all Zamiaceae species are excellent ornamental plants. This situation also affects their natural populations. The rare species of Cycadophyta were included in Appendix 2 of CITES in 1975.

7

Oceania

280 The largest group of islands on earth is located in the western and central parts of the Pacific Ocean. A large number of these islands are grouped into archipelagos and are combined under the common name Oceania. The islands differ in size and origin and are located in the northern and southern hemispheres. Their northern border is Hawaiian island, southern border Campbell Island, western border Misool Island, and eastern border Salay Gomez Island. Historically, the Oceania is divided into Melanesia, Micronesia, Polynesia, and New Zealand. Oceania is included in the Paleotropical floristic region. Besides tropical elements of the flora, Australian, Antarctic, and American elements are also found on the islands. The flora of Oceania is very poor in comparison with the continental floras, but it is characterized by a high degree of generic and specific endemism. Oceania flora becomes increasingly poor towards the southeast. The Melanesian flora includes 13,000 species, including 8000 in New Guinea, and the endemic plants constitute 85 percent of the flora here. The flora of the Hawaiian Islands consists of 2200 species, 75 percent of which are endemic. The New Zealand flora comprises 1850 species, of which 79 percent are endemic. The flora of Pashi Island has only a few dozen species.

The flora and vegetation of Oceania have suffered and continue to suffer from commercial activity. The natural vegetation is completely destroyed in some islands and plantations of agricultural and commercial crops have been developed in their place. In the western islands, the ancient forests have been cut or changed by selective felling, at some places they have been replaced by artificial forest plantations. The introduced or randomly transported plants to the island have played a significant role in changing and destroying the ancient flora. In several cases, the local species were less competitive than the new introductions and surrendered their positions to them. Collection and digging of ornamental plants for private collections and for sale, and introduction of domestic cattle and other animals on the islands, including rats, which eat away or destroy the fruits and seeds and prevent regeneration, are among the specific causes of destruction of one or another plant species or their groups. As a result, the condition of the ancient island flora became catastrophic. The list of Hawaiian plants, for instance, includes 1113 species, subspecies, and varieties, i.e. 50.6 percent of the total flora.

The situation in New Zealand is the best. The tentative list of rare, endangered, and extinct plants includes almost 360 taxa. The published Red Data Book includes 66 species.

FAMILY AMARANTHACEAE

In the family Amaranthaceae, the **Mangareva chaff tree** (*Achyranthes mangarevica* Suesseng) is included among the endangered or extinct species. It is a 5–7 m tall tree with greenish-gray bark and spreading leaves up to 8 cm long. The flowers are borne in terminal racemes. The plant was found in 1934 in a small area of a natural forest (on the southern slopes of the Mokoto mountains at an altitude of 290 m), surrounded by burnt arboreal vegetation or lands where goat grazing was common. Even at that time it was on the verge of extinction, but many repeated fires since then have, most probably, made it extinct. The only location at present is on Nabgereva island (1300 ha), the largest island in the group of volcanic Gambier islands in the southern Pacific Ocean. This island is surrounded by coral reefs and is distinguished by poor floristic composition. The island flora in this part of the Pacific Ocean includes only 29 angiosperms, among which are the endemic species, for example, the **Mangereva gouania** (*Gouania mangarevica* Fosberg) of family Rhamnaceae, which probably has also disappeared.

The Mangereva chaff tree and two other ancient species, the **arboreal chaff flower** (*A. arborescens* R. Br.) from Norfolk Island and **Marquesas chaff flower** (*A. marchionica* Forest Brown) from Marquesas Island, constitute the relicts of the ancient Pacific flora. At present, the habitat of Mangereva chaff tree on Mokoto Mountain is protected from goats and measures are being taken to prevent fire. The chaff tree is highly ornamental and has practical value for gardening in tropical countries; this species is included in the Red Data Book of IUCN.

FAMILY AMBORELLACEAE

The family Amborellaceae is represented by the monotypic genus *Amborella*. This is an endemic family of the flora of the New Caledonia. Its only species, **hairy-pedicel amborella** (*A. trichopoda* H. Bn.), is an evergreen shrub, up to 8 m tall, with pinnatinervate leaves. The flowers are unisexual, borne in axillary inflorescences. Perianths number 5–8; stamens are broad, and peduncles [sic] have 2 broad processes; the fruits are stones. Amborella has typical characters of nonvascular flowering plants. It grows in a tropical forest in New Caledonia, which is known for its unique and primitive plants.

FAMILY APOCYNACEAE

The endemic Hawaiian genus *Pteralyxia* with a few species belongs to this family. All pteralyxias are trees with milky latex; their thick branches are covered with large lenticels discharging latex, and the flowers are waxy yellow. The species *P. caumiana* Degener is among plants which have become completely extinct. It is a tree up to 15 m tall with grayish bark and quite distinct lenticels on the branches. The leaves are dark green, shining, leathery, 15–25 cm long and 6–9 cm broad; midrib is broad, lateral veins parallel to each other originating from it. The flowers are regular, numerous, in a panicle; corolla is greenish-yellow. The fruits are ellipsoid; immature fruits are orange, mature red. In March, when pteralyxia flowers, the tree still has maturing and ripe fruits. The seeds of even the immature fruits falling on the ground are eaten by rats which have got into the Hawaiian Islands along with humans. The absence of seeds, thus, precludes its regeneration. The plants of this species were found in a dense forest at the northernmost tip of Oahu island (Hawaii). A total of 12 plants and a few more trees grew here along the road connecting Pupukea and Kahuku. It is believed that the last specimen of *Pteralyxia caumiana* died in 1933.

FAMILY ARALIACEAE

Lydgat's tetraplasandra [*Tetraplasandra lydgatei* (Hilled.) Harms], a few specimens of which were found on Oahu island in a dry and somewhat closed forest in the Wailupe valley, is included among the endangered plants of Hawaii. This is a tree with a symmetrical crown, 7–10 m tall. The leaves are 20–45 cm long, yellowish-green, papery and thin. The inflorescence is umbellate; calyx is ovoid-conical, 5 mm long; petals form in a cap. The fruits are juicy and ovoid.

Among the endangered plants of the same family is included the **tripistillate cheirodendron** (*Cheirodendron trigynum* A.A. Heller), a medium-sized tree, growing in the humid forests of Hawaii island. It has opposite, 3–5 times divided leaves; the racemose inflorescences of small green flowers are terminal on branches. The fruits are black (Plate 56). A few varieties are distinguished on the basis of their leaf shape, of which *C. trigynum* var. *rochii* and *C. trigynum* var. *subcordatum* are endangered.

Stilbocarpa robusta (T. Kirk) Cockayne is an endangered plant of New Zealand flora, which has been included in the Red Data Book of the country. This is a large herb with short but strong, sometimes lodging stem. Leaves are on long petioles; inflorescence is umbellate and has greenish-yellow flowers; fruits are black. The plant grows in forest glades, cleared areas, in shrubby thickets on wet areas. The distribution of this species is limited to the Snares Islands. The area of its habitat does not exceed 260 ha, and population strength is a few hundred plants. The population is decreasing as a result of grazing of the young plants by the animals introduced into the islands, including rats, although a reserve formally exists on Snares Island.

FAMILY ASTERACEAE

The genus *Celmisia*, including about 60 species widely distributed throughout New Zealand, belongs to the family Asteraceae. These are perennial herbs with whorled, simple, silky-pubescent leaves. The stems bear one inflorescence each; outer (marginal) florets are white. The plants mainly grow in the alpine belt of mountains. Many of them are ornamental and cultivated in botanical gardens and private collections.

The list of rare and endangered plants of New Zealand includes 9 species and 6 varieties of this genus. *Celmisia morganii* Cheeseman is represented by a single population, which includes a few hundred plants concentrated on the rocks along the Ngakavau River near Westport (South Island). It is a small plant with lanceolate leaves up to 45 cm long and with hairy lower surface. The flowering shoot is slender, 35 cm long, hairy, bearing a single head up to 3 cm in diameter. The marginal florets are white, 25 mm long; disc florets are yellow. This species grows almost on suspended rocks together with other herbaceous plants. The reduction in the strength of this population and disappearance of another population in the lower reaches of the Buller River (environs of Nelson) is associated with the collection of plants and change in their habitat conditions (laying out of tourist routes). The proposed construction of a thermal power station near Westport may destroy the remaining population of *C. morganii*. However, there is a project to organize a scientific reserve in the lower reaches of the Ngakavau River, which should include a larger part of the distribution area of this species. It is proposed to develop a biological reserve in the upper reaches of the same river where this plant can grow.

A less rare but taxonomically more enigmatic species is a shrub forming hemispherical pads up to 70 cm in diameter. This is *C. philocremna* Given, forming 5 populations with a total strength of a few hundred plants. All the populations are concentrated on a small territory on Eyre Mountain (South Island), on the southern side of Lake Wakatipu. The plant grows on rocks where the soil cover is poorly developed, in a community of dense-flowered and branched celmisia and Biggard's hebe at an altitude of 900–1800 m. The normal existence of the species is endangered mainly by plant collection, although they reproduce well by seeds as well as vegetatively. The other cause for reduction in the population of *C. philocremna* is grazing by wild herbivores. At present, the animal density is regulated and this cause has been eliminated. Work has also been carried out on mapping the area of distribution of the species and its condition is under constant control.

Helichrysum dimorphum Cockayne differs from other species of the genus by its morphology. It is a creeper or prostrate plant up to 8 m long with lignified base; the leaves are white-tomentose, small. Its primary distribution is not yet established. Now a few populations are known, one of which was discovered in 1920. Since then, its population strength has sharply declined due to fire and cattle grazing. The plant grew on dry rocky places among communities of herbs and shrubs of coprosma, corokia, and discaria. Further survival of the species is threatened by the introduction of weeds and taller shady plants. The area of its distribution is on South Island, in the Waimakariri River valley between the Broken Thomas and Poulter rivers (New Zealand). Particularly here, it is proposed to develop the Waimakariri valley reserve for the conservation of dimorphic helichrysum and the characteristic shrubby scrub population of this region. Helichrysum cultivation has been started in Wellington in the Otari Botanical Museum (New Zealand). The plant is included in the Red Data Book of IUCN and New Zealand.

The **Henderson bur marigold** (*Bidens hendersonensis* Sherff.) belongs to the category of endangered plants, its distribution is limited to Henderson Island and Oeno atoll, included in Pitcairn Island, in the southern parts of the Pacific Ocean. The flora of the Henderson coral island (area about 300 ha) comprises 55 species of flowering plants, including 10 endemics, to which the Henderson bur marigold with 3 varieties, also belongs. Two of them grow on Henderson Island and are distinguished by the shape of leaves and floral heads; the third (smaller in size, with fewer heads and larger leaves) has been found on the small Oeno atoll (4 km across, altitude 3.6 m, unpopulated). Here the natural population has been partly destroyed and a coconut plantation has been developed at its place. The flora of the atoll includes 15 higher plants.

Morphologically, the bur marigold is a tall shrub or a short tree, up to 4 m tall, with a stem 2.5 cm thick at the base. It grows in a dense forest of low plants with endemic sandalwood and portia trees, on the flat peak of the island, 3 m above the sea level. This forest is a remnant of the vegetational cover occupying the entire island. The vegetation has experienced the effect of goat grazing. The goats were brought here and possibly died in the absence of sweet water. Besides, the local population regularly visits the island and collects sandalwood, portia wood, and other trees, which has affected the present condition of bur marigold.

At present, it is proposed to declare Henderson Island a natural reserve so as to conserve the endemic flora of the island as well as the vegetation which is considered to be the most typical for the coral island of this region of the Pacific Ocean.

Poplar-leaved bur marigold (*B. populifolia* Sherff.) is an endangered plant of Hawaii. Till now, there is no uniform opinion about its life span; some specialists consider this plant an annual, others consider it a perennial. It has many-branched herbaceous tetraquetrous stems up to 80 cm tall. The leaves are pale green, deltoid-cordate, covered with fine pubescence on both sides, serrate, and on petioles; the lamina, together with petiole, is 20 cm long. The heads are few, forming a loose bunch. The ray florets are yellow with reddish streaks. The poplar-leaved bur marigold is endemic to Oahu island, where it grows on grassy slopes of the Kahana, Kaawa, and Kaipapau valleys. It has been included in the category of endangered plants of the Hawaiian Islands.

FAMILY BALANOPHORACEAE

Taylor's balanophore (*Dactylanthus taylori* Hook. f.) is the only member of the tropical family Balanophoraceae in New Zealand. It parasitizes on the roots of other plants, in particular, on the roots of cheesewood and southern beech, deforming them. It is a small plant without chlorophyll, with a short stem; the upper part of the stem consists of fleshy, dark chocolate-brown, clustered scales in the form of a head. Dirty-white or chocolate-brown flowers are borne in the center of the scales. This balanophore emits an unpleasant smell. It grows in the forests on North Island, and is found extremely rarely. The extinction of this species is related not only to cutting of forests but also to its use as a source of balanophorin, a waxy substance. It is included in the Red Data Book of New Zealand and is conserved in a few national parks.

FAMILY BIGNONIACEAE

Bignoniaceae is a tropical family, of which only a few species are found in other climatic zones. The genus *Tecomanthe*, with only one species, *T. speciosa* W.R.B. Oliver, found in New Zealand, is an example of this. It is a woody evergreen creeper with the stem up to 10 m long, pinnate dark green leaves, and bunches of cream-white symmetrical flowers. It is known in nature from a single specimen found in 1945 on Great Island northwest of the northern end of New Zealand. The primary flora of this island is highly depleted, but 12 species of plants, endemic to the Three Kings Islands, are still conserved here, among them the unique specimen of **Baylis' pennantia** [*Pennantia baylisiana* (W.R.B. Oliver) Baylis] of family Icacinaceae and 12 trees of **Johnson's elingamita** (*Elingamita johnsonii* Baylis) of family Myrsinaceae. Both species are members of monotypic genera. The reserve has been established on the islands since 1946. Now the islands are not populated and observations on the restoration of island vegetation are recorded only periodically. It is assumed that the basis of natural vegetation on Great Island is the creation of coastal mixed forests. According to the observations of 1951 and 1963, the **tea tree** or **kanuka** (*Leptospermum ericoides* A. Rich.) now plays dominant role here and the population of many endemic species has increased; nothing has changed in respect of tecomanthe and pennantia. The only tecomanthe plant climbs along the trunk of kanuka in a marshy area at an altitude of 150 m. It produces flowers and fruits, but not a single sapling has been found so far. Experiments on the germination of its seeds and vegetative propagation did not succeed. Tecomanthe cultivation has been started in several regions of New Zealand and is of great interest for horticulturists. This species is included in the Red Data Books of IUCN and New Zealand.

FAMILY BORAGINACEAE

The genus *Myosotidium* is represented by a single species in New Zealand, the **Chatham island forget-me-not** [*M. hortensia* (Decaisne) Baill.]. It is a succulent perennial herb with large bright green leaves on long petioles arising from a fleshy rhizome. The flowers are pale, dove-blue, borne in dense branched pyramidal cymes (Plate 56). There are 2 separate populations of Chatham forget-me-not. The plant is widely cultivated in the gardens of New Zealand and has been included in the Red Data Book of that country and conserved in the natural reserve at the Chatham Islands.

FAMILY CUPRESSACEAE

The genus *Neocallitropsis*, close to the American cypress genus (*Callitrix*), is an endemic monotypic genus of New Caledonia. Its only species, *N. araucarioides* (R.H. Compton) Florin, grows in wet coniferous forests. This is a small tree with grayish, hard bark and conical crown. Its slender branches, resembling those of araucaria (on the basis of which the species is named), carries minute, bent, scaly leaves, in regular rows. The present condition of the species requires detailed study and constant control.

FAMILY EUPHORBIACEAE

A member of the family Euphorbiaceae, *Neowawraea phyllanthoides* Rock, is the tallest tree of the Hawaiian Islands: it grows up to a height of 10–30 m. It has hard wood of reddish tinge and black heartwood. The plant is dioecious. The male flowers are small, greenish-yellow, and clustered; the female flowers have a cup-shaped bract. The flowers are pollinated by insects. The seed productivity of this plant is low because there are few flowering plants. Besides, trees of various ages die every year for unknown reasons. *Neowawraea* grows in dry forests on lava formation, along dry slopes of valleys in the communities of nigger's cord, cheesewood, Cooper's wood, snakewood, and other endemic plants. Its present population is too small for the continued survival of the plants. Most trees are found along the valleys of the Bayana Mountains on Oahu. One specimen each (in most cases already dead) is known from the Molokai, Kapua, Southern Kona, Hawaii, and Kauai islands. In order to conserve *Neowawraea*, it is proposed to bring under cultivation its male and female plants so as to multiply them vegetatively and then transfer their seeds or plants to their earlier habitats. Experiments on its cultivation showed that this is a difficult task because of limited seed productivity of *Neowawraea* and difficulties in vegetative propagation.

FAMILY FABACEAE

The New Zealand broom genus (*Carmichaelia*) including 39 species is endemic to New Zealand. The list of endangered plants includes 12 of its species, of which *C. prona* T. Kirk may already be extinct.

Spreading New Zealand broom (*C. exsul* F. Muell.) was found on Lord Howe Island in the Pacific Ocean. It is a perennial shrub, up to 5 m tall. Its leaves are often replaced by phylloclades (leafy cauline structures). The pale cream-colored flowers are borne in racemes. About 10 habitats of New Zealand broom are known to be confined to the mountains in the southern part of the islands. It grows on the upper slopes and peaks of the Lidgeberg and Gower mountains at an altitude of 600 m. The majority of the plants were found at the foothills or on the basaltic terraces; a few plants were found in a rainforest.

Besides New Zealand broom, 74 endemic plants have also been found on the island, including 4 species each of palms and tree ferns. Such high endemism of the island flora, and relatively well-preserved tropical and subtropical vegetation, is of great scientific interest and also attracts the attention of tourists. Earlier the vegetation suffered due to goats and sheep brought on the island, reduction in whose population during recent years have had a favorable effect on the condition of endemic biota. The tourist pressure is now increasing. It is proposed to convert a part of the island into a reserve. The plant is cultivated in botanical gardens and private collections. It is included in the Red Data Book of IUCN.

Chordospartium stevensonii Cheeseman is a member of the endemic monotypic genus of New Zealand. This is the largest broom in the world, up to 8 m tall with numerous branches and large inflorescences. The plant was discovered relatively recently in the Clarence River valley (South Islands) on alluvial soils. According to the latest counts, about 100 plants have survived in this region which do not stand competition with either the agricultural crops, which are almost adjoining their thickets, or with the weeds. The herbicides are causing serious damage to the chordospartium populations. No measures are being taken to protect this species. It is included in the category of endangered plants of New Zealand and also in the National Red Data Book.

One of the most attractive garden plants for a long time not only in New Zealand but far beyond its borders was the **Parrot's bill** [*Clianthus puniceus* (G. Don) Solander ex Lindl.]. It was highly valued for its large, bright red flowers and was cultivated even by Maoris as it grew fast. It is a beautiful plant but has a short life and often suffered from diseases. This was the main reason for a universal reduction in its cultivation. Now the danger of its extinction from natural conditions has become real. *Streblorrhiza speciosa* Endl. which was found on Philip Island in the group of Norfolk Islands and cultivated in Europe in the 19th century for its large pink flowers, has also suffered the same fate. When it stopped flowering, it was completely forgotten and disappeared.

The parrot's bill is a shrub up to 2 m tall with spreading branches on herbaceous silky-pubescent twigs, bearing 15 pairs of leaflets. The racemes have up to 15 large flowers, with a length of 8 cm (Plate 56). At present, only 6 populations of this plant are known from North Island. In 3 of these populations, only 2 plants have survived, and the population strength of others is also low; 5 populations are located around Lake Waikaremoana (eastern coast of the island) in the forests together with *Metrosideros, Dacrydium,* and *Beilschmiedia,* and the other population in the coastal shrubby thickets near Kailara Harbor. Multiplication of parrot's bill by seeds is very poor (only a few seedlings have been found), vegetative propagation is minimal, found only in the lodged plants. The role of parrot's bill in the forest communities and its natural distribution in the past still remains unexplained. There is a suggestion that this plant could be found not only in the coastal areas but also in the interior of the island. At present, the populations around Lake Waikaremoana are on the territory of Wrever National Park. It has been decided to continuously record observations on its status here, organize search for new habitats of parrot's bill, and impose strict control on the populations of introduced animals, mainly the Australian opossum, which gnaws the plants.

Members of the genus *Mezonevron* are almost all herbaceous plants, rarely shrubs; and only *Mezonevron kavaiensis* (H. Mann) Hilleb. is a tree. It has a black trunk up to 10 m tall and widely spreading branches; leaves are bipinnate, 4–9 cm long. The umbellate inflorescences bear numerous pink-purple or pinkish red flowers on long pedicels. The fruits are flat, young fruits are light glaucous, and mature fruits pale pink. Initially the arboreal mezonebron grew on several Hawaiian islands: Kauai, Oahu, and Hawaii. As a result of the introduction of goats on the island and commercial activity, almost all its habitats were destroyed and this plant now remains only on Hawaii island.

The pink broom genus (*Notospartium*) is an endemic genus of New Zealand. It comprises 3 species. **Pink broom** (*N. carmichaeliae* Hook. f.) is a small tree with slender branches and drooping twigs. The flowers are small, pink, and terminal on branches. The pods are linear, compressed, multi-seeded (Plate 57). The plant was found on sandy and rocky areas in the Waihopai River valley (Middle Island). The **beaded pink broom** (*N. torulosum* Kirk) is a leafless shrub, up to 4 m tall with delicate drooping branches and racemose inflorescences of purple flowers. It grows in the shrubby riverine thickets. Its largest population (about 100 plants) was found on South Island near Peel Forest in Canterbury.

The population of pink broom is declining mainly as a result of herbicide applications to the agricultural fields as well as fire. All the species of the genus are cultivated in New Zealand. The beaded pink broom is included in the National Red Data Book.

The flora of the Mariana Archipelago in the Pacific Ocean consists of 54 endemic species of ferns and flowering plants. The largest and the southernmost island in this group is Guam (USA), to the north of which Rota Island is located. A few plants of large tree species up to 20 m tall have been conserved on both these islands. These are **Nelson's serianthes** (*Serianthes nelsonii* Merr.), a plant discovered on Guam in 1916 by Nelson and named for him. Subsequently, botanical studies on the island confirmed the physical state of serianthes: only 4 profusely fruiting trees were found. However, the young plants developing from seeds were avidly grazed by the Asian deer introduced into the island. At present, 2 trees are growing in the northern part of the island on recrystallized coral limestone rock typical for western Micronesia, covered in the past with dense forests with a rich floristic composition (now almost all the forests have been replaced by serianthes plantations). Two more trees grow on volcanic soil in the southern part of the island in a valley forest. Such a wide separation of the habitats of this species indicate its wide distribution in the past. The fifth tree, found on Rota Island, grows in the forests together with other species endemic to Mariana Island: *Hernandia labyrinthica*, *Boerlagiodendron* sp., *Heritiera longipetiolata* Kanehira. At present, only the northern part of Guam is declared a protected territory. A few young plants were planted in the garden of Agana Town. Serianthes is cultivated in the Waimea arboretum in Hawaii.

The trees of the genus *Sophora* belonging to 6–7 species, are distributed in the tropics and subtropics of America and Asia. One of them—the **Golden-leaved sophora** (*S. chrysophylla* Seem.)—grows on the Hawaiian Islands and is endemic to the island flora. Like other plants of these islands, the sophora is distinguished by a great variety of forms. Several varieties of pagoda tree are distinguishable, of which 2 are already believed to be extinct, and are endangered and have been included in the National List of Endangered and Vulnerable Plants of Hawaii. The sophora populations are declining because its leaves and young plants served as fodder for domestic and semi-wild cattle, especially goats and sheep grazing on the islands. Eating of foliage and bark prevents the growth and development of trees, and reduces fruiting and seed-setting. As with many other species typical to the amazing and unique vegetation of Hawaii, the only way to conserve this endemic plant is to eliminate the introduced animals and plants, and impose strict control on the aboriginal flora and fauna.

The golden-leaved sophora is a short shrub or tree up to 12 m tall; leaves are imparipinnate; the young leaves are yellowish-green (hence the name of the species), and adult leaves grayish-green. Bright yellow flowers are borne in racemes at the tip of branches (Plate 57). The plant is in full bloom in the middle of winter. It grows at an altitude of 450 m up to the upper forest border (2500 m), and is found on the Kipuha Heke, Kipuha Paulu, and Hawaii islands.

The **toromiro sophora** [*S. toromiro* (Phil.) Skottsb.] is an endemic species of the Pashi Islands (Rapa Nui and Easter Islands). By the time of discovery of the islands in 1774, it was the only tree of the island. Its plantations were found as small thickets on volcanic slopes. During the visits of investigators to the Pashi Islands in 1917 and 1935, one tree each was sighted, a few trees were found at the same place during 1955–1956, and in 1962 all searches for toromiro sophora were unsuccessful. Since then, the plant has been listed among the extinct species and included in the Red Data Book of IUCN. Toromiro sophora is a shrub or a small tree, up to 3 m tall, with pale green pinnate leaves covered with silky white hairs. The flowers are yellow; fruits are long and narrow.

At one time toromiro sophora was a source of wood for the natives of the Pashi Islands. The Europeans introduced sheep on the islands in the 18th century. They ate the bark and the foliage of young plants, and the plants died because of the damage. In spite of the tropical climate, the island is presently covered with herbaceous vegetation, which is poor in species composition (only about 30 species of higher plants) and less endemic (only sophora and 3 grass species are endemic to the islands). It is beyond doubt that the flora of the island, especially the tree flora, was slightly different before it was occupied by Polynesians. Conservation of the living plants of toromiro sophora had a great significance for understanding the botanical-geographic relationships of the islands and evolution of the widely distributed sophora genus on the islands of the Pacific Ocean. It seems Tur Heyerdahl once brought from Pashi Island a few seeds of one surviving plant and handed them over to the botanical garden in Goteborg (Sweden). In 1979, the workers at the botanical garden noticed the young shoots. The seeds of the toromiro sophora tree, which did not exist any more in nature, germinated under controlled conditions. Now scientists are hoping to reintroduce it on the Pashi Islands.

Streblorrhiza speciosa Endl., a perennial woody creeper with pinnate leaves and large (up to 25 mm long) pink-red flowers arranged in racemes, was considered to be the most modern greenhouse plant of Europe in the beginning of the 19th century. This plant was discovered during the visit of Cook to Philip Island in the Indian Ocean (in 1774). The island was then covered with dense forest and was uninhabited. With the beginning of colonization of Norfolk Island, goats and pigs were brought on to Philip Island. Even the beginning of the 19th century, the natural vegetation was confined to the valleys and erosion processes could be seen everywhere. The remaining vegetation completely disappeared with the introduction of rabbits, and with that partly or completely disappeared 3 endemic plants of the island: the **king couchgrass** [*Agropyron kingianum* (Endl.) Laing], which was last seen in 1912, **island mallow rose** (*Hibiscus insularis* Endl.), now endangered, and the already extinct *Streblorrhiza speciosa*.

There is no hope of preserving *Streblorrhiza speciosa* under cultivation. This species has been included in the Red Data Book of IUCN.

Among the very rare, almost extinct plants is the **Oahu cowpea** (*Vigna owahuensis* Vog), a delicate biennial creeping on the soil and sometimes climbing around the lower shrubs and grasses. It has an angular stem with thin, ternate leaves on long petioles, pubescent all over. Flowers up to 2 cm long, solitary, lemon yellow; fruit is narrow, 4–5 cm long, and 4–5 mm across. The plant was first collected on Oahu island, in the Makaleha valley and in the Kaala Mountains, in the 19th century. Now it does not grow there. Somewhat later, this species was collected on the same island from a different place on the limestone outcrops under rocks as well as on the Molokai Islands. The plants from the surviving habitats have small differences and strongly differ from the other endemic species of Hawaii: the **Sandwich cowpea** (*Vigna sandwicensis* A. Gray), already endangered.

FAMILY GERANIACEAE

The family Geraniaceae includes **wedge-shaped cranesbill** (*Geranium cuneatum* Hook.)—a shrub up to 1 m tall with beautiful white pubescence, reflecting sunlight; flowers are white with purple veins (Plate 57). It grows in the mountains of Hawaii island at an altitude of 2100–2400 m. This rare endemic of Hawaii is included among the vulnerable and endangered plant species.

FAMILY GOODENIACEAE

The **Kilauea fan flower** (*Scaevola kilaueae* Degener) is endemic to the Hawaiian Islands. All of its few locations are confined to the crater of Kilauea volcano, where it grows on the lava and ash deposits of the southwestern slopes. It is a short, xerophytic shrub, up to 80 cm tall, with thick stems covered with black, rough bark. The leaves are leathery, pale green, narrowed towards the base, 4 cm long and 1.5 cm broad. Flowers are in clusters of 3: central flower is sessile, and the lateral ones are on short pedicels. The flowers open at the lower end on one side; this gives the impression that it is only half of the flower; calyx has yellowish coatings and corolla is light chocolate-brown to yellow. The fruits are black oblong berries (Plate 57). The Kilauea fan flower is an element of the Australian flora. At present, it is endangered and included in the National List of Endangered Plants of the Hawaiian Islands.

FAMILY GUNNERACEAE

The family Gunneraceae is distributed in the southern hemisphere, mainly Australia and South America. **Hamilton's prickly rhubarb** (*Gunnera hamiltonii* T. Kirk) is known only from New Zealand, where only one of its habitats exists now. This is a small herbaceous plant, forming compact pads up to 1 m across, from which scapes with a large number of small flowers arise. This prickly rhubarb grows in the zone of coastal sandy dunes, occupying a small area (100 m long, 2.5 m wide) on Stewart Island, which is south of South Island. The second habitat of Hamilton's prickly rhubarb, south of the Oreti River (South Island), does not exist any more because of sand and gravel mining, introduction of various weeds in the area, including lupine, and cattle grazing. The area on Stewart Island is also experiencing the pressure of weeds and grazing, which is endangering the survival of this species. Attempts to find this plant in similar locations are not likely to yield positive results. The only possible way to conserve it is its propagation and transplantation to locations closely similar to the natural. Definite success has already been achieved in raising the species in several botanical gardens. Each plant easily reproduces with seeds and cuttings, and grows fast. Hamilton's prickly rhubarb is included in the Red Data Book of IUCN and New Zealand.

FAMILY ICACINACEAE

The family Icacinaceae includes 45 genera, most of which are monotypic. The members of Icacinaceae are mostly distributed in the tropical regions of Asia, Africa, and America.

Pennantia baylisiana (W.R.B. Oliver) Baylis, the only member of the genus, grows on the Great Islands. This is a small tree; leaves are leathery, dark, crowded near the tips of the branches; flowers are in panicles. Only a single plant of Baylis' pennant (*P. baylisiana*) has survived under natural conditions; a few plants are found in state and private gardens. It is being conserved in the reserved forests on the islands, and is included in the Red Data Book of New Zealand.

FAMILY LAMIACEAE

The family Lamiaceae includes *Phyllostegia hirsuta* Benth., a perennial herb with a height of 90–120 cm. All the plant parts are covered with stiff hairs, projecting on all the sides. The leaves are pale green, thick, 9–18 cm long and 5–12 cm broad. The flowers are borne on short pubescent peduncles in whorls of 10–15 flowers, forming a raceme or panicle at the end of the stem. Corolla is white with pinkish tinge. The hirsute phyllostegia grows in hill range forests in the Kulau and Waianae ranges on Oahu island; it is endemic to the islands and forms small thickets. Two varieties of hirsute phyllostegia are included in the National List of Endangered Plants of the Hawaiian Islands: the sparse and hirsute one has more sparse verticils, and the other has more dense pubescence.

The List of Rare and Endangered Plants of the Hawaiian Islands includes 22 species and varieties of the genus *Phyllostegia* of which 9 have already disappeared, and 7 are endangered.

The genus *Stenogyne* is richly represented by species and varieties which are endemic to the Hawaiian Islands. The **crenate stenogyne**

PLATE 57. 1—Pink broom; 2—Kilauea fan flower; 3—golden-leaved sophora; 4—wedge-shaped cranesbill.

PLATE 58. 1—Kokioid lebronnecia; 2—island rose mallow; 3—beautifully staminal xeronema; 4—ribbon wood.

(*S. crenata* A. Gray) is a herbaceous perennial with branched and twining tetraquetrous hairy stem. The leaves are crenate, petiolate. The flowers are on short pedicels with campanulate calyx and greenish-yellow corolla with uniform petals. It grows on rocks in a crater up to the elevation of 1800 m (Maui Island). A total of 37 taxa of stenogyne are included among the Rare and Endangered Species of Hawaii, among which are 4 varieties of crenate stenogyne; 17 species and varieties of this genus have already disappeared from the island flora.

FAMILY LILIACEAE

The arboreal lily of New Zealand—**kaspar club palm** (*Cordyline kaspar* W.R.B. Oliver)—an endemic plant of the Three Kings Islands, belongs to the family Liliaceae. This is a small tree up to 5–7 m tall, with a crown of leaves at the apex and racemose inflorescence of 12 yellow flowers. It grows on the islands whose flora is distinguished by a high degree of endemism. In 1946, goats were removed from the Great Island and young plants of club palm started appearing. It was fairly widespread by 1963, forming large concentration at isolated places. A few plants have been conserved on 2 other smaller islands. It grows in the coastal forests, having undergone repeated changes over the period of their inhabitation; the vegetation is being gradually restored there with the organization of a floral-faunal reserve on the islands. The kaspar club palm is cultivated in Auckland; a few young plants have been transferred for cultivation in gardens. The present condition of the species has led to its inclusion among the rare plants.

An unusually attractive plant at the time of flowering is a member of the New Zealand flora, *Xeronema callistemon* W.R.B. Oliver. Its flowering shoot, 120 cm long, first grows vertically, then bends above the middle. As a result the flowering bunch (up to 30 cm long) is horizontal. It consists of numerous erect stamens and pistils and very much resembles a bright red bract (Plate 58). Such an inflorescence is also found in another species of the same genus, *X. moorei* Brongn. et Gris., growing in New Caledonia. Both the plants are rich in nectar, and are pollinated by birds.

The area of distribution of this xeronema is confined to 2 groups of islands: Poor Knights and Hen and Chickens (these are rocky formations, consisting of crystalline rocks). Xeronema grows in open areas with rich and wet soils, sometimes found in shady places under the canopy of ironwood and tea tree. At present, both groups of islands are included in the floral-faunal reserves and entry to them is restricted. Xeronema is cultivated in the gardens of New Zealand, propagated by seeds and rhizomes. The present status of this xeronema and the measures taken for its conservation make it possible to include it under the category of rare species. It is included in the Red Data Book of IUCN.

FAMILY LORANTHACEAE

The family Loranthaceae includes the only member of the monotypic genus *Trilepidia* of New Zealand, **Adam's trilepidia** [*Trilepidia (Elytranthe)* adamsii (Cheeseman) van Tieghem]. It is a small plant, up to 1 m tall, with green, thick, rhombic leaves and inflorescences of 2–4 reddish flowers arising from leaf axil. It parasitizes on woody coprosma and Australian myrsine, penetrating the wood and feeding with the help of haustoria. According to the published reports and herbarium specimens, Adam's trilepidia grows in northern Auckland (Waipawa and Kaipara) and southern Auckland (the Great Barrier Island, Waiheke Island, Cape Colville, Hunua and Maungakawa peninsula) in New Zealand. It was last seen under natural conditions in 1954, and now this plant does not exist even under cultivation. There is only the hope that this trilepidia may survive on the Coromandel peninsula or among the fragments of forest marshes in northern Auckland. This plant has disappeared mainly due to forest felling and subsequent reclamation of the territory, and to a lesser extent because of collection and damage caused by the Australian opossum. It has been included in the Red Data Book of New Zealand.

FAMILY MALVACEAE

The family Malvaceae includes the **Hawaiian cotton** (*Gossypium tomentosum* Nutt.), a 80–140 cm tall shrub with small grayish pubescence. The leaves are densely pubescent, glandular-dotted; petiole is 2–4 cm and lamina 3–6 cm long. Flowers are on long pedicels in the axils of terminal leaves; calyx is purple to chocolate-brown; corolla is sulfur-yellow. The fruits are leathery and ovoid. This endemic of the Hawaiian Islands grows in arid parts of stony or clay plains of the large islands: Oahu, Molokai, Maui, and Hawaii. Its present population strength is small and it has been included in the category of endangered plants of the Hawaiian Archipelago.

Among the endemic plants of the Hawaiian Islands, the genus *Hibiscadelphus*, including 5 species, is endangered. The area of distribution of the genus includes the Hawaii, Kauai and Maui islands. Its evolution is an example of close relationship between the endemic plants and animals. While the flowers of most malvaceous plants are fully open, the corolla of hibiscadelphus forms a tube, which is adapted for pollination by birds (belonging to the endemic group Drepanididae, feeding upon nectar).

Giffard's hibiscadelphus (*H. giffardianus* Rock) is a tree up to 7 m tall with highly branched, roundish crown. The leaves are cordate, 9–15 cm broad; flowers are 5–7 cm long forming a tube with a wide mouth. The species has been described from the lone specimen from the eastern slope of Mauna Loa, a few miles from the Kilauea volcano. This tree died in 1940. By that time the plant was brought under cultivation and subsequently reintroduced in its earlier habitats. By 1968, there were 10 adult and a few young trees. The most probable cause of the rarity of the species in the past was volcanic eruption. At present, the habitat of the Giffard hibiscadelphus is included in the Volcanoes National Park on Hawaii Island. It is being cultivated successfully in the Waimea arboretum (freshly collected seeds germinate easily).

Spreading hibiscadelphus (*H. distans* L.E. Bishop et Herbst) was discovered only recently on Kauai island in the valley of the Waianae River, where it grew on dry areas. Attempts to introduce it in cultivation were successful: the plants raised from seeds flowered in the third year. They are not distinguished particularly by their beauty, but are of special interest (their flowers are greenish and young leaves silvery). The species have been maintained in the arboretum together with another species: the **Hualalai hibiscadelphus** (*H. hualalaiensis* Rock).

Wilderian hibiscadelphus (*H. wilderianus* Rock) has become extinct in nature as well as in cultivation. Its only specimen was found in 1910 on the southern slope of the Haleakala Mountain on Maui island, in dry forests, on Kapukas (islands isolated from lava). In 1912, it was found already dried and did not survive. Wilderian and Giffard's hibiscadelphus are included in the Red Data Book of IUCN.

Island rose mallow (*Hibiscus insularis* Endl.) is a dense shrub up to 2.5 m tall. Its flowers are solitary, up to 8 cm in diameter, pale lemon-yellow with purple veins; corolla becomes purple when in full bloom (Plate 58). From the time of the discovery of the species to the 1960s only 4 plants have survived. The habitats of this plant are confined to the shrubby thickets on volcanic fragment. In 1967, when the island rose mallow was last seen, it was growing in a community of paniculate hackberry, Peterson's sugarplum tree, apetalous nestegis, and varied-leaf araucaria. Even at that time it was proposed to establish reserves on the island for restoration of its flora and fauna. Since 1964, the island rose mallow is being cultivated successfully on Norfolk Island, in Australia and in the botanical garden in Honolulu (Hawaii island). The plant is included in the Red Data Book of IUCN.

Cook's kokia (*Kokia cookei* Degener), discovered at the end of the 19th century, is included among the extinct Hawaiian plants. It is a 3.5–4.5 m tall tree. Its leaves are thin, leathery, pale green with reddish spots, tri-quadrangular, 9–13 cm broad on long petioles, and hairy on the lower surface at the point of branching of the veins. Flowers are solitary in the axils of upper leaves on pedicels; calyx is thin, leathery, with dark spots. Corolla is orange-red with divergent petals up to 8 cm long and 6 cm broad; stamens are reddish orange; ovary is conical, 8 mm long. The fruits are ovoid-flattened, up to 3 cm across. Only 3 plants of Cook's kokia were found on Molokai island, on the basis of which the species was described. The trees gradually died, but 3 seedlings were obtained from natural seeds and planted on the island, of which only one survived till adult stage. It flowered and produced fruits every year, and this saved kokia from complete extinction. This plant flowers from February to June–July, flowers remaining fresh for many days; flowers and fruits are found together on the same plant. The leaves turn red by the end of the vegetative period. Dormancy begins after the leaves drop, but the fruits remain till the next season. It is proposed to sow the seeds in its natural habitats so as to restore its natural population.

Kauai kokia [*K. kauaiensis* (Rock) Degener et Duval] is an endangered plant of the Hawaiian flora. This is a tree up to 5–10 m tall, with thick branches. The leaves, at the end of branches, are large, dark green, with numerous black spots and yellow-chocolate-brown hairs at the end of petioles. The flowers are 22 cm in diameter, solitary, on long pedicels in leaf axils; pedicels are densely pubescent; corolla is brick-red. The Kauai kokia is endemic to Kauai island. Its population is extremely low: one tree grows in the Koaloha canyon and a small grove of 15 trees is maintained in the Paaika valley. The plant is included in the List of Rare and Endangered Species of the Hawaiian Islands.

Lebronnecia kokioides Fosberg is a recently discovered species. The only tree, on the basis of which the species was described, was found in 1966 on Tahuata Island (group of Marquesas Islands in the South Pacific). It was later established that the lebronnecia also grew on Mohotan Island, northwest of the Marquesas Islands. This plant is of great interest for botanists primarily because of its geographic isolation as well as relationship to the cotton genus (*Gossypium*); it has a great significance for understanding the evolution of this genus. Lebronnecia is also a part of the endemic flora of the Marquesas Islands, which comprises about 80 species of higher plants. It is assumed that 18 of them are endangered or have already become extinct. Such critical conditions of the flora result from several factors, including cattle grazing as well as forest fellings and use of the cleared area for agriculture. Therefore, lebronnecia has practically disappeared from Tahuata Island and has survived on Mohotan Island, which is not yet populated; the forest vegetation in its center remains practically free from external disturbance and is represented by the plantations of pisonia. Nevertheless, the threat persists even here: because of grazing by sheep, xerophytic shrubby thickets are already destroyed and their areas have been converted into stony deserts.

On Tahuata Island, lebronnecia grows on a dry hilly slope with remains of tree vegetation. On Mohotan Island, this plant was found in a dry forest and outside it at an altitude up to 200 m. Lebronnecia reaches its normal height of 10 m in the forests along ravines, flowers (it has white, wide open flowers with dark centers), produces fruits, and reproduces. The fruits of lebronnecia are capsules, as in cotton, 2–3 cm long, with 3 internal locules, each with a single seed and reddish chocolate-brown hairs up to 1 cm long (Plate 58). In a test on its fibers with sulfuric acid, this species showed greater resistance than cotton.

A natural reserve has been established on an area of 1554 ha on Mohotan Island since 1971 to conserve the natural habitats of lebronnecia and other species. The number of sheep grazing on the island is being constantly curtailed. Kokioid lebronnecia is cultivated in tropical botanical gardens on the Hawaiian Islands, Tahiti, and in the Kew Botanical Garden. It is included in the Red Data Book of IUCN.

The **Chatham** variety of **ribbon wood** (*Plagianthus betulinus* A. Cunn. var. *chatamica*) is distributed on Chatham Island (east of New Zealand). The young plants are shrubs, and adult ones are trees 12–20 m tall. Depending on the age, differences appear in the shape, size, and texture of leaves. The leaves of young plants are tender, 6–12 mm long, ovate-orbicular, unevenly serrate; in adult plants, hard, 2.5–5 cm long, ovate-lanceolate, pointed, dentate-serrate. The lamina is hairy on both sides, and the petiole is thin. It is a dioecious plant. The male flowers are whitish-yellow, arranged in dense panicles; female flowers are greenish, forming a lax inflorescence (Plate 58). It grows on volcanic soils in forests (the local population uses its bark for making mats, fishing nets, and other items). The plant is included in the list of protected species of New Zealand under the category of endangered plants.

FAMILY MARATTIACEAE

The family Marattiaceae includes tree ferns, occupying large areas on the earth during the Carboniferous and Permian periods of the Paleozoic Era. Fossil remains of their trunks have been found in the deposits of all the continents right up to 70° N latitude. At present, these plants have been conserved only in the tropical forests. The family includes 7 genera. But 60 species of the genus *Marattia*, growing in the tropics of both hemispheres, are among the largest ferns on earth. Their large size is not due to the stem (up to 1 m tall), but due to the leaves, up to 6 m long.

Willow-like marattia (*M. salicina* Smith.) is a fern with large, tuber-like, thick rhizomes and 3–4 m long leaves. The sporangia are borne along the leaf veins closer to the margins of the segments of repeatedly divided leaves. On maturation of the spores, longitudinal rupture of the fused sporangia occurs (along the veins) and the two walls separate.

The willow-like marattia grows on the small Lord Howe Island in the Pacific Ocean, in a dense tropical forest at the peak of the Lidgeberg and Gower Mountains with an altitude of 776 and 803 m, respectively. Besides this endemic fern species, 75 species and subspecies of flowering and spore-producing endemic plants have been found on the island. The habitats of marattia are confined to the wet areas under the canopy of rainforests. Only 3 habitats of this species are known, one of which was discovered in 1977. At present, the population strength of the species is highly reduced; only a few plants are known. The most important reasons for the catastrophic position of this primitive fern as well as other endemic plants of Lord Howe Island are the gradual reduction in the area of tropical forests, introduction of goats and pigs on the island, and ever-increasing tourist pressure. This species is included in the Red Data Book of IUCN as an endangered plant.

FAMILY MONIMIACEAE

Taxonomically, the family Monimiaceae is closest to the endemic Australian family Austrobaileyaceae.

The native mulberry or smooth holly genus (*Hedycarya*) was discovered towards the end of the 18th century. It includes 37 species, distributed in eastern Australia, New Zealand, Fiji, and New Caledonia. On New Caledonia alone, 10 endemic species are found, the most rare among them is **riverside smooth holly** (*H. rivularis* Guillaumin). This is a tree up to 2 m tall, with opposite, lanceolate, aromatic leaves. The inflorescences have 3–5 flowers with a common thalamus, in which the mature seeds are submerged. The sweet holly grows in humid gallery forests in the central part of the island, confined to riverbanks at small heights. A few habitats with a small number of plants are known. The present status of this genus and its conservation measures are unknown.

FAMILY MYRSINACEAE

The family Myrsinaceae is distributed in the tropical and subtropical regions of both hemispheres. Most species are evergreen shrubs or small trees with alternate or whorled leathery leaves, with glandular dots. The small flowers are white, pink, rarely red, yellow, or green, arranged in panicles or racemes. The family includes about 1000 species distributed over 35 genera.

The distribution of the monotypic genus *Elingamita* is limited to the west island of the Three Kings Islands. **Johnson's elingamita** (*E. johnsonii* C.T.S. Baylis) is a small tree with leathery, glandular leaves, 18 cm long and 9 cm broad, and racemose inflorescences of yellowish flowers. The 12 surviving plants of elingamita form a small population on the rocky slopes facing the sea. A few plants have been conserved under cultivation; their seeds germinate easily, but the young plants are very sensitive to even mild frosts. Elingamita is being conserved in the Three Kings National Reserves and included in the Red Data Book of New Zealand.

FAMILY MYRTACEAE

The wild relatives of eucalyptus of the ironwood genus (*Metrosideros*) are found in the tropics and subtropics. They are about 20 species, half of which grow in New Zealand, one in South Africa and others in the Pacific Islands. The plants of this genus are called 'Christmas trees' in New Zealand, because they are in full bloom before the new year, covered with a large number of bright flowers. These plants are also famous for their roots. They have the usual long, horizontally spreading roots, which twine around stones fixed in the cracks and crevices of rocks, spread on coastal sands, and even stretch into the swirling wave front. But they also produce aerial roots whose functions are still not clear. Sometimes one or another tree produces a large number of fibrous roots. They droop from the branches, without reaching the ground, and surround the trunk in the form of long plaits. Some specialists believed that formation of such roots is a symptom of old age of the tree, others are of the opinion that these roots absorb moisture from the air if the underground roots are clogged by the salts from sea water.

Among the endangered plants of New Zealand is the **carmine ironwood** (*M. carminea* W.R.B. Oliv.), a creeper twining along the trunk of Kauri dammar pine, with short-petiolate, leathery leaves and numerous carmine red flowers in racemes. It grows in coniferous forests of Kauri in the lowlands. A few habitats are known in northern and southern Auckland and on many other islands, but the total population is not very large. The main cause of the disappearance of the plant is forest felling. The carmine ironwood is cultivated in the botanical gardens and included in the Red Data Book of New Zealand.

FAMILY NYCTAGINACEAE

The **bird-catching plant** [*Pisonia brunoniana* Endl. (*Heimerliodendron brunonianum* (Endl.) Scottsb.)], a 3.5–6 m tall tree, is a rare species of the New Zealand flora. Its leaves are opposite or whorled, 10–30 cm long, ovate. The flowers are small, green or reddish, hairy, in umbellate inflorescences (Plate 59). The fruit wall is sticky and is attached easily to the wings of birds, which disseminate them. Flowering in the bird-catching plant can be observed round the year. The plant has been noticed on the eastern coasts of North Island, in the coastal zone near Ngunguru (Auckland province). The species is included in the List of Endangered Plants of New Zealand.

FAMILY ONAGRACEAE

The area of distribution of the *Fuchsia* genus is in South America, mainly in the Andes. A few of its endemic species grow in New Zealand. The **basket fuchsia** (*F. procumbens* R. Cunn.) is most endangered: it is disappearing from the natural habitats. This plant has almost prostrate, 15–45 cm long stem with small, cordate, petiolate leaves. Its beautiful yellow flowers are in sharp contrast with the pure dove-blue color of the pollen. The fruits are sticky (Plate 59). The plant flowers from November to February. It grows in moist forests along the eastern coast of North Island. The disappearance of the species in nature is related to change in its habitats: aridification, felling of forests, and collection of the plants for private collections. The basket fuchsia is fairly well known as a cultivated plant and is in great demand under the name 'Kirk's fuchsia'.

FAMILY ORCHIDACEAE

The **Australian yoania** (*Yoania australis* Hatch.), which spends a large part of its life under the soil as a branching whitish rhizome with a diameter of 4 mm, is a saprophytic orchid. The dark aerial shoots developing from the dormant buds are covered with almost colorless, 1 cm long, scaly leaves. The stem terminates in an inflorescence of 1–5 flowers, which is chocolate-brown to pink in the lower part and white above. It is not yet known how this plant is pollinated. On the one hand, it is capable of self-pollination, and on the other the lip of its flower is so close to the upper petal that it almost forms a tube which can attract insects. The plant does not have chlorophyll and its viability depends on the availability of nutrition from the humus directly or with the help of mycorrhiza (*Lycoperdon perlatum*). The Australian yoania grows on highly humus-rich soils in close middle-age plantations of *Beilschmiedia tarairi* from the family Lauraceae, which is an endemic tree of North Island (New Zealand). The orchid as well as *B. tarairi* form mycorrhiza with the same fungus. The orchid needs 2 components (the tree and fungus), but yoania is not always found where *B. tarairi* grows. Australian yoania is distributed at several places on North Island. Three populations are known in the Atuanui forests: the largest comprises 20 plants over an area of 5 ha. This area is located along both sides of a small stream. However, precise information on the yoania population is not available, as this plant is subterranean and does not always form aerial shoots. The species is also disappearing from natural habitats as a result of forest felling with frequent reclamation of the vacant lands for agriculture. There is intensive drying of branches and drying of the *B. tarairi* trees as a result of a series of drought years, and the number of flowering plants of yoania in Atuanui has decreased. At present, the habitats of yoania concentration are being conserved in the Waipua forest reserve, in Atuanui and in the Centennial and Kirks-Bash parks. Observations are being recorded on its condition. It is difficult to cultivate yoania because of its unique requirements of nitrogen nutrition. It has been included in the Red Data Book of IUCN and New Zealand.

Small caleana (*Caleana minor* R. Br.) is a delicate plant, 5–20 cm tall. Its leaves are very narrow, linear, and shorter than the stems. The inflorescence is racemose with a small number of flowers, sometimes a solitary flower. The flowers are greenish, lip is bright red. The plant grows in open glades, in shrubby communities of leptospermum at the northwestern tip of North Island (Kaitaia, Rotorua, Waiotapu). Recently, the caleana population has been reduced drastically as a result of construction of the Rotorua Aerodrome. The plant is included in the category of endangered species.

Chiloglottis formicifera Fitz. belongs to the category of plants which have, most probably, disappeared from the New Zealand flora. Its only habitat was found in Chisman in 1901. It was located near Kaitaia on North Island. Since then, this orchid has been preserved only in herbaria. The species is included in the List of Rare and Endangered Plants of New Zealand.

FAMILY PALMAE

The flora of New Caledonia is characterized by a large variety of palms and a high degree of their endemism. Thirty-one species belong to 17 genera have been listed on its islands, including the Loyalty Islands; one species has already become extinct, 4 are endangered, the population of 3 species is declining, and 18 species have become rare. A large number of rare palm species is concentrated on Mont Panier, which has been declared a botanical reserve.

The palm tree *Burretiokentia hapala* H.E. Moore is endangered in New Caledonia. This plant was discovered only in 1964, and only 2 of its habitats are known at present. One is located in the Parari Pass Mountains at an altitude of 300 m, and another in Palm valley near the Newe River, 30 km west of the first habitat, at an altitude of 100 m. In both these areas, *B. hapala* together with other palm species is included in the formation of a gallery forest, which is characterized by smooth-trunk trees and poor development of the lower tiers.

The delicate *B. hapala* grows up to a height of 12 m. Its crown consists of 10 pinnate leaves, which are bright green on the upper side, and pale and covered with scales on the lower surface. The inflorescences are axillary. It is planned to organize a reserve combining the Parari Pass and a part of the Pani Iguambi Range with Mont Panier to conserve this palm as well as other rare species of the family. At present, no measures are being taken to protect the habitats of *B. hapala*. Both of its habitats are frequently visited by tourists, which is endangering the very survival of this species.

One more species of New Caledonia—*Cyphophoenix nucele* H.E. Moore—is endangered on the Loyalty Islands. This palm was discovered in 1925 and included under the genus *Microkentia*. In 1976 it was described as a new species of the endemic genus *Cyphophoenix*. This palm grows up to a height of 12 m, its 8 pinnate leaves forming a crown, under which long large inflorescences, up to 60 cm long with a span of 90 cm. The palm produces abundant fruits. Like the coconut palm, it grows on coral formations in wet forests at an altitude up to 60 m. At present, only one of its habitats has remained in the central part of the northeastern Lifu coast. It is believed that a small natural reserve must be created to save this species. It can also serve as a source of seeds for cultivating the palm in botanical gardens. Cyphophoenix has been introduced into southern Florida (USA). Both the species described above are included in the Red Data Book of IUCN.

The **elegant cyphophoenix** [*C. elegans* (Brongn. et Gris.) H. Wendl. ex Salomon], a member of the endemic genus of New Caledonia, also belongs to the category of endangered plants. It grows in the northeastern part of the Grand Terre region and forms a small population, which is endangered by forest fires.

The flora of the Fiji Islands, like that of other islands of Oceania, is rich in endemic species. They include a member of the monotypic genus from family Palmae, **Storck's neoveitchia** [*Neoveitchia storckii* (H. Wendl.) Becc.]. At present, one population of this species has survived, which has 200–300 adult trees over an area of less than 2 ha. Three young plants up to 1 m height were found in this region. It appeared that with such an abundant undergrowth there should be no cause for concern about the fate of this species. Unfortunately, this was not so, as surrounding banana plantations are spreading as a result of the clearing of the neighboring territories. As late as in 1971, neoveitchia was seen at 12 km from the protected area. Now it does not exist there. The introduction of rhinoceros (*Oryctes rhinoceros*) on the Fiji Islands is also endangering the survival of this species. It has been noted that rhinoceros has already caused severe damage to 2 adult trees in 1972–1973. The last habitat of neoveitchia is located in the southeastern part of Viti Levu Island, where the palm was initially widespread along the banks of the Reva River. Now only secondary forests including neoveitchia in the upper tier and undergrowth are growing here.

This palm grows up to a height of 10 m, its crown being formed by 14 variegated leaves up to 4.5 m long. The inflorescence is on short peduncles. The fruits are yellowish or reddish-yellow, edible. No measures are being taken for protecting the natural habitat of this palm, although the area definitely requires immediate protection from exploitation. It is also proposed to establish a seed bank of this palm. In 1972–1973, 4 young plants of neoveitchia were transplanted in the botanical garden of Suva University (Viti Levu Island), where they have survived well. One specimen is growing in Florida (USA) in the Fairchild tropical garden. Storck's neoveitchia is included in the Red Data Book of IUCN.

Henry's pelagodoxa (*Pelagodoxa henryana* Becc) is an endemic species of Marquesas Island (southern Oceania). In the Red Data Book of IUCN, it is included under the category of endangered plants. Till 1970, only one of its populations was known over an area of 0.5 ha with 30 plants, 2 adult, and others young. The natural origin of this population was doubtful, as all the trees were located close to old constructions and could be remnants of cultivated plants. The area is located in the center of Nukuhiva Island, the largest island of the Archipelago. A few more trees were found in 1978 in the northeast of Nukuhiva, but this discovery is yet to be confirmed. There are also reports about the growth of pelagodoxa on the Hivaoa, Tahuata, and Austral islands. Most probably, these are introduced plants. The palms on the southern coast of San Cristobal (Solomon Islands) developed from the fruits thrown on the coast.

Henry's pelegodoxa is a very beautiful palm with an erect stem up to 8 m tall and large, 2 m long, entire leaves, which become fan-shaped with age. The leaves are bright green on the upper surface and silvery gray on the lower surface. The fruits are up to 15 cm in diameter, resembling coconut, and are edible when unripe (aqueous extracts of the endosperm are used as medicine). This pelagodoxa grows in humid tropical forests together with *Inocarpus edulus* and *Hibiscus tiliaceus* on a slightly slanted area at an altitude of 40 m, 500 m above the waterfalls. The main reason for the destruction of natural vegetation is sheep and cattle grazing. Unfortunately, the area with this palm is not being protected. Henry's pelagodoxa is being cultivated in several botanical gardens, including the garden on Papeari (Tahiti) and Bogor (Lava).

Only one palm genus, *Pritchardia*, grows on the Hawaiian Islands. It includes 33 species, of which 11 are endangered and one, **large-fruited pritchardia** (*P. macrocarpa* Linden), has most probably already become extinct. At present, the only plant of this species is being conserved in the Honolulu Botanical Garden (Hawaii island). In spite of the healthy condition of the palm, there is no certainty that it will produce mature seeds.

The height of this palm is 3 m, the petioles and laminas are 1 m each, leaf is fan-shaped up to one-half or one-third. The inflorescences are weakly branched, 1 m long. It is known that the palm grew in the Nuuanu Valley near Honolulu, and on Oahu island. Two of its habitats were known even in 1888: in the upper part of the valley and on a suspended rock; now they do not exist. The valley region is characterized by a high rainfall and frequent strong winds. The most probable cause for the disappearance of large-fruited pritchardia is the collection of the plant for commercial trade. It is still not clear why only one plant has survived under cultivation (the seeds were once sent to Europe).

Munro's pritchardia (*P. munroi* Rock) is among the endangered plants. It was discovered in 1920 on Molokai island in a dry valley at an altitude of 550 m in a community of shrubby vegetation together with black persimmon, narrow-winged and sticky dodoneas, and golden pleomela, one or two plants of pritchardia have survived in the same area till now. These are short, 3–5 m tall plants with numerous deeply divided small leaves and short paniculate inflorescences. For many years, neither falling fruits nor young plants have been seen in the area, only a few seeds could be obtained in 1975. Seedlings were obtained under cultivation. It is assumed that most seeds are eaten by rats. The seedling and young plants are destroyed here by goat grazing. At present, it is planned to reintroduce young saplings of the cultivated plants and construct fences to protect them from goats in future. Plants of Munro's pritchardia are present in the Waimea Arboretum on the Kaala Mountains on Hawaii island since 1975. Also growing here is the kaala pritchardia, which was found on the Kaala Mountains (Oahu), where it forms a small plantation. It is proposed to establish a collection of all the pritchardias of the Hawaiian Islands. Two pritchardia species—large-fruited and Munro's pritchardia—are included in the Red Data Book of IUCN.

FAMILY PASSIFLORACEAE

Herbert's passion flower (*Passiflora herbertiana* Ker-Gawl. ssp. *insulae-howei* P.S. Green), subspecies of Lord Howe Island is one of the 75 endemic species and subspecies of the island. It is a tendril climber. Its leaves are 5–8 cm long and 5–8 cm broad with 3 lobes and glandular petioles. The flowers are solitary, cream-yellow, pale pink or red on the inner side. It grows in the lower tier of rainforests, where *Drypetes australasica* and *Cryptocarya triplinervis* or *Cleistocalyx fullageri* and *Linociera quadristaminea* are predominant. Initially, when the rainforests were widely distributed on the island and vegetation was not affected by cattle grazing and urbanization, about 10 habitats of passion flower were known. Now only 2 remain, and their condition is critical. In one area, in the northeast of the island, cattle grazing is prevalent and it is proposed to construct houses, the other habitat is far away from the former and is also influenced by cattle grazing; besides, reconstruction of a road is planned here, which will undoubtedly destroy its habitat. In order to save the last habitat of Herbert's passion flower, it is proposed to organize reserves in both areas, re-examine the house construction plan, impose strict control on cattle grazing, and restrict tourism. The passion flower subspecies of Lord Howe Island is included in the Red Data Book of IUCN as an endangered plant.

FAMILY PIPERACEAE

The family Piperaceae includes 8–10 genera and about 3000 species. Mainly 3 genera are represented in the flora of Oceania: pepper elder (*Peperomia*), pepper (*Piper*) and pittosporum* (*Pittosporum*). The pepper elders are succulent terrestrials or epiphytes, glabrous or pubescent, erect or prostrate herbs. Their flowers are usually arranged in racemose inflorescences; buds, flowers, and fruits can be found simultaneously in a single bunch.

Short-branched pepper elder (*P. breviramula* DC.) is an epiphytic herbaceous perennial with creeping, rooting, and strongly hairy stems. The leaves are alternate, 1–2 cm long, 1.6 cm broad, ovate-elliptical, light-colored on the lower surface due to pubescence. The inflorescence is 1–1.8 cm long. This endemic plant of Ponape, one of the Caroline Islands, is known only from the type habitat—Nipit Mountain, at an altitude of 350 m in hilly forests. **Kremer's pepper elder** (*P. kraemeri* DC.) is morphologically similar, but it is completely glabrous with the exception of fine pubescence along the leaf margins. This is endemic to the volcanic

---

*Pittosporum is now separated under a separate family, Pittosporaceae.

Palau Islands, where it grows on trees in forests. The **Kusaie pepper elder** (*P. kusaiensis* Hosokawa) is a succulent epiphyte with succulent creeping stem, covered with long hairs. The leaves are broadly elliptical, puberulent, on short petioles; inflorescence is 1–2 cm long. This endemic plant of Kusaie (Caroline Islands) has been found on moss-covered tree trunks in hilly forests in the zone of permanent fog.

The pepper genus (*Piper*) comprises shrubs growing under forest canopy and climbers twining around trunks or creeping on the ground. Their flowers are arranged in racemes or spicate inflorescences. Some species have commercial importance and are cultivated. The highly endemic plants are included under rare species.

Guaham pepper (*P. guahamensis* DC.) is a shrub up to 3 m tall with ovate-cordate leaves on long petioles. The flowers are dioecious, borne in spicate inflorescences. This species is endemic to the Marianas (Micronesia), where it grows under the canopy of wet forests. Three forms of this species are distinguished on the basis of pubescence on the lower surface of leaves: one with glabrous leaves, and the other pubescent. The roots of this plant are used for medicinal purposes.

Hosokawa pepper (*P. hosokawae* Fosberg) is a creeper with ovate-cordate, petiolate leaves. The inflorescences are longer than the leaves. It is endemic to the volcanic Palau Islands, growing in forests, sometimes forming small clusters, but found rarely.

Dall's white-wood (*Pittosporum dallii* Cheeseman) is an endemic plant of South Island in New Zealand. This very rare plant forms a few populations with a total strength of 12 specimens. They are confined to the northwestern end of the island. Dall's white-wood is a 4–6 m tall tree with spreading branches; the leaves are closely placed at the apices of twigs; white flowers form dense umbellate inflorescences (Plate 59). The plant grows in forests on rocky slopes, in the forests among large granite boulders, and in communities of southern beech at an altitude of 600–1000 m; the young trees prefer open areas. Some of the habitats in the Tasman Mountains are in the forest part, while others are located in various protected territories. A sharp reduction in the population of this species in the past was caused by destruction of its habitats in the course of mining. This threat persists even now, and may destroy several populations in future. Construction of a road in 1950 caused serious damage to this species. At present, attempts are being made to reduce mining of minerals so as to conserve the last habitats of white-wood. The plant is cultivated in New Zealand and in England, and included in the National Red Data Book together with 4 other species of the same genus.

FAMILY PODOCARPACEAE

The genus *Acmopyle* includes 2 species, which are characterized by the presence of dimorphic leaves: foliage and scaly. Study of the anatomical structure of the foliage leaves of acmopyle, which do not differ in external morphology from the usual linear or lanceolate leaves, revealed an unusual internal distribution of vessels of the midrib: the vascular bundles are directed towards leaf margins horizontally and not vertically, as normally happens, as a result of flattening of the leaves on the sides. These changes occurred in ancient times, when conifers were predominant on the earth surface. Such a structure of leaves has also been noted in 2 *Falcatifolium* species and a few *Podocarpus* species.

Pancher's acmopyle (*A. pancheri* Pilg.) is a tree up to 25 m tall, growing in humid mixed hilly forests at an altitude up to 1000–1200 m in New Caledonia. **Sahni's acmopyle** (*A. sahniana* Buchholz et N.E. Gray) is a tree not taller than 5 m. It grows under conditions similar to those of the previous species: in humid forests on the Viti Levu Islands (Fiji). Such a limited distribution of the acmopyle species and their adaptation to the island floras, which have undergone maximum changes, undoubtedly demand the most stringent check on their status.

The genus *Falcatifolium* includes 4 species. Its name indicates the peculiarities of its leaf structure, which are sickle-shaped, bent at the base.

Yew-like falcatifolium [*F. taxoides* (Brongn. et Gris) de Laubenf.] is an endemic of New Caledonia. It is a shrub or 2–15 m tall tree with conical crown. The branches are spreading; young branches are purple. It grows in the hilly coniferous forests at an altitude of 200–900 (sometimes up to 1400) m together with other members of Podocarpaceae: *Dacrydium* and Pancher's acmopyle. It is interesting not only as an endemic species of the island flora, but also as a host plant for a more rare species, *Parasitaxus ustus*.

The monotypic genus *Parasitaxus* of the family Podocarpaceae is the smallest genus. *Parasitaxus ustus* was discovered in the middle of the 19th century when colonization of the Pacific islands began. This parasitaxus was found on New Caledonia, but subsequently was considered extinct for a long time, as the local population, attributing some miraculous properties to it, protected it from outsiders. Only during the 1960s was this rare plant rediscovered in the forests and studied. Parasitaxus is a branched monoecious shrub, 25 cm tall, sometimes growing up to 1.5 m, with variable colors: reddish, honey-red, purple, rusty. The leaves are 1–2 mm long and 1.5 mm broad, scaly, and somewhat fleshy. The seeds are small, spherical. Because of the unusual color, this plant received the species epithet *ustus,* meaning scorched.

This is the only true parasite among conifers: it grows on the roots and trunks of another member of Podocarpaceae, the yew-like falcatifolium. The roots of the parasite first penetrate the bark of the host trunk and are then located between the bark and the wood, spreading upward and not downward along the trunk over a considerable height. Parasitaxus is mostly found in dense shady forests on the Black Mountains and mountain slopes at an altitude of 500–800 m. The present status of the species is unknown.

FAMILY RANUNCULACEAE

Six crowsfoot species, including the **pauciflorous crowsfoot** (*Ranunculus pauciflorus* T. Kirk.), are among the endangered plants of New Zealand. Pauciflorous crowsfoot is a small perennial herb with a basal rosette of leaves and bright yellow pentamerous flowers on short stems. Forty years ago, this rare crowsfoot species was on the verge of extinction. There were only 14 plants in the only habitat of Canterbury (South Island). Measures for its conservation were taken up immediately: the area was fenced to prevent trampling and collection, cattle grazing was prohibited, growth of weeds was checked, and limestone mining and construction of houses near that area were prohibited. These measures as well as artificial sowing of its seeds and constant check on its conditions not only helped in saving the plant from extinction but also have increased its population to a few hundred plants.

Pauciflorous crowsfoot grows on the limestone rocks at an altitude of 600 m, and reproduces well if not suppressed by other plants. The habitat of this endemic crowsfoot is characterized by the wealth and variability of flora. A few more endemic plants have been found here, including Brock's bluebell *Wahlenbergia brockiei* J.A. Hay (family Campanulaceae, Plate 56), *Myosotis colensoi* (T. Kirk) J.F. Macbride, and *Myosotis traversei* Hook. f. var. *cinerascens* (Petrie) L.B. Moore (family Boraginaceae). All these species and varieties are included in the List of Endangered Plants of New Zealand. The habitats of crowsfoot are under protection since 1954.

The status of **godley crowsfoot** (*R. godleyanus* Hook. f.) remains insufficiently known. A large part of its habitat is located in the center of the Southern Alp Mountains (South Island, New Zealand) from Mount Hunt in the south to the Sily Range. The total length of the habitat is 160 km. Crowsfoot grows mainly on the moraine areas or segmented placers at an altitude of 1200–2100 m at the places where water flows out from the melting glaciers. It is found quite commonly but all its habitats have only a few plants. The plant suffers from grazing by domestic as well as wild, mainly introduced, animals and also from mass collections of fresh

PLATE 59. 1—Basket fuchsia; 2—bird-catching plant; 3—Dall's white-wood; 4—godley crowsfoot.

PLATE 60. 1—Woolly dodonea; 2—ornamental hebe; 3—Fijian degeneria; 4—giant dorstenia.

PLATE 61. 1—Pine-shaped viper's bugloss; 2—Canary strawberry tree; 3—royal poinciana; 4—Mauritian lily.

PLATE 62. 1—Cotyledonous stork's bill; 2—Berthelot's bird's-foot trefoil; 3—dragon tree; 4—opposite-leaved medusa tree.

specimens for sale or for private collections. Most populations that have been conserved do not exceed 50 plants, and the largest occupy an area of 0.5 ha. At present, this crowsfoot species is being conserved in the Westland, Mount Cook, and Arthur Pass National Parks as well as on the lands controlled by the forest department. Because of its large and attractive golden-yellow flowers, it is often cultivated in the gardens on alpine mountains (Plate 59).

Both pauciflorous crowsfoot and godley crowsfoot are included in the National Red Data Book, and godley crowsfoot also in the Red Data Book of IUCN.

FAMILY RESTIONACEAE

The family Restionaceae includes the endemic monotypic genus *Sporodanthus*, which is distributed on North Island in New Zealand and on Chatham Island. **Traver's sporodanthus** [*S. traversii* (F. Muell.) F. Muell. ex Kirk] is a herb 1–2 (3) m tall with numerous shoots arising from the horizontal rhizome. The stem is 3–10 mm in diameter, very sparingly branched; leaves appressed to the stem. The inflorescences are paniculate, unisexual; flowers are yellowish-brown, trimerous; fruit is an achene. It grows on marshes and marshy lands. Initially, it formed dense thickets in the environs of Kaitaia, but became extinct after reclamation of the area. At present, isolated habitats of this plant, which may become extinct very fast without special conservation measures, are known as the Hauraki and Waikato plains. Sporodanthus is included in the List of Rare and Endangered Plants of New Zealand.

FAMILY RUBIACEAE

Trees and shrubs of the family Rubiaceae are found extremely rarely in the northern hemisphere, but they are common in the southern hemisphere. The Tasmanian native currant genus (*Coprosma*), comprising evergreen trees and shrubs with unisexual flowers on different trees, is characterized by great species variability. The members of the genus are widespread in New Zealand and on the islands of the Pacific Ocean, as well as in Australia.

Pointed-leaf native currant (*C. acutifolia* A. Cunn.) is an endangered plant of New Zealand. This is a small, entirely glabrous tree with slender branches and pale bark. The leaves are long, broad, pointed, on thin petioles; stipules are broad and deciduous. Inflorescence is branched, with 3 flowers, at the end of branches. The plant is dioecious. The native currant grows on the Kermadec Islands from the coast to the mountain peaks. It is included in the List of Rare and Endangered Plants of New Zealand.

The tall Rubiaceae plants are found in the genus *Mastixiodendron*. The **robust mastixiodendron** (*M. robustum* A.C. Sm.) is a tree up to 25 m tall and 80 cm in diameter. Its crown is bent and roundish; bark is naked and cracked; wood is light-colored. The leaves are petiolate, leathery, elliptical, 9–17 cm long and 3.5–8 cm broad, often recurved along the margins. Inflorescence is strong, bearing 7 to 50 yellowish-white flowers. This is an endemic plant of Fiji Levu and Vanua Levu of the Fiji Islands. It grows from the coast to an altitude of 120 m. Since ancient times, this tree has been a source of wood, which has reduced its population to a great extent.

Another endemic member of the genus found on these same islands is the **hairy mastixiodendron** (*M. pilosum* A.C. Sm.), whose height does not exceed 10 m. It differs from the previous species by the pubescence on all plant parts, thicker leaves, and multiflorate inflorescence (with 25–55 flowers). The flowers and fruits form from September to January. Hairy mastixiodendron grows in dense forests up to an altitude of 250 m. It has a very hard wood, which is used for construction and other purposes.

FAMILY SANTALACEAE

The cherry tree genus (*Exocarpos*) of the family Santalaceae includes about 20 species, which are mainly distributed in Australia. These are trees and shrubs with opposite or alternate, sometimes scaly leaves and spicate or racemose inflorescences. An endemic species of the Hawaii is the **yellow cherry tree** (*Exocarpos luteolus* C.N. Forbes), known from one habitat on Kauai Island. This is a shrub, 60–180 cm tall, with alternate, leathery, yellowish-green, 2.5–5.5 cm long and 1.5–2 cm broad leaves which are pointed towards the base. Flowers in groups of 4–9 form a loose spike. The cherry tree grows in humid places along the edges of the Wahiawa marsh on Kauai Island. It belongs to the category of endangered plants.

FAMILY SAPINDANACEAE

The family Sapindanaceae includes **woolly dodonea** (*Dodonea eriocarpa* Sm.), with a great variety of forms. At the sea coast, they are small shrubs with hairy leaves. At an altitude of 1500–1800 m, they are 6–8 m tall trees. At still higher altitudes the woolly dodonea is modified into a shrub. Its leaves are oblong, narrow. Flowers are in cymes (Plate 60). This plant is usually dioecious, but bisexual flowers, which are capable of forming fruits, sometimes appear on the male plants at the end of flowering season. Thus, some plants bear many fruits, while others have a few and small fruits. The fruits are red, yellowish-green or chocolate-brown to green with broad leaves. The woolly dodonea grows on Hawaii Island from the foothills to the mountain tops. Nine varieties of woolly dodonea are included in the List of Endangered Plants of the Hawaiian Islands; all of them are endangered.

FAMILY SCROPHULARIACEAE

The family Scrophulariaceae includes the very interesting and ornamental genus *Hebe*, which is close to genus *Veronica* and at the same time has several significant differences. The genus *Hebe* comprises about 100 species, which are mainly distributed in the temperate zone of the southern hemisphere. These are shrubs, rarely trees with oppositely arranged evergreen leaves. The flowers are dove-blue, purple, or white, borne in racemose inflorescences. Hebe grows from the sea coast to the alpine meadows, with the exception of dry open plains occupied by senior communities. The List of Rare and Endangered Plants of New Zealand includes 17 species, of which *Hebe breviracemosa* (W.R.B. Oliver) Cockayne et Allan has already become extinct. Not long ago, this plant was common on Raoul Island (Kermadec Islands), but it was not to be found there recently; it is also not known under cultivation. It is included in the Red Data Book of New Zealand.

Ornamental hebe (*H. speciosa* R. Cunn.) is a glabrous shrub with thick, angular branches. The leaves are sessile or on very short petioles, ovate-oblong, roundish at the apex, leathery, glossy. The racemose inflorescences are multiflorate, erect; corolla is red, 8 mm in diameter, anthers and pistils are long (Plate 60). The shrub flowers from February to October when the air is saturated with humidity from the sea. The earlier habitat of the ornamental hebe along the coast near Hokianga on North Island is at present destroyed; the plant has survived on the Middle Island in the outskirts of Ship Coup and Port Nicholson. It is cultivated in the botanical and private gardens.

Dieffenbachi's hebe (*H. dieffenbachii* Benth.) is a strong, glabrous shrub. Its leaves are sessile, 7.5 cm long, leathery, and pointed. The length of floral bunches is greater than that of the leaves, width 18 mm; peduncles are thin, and corolla 6 mm in diameter. The shrub grows on Chatham Island. It is included in the category of Rare Plants of New Zealand.

Cypress-like hebe (*H. cupressoides* Hook. f.) is a bent, many-branched shrub, up to 2 m tall. The branches are erect, equal in length, thin and glabrous. Leaves are small, fleshy, and smooth; flowers are very small, in groups of 3–4 at the ends of thin branches; bracts are larger than sepals; corolla is violet. The shrub grows at an altitude of 1200 m on the Middle Island. It is included in the category of endangered plants.

FAMILY STERCULIACEAE

The family Sterculiaceae comprises 660 species, distributed exclusively in the tropics. It includes the chocolate tree, from the seeds of which cocoa powder and cocoa oil are obtained, the cola tree with seeds rich in theine and theobromine, as well as the heritiera, and fremontodendron.

The genus *Heritiera* includes 4 species, growing along the Asian coast in Africa, in Asia, and on the Pacific islands. **Long-petiolate heritiera** (*H. longipetiolata* Kanehira) is a plant of Guam and the islands of the Mariana: Rota, Saipan, and Tinian. This is a rare tree up to 15 m tall with large, up to 30 cm long, white-pubescent leaves on the lower surface, on 4–6 cm long petioles; inflorescence is paniculate, of bisexual flowers. The species is included in the Red Data Book of IUCN, among the endangered species. Its population strength has come to a critical stage on Guam, which is its largest area of distribution. This plant grows on the standard limestone rocks and plateau, on areas always exposed to wind. The species is partly protected on Guam because of the limited access to the island territory. As proposed by the specialists, a more strict regime for conservation as well as reintroduction of this plant into the earlier habitats from the plantation established on Guam and in the Waimea arboretum (Hawaii Island) is necessary to restore its population.

FAMILY TAXACEAE

In the southern hemisphere, the family Taxaceae is represented by the genus *Austrotaxus*, with its only species, **spicate austrotaxus** (*A. spicata* R.H. Compton). This is an evergreen, up to 25 m tall tree with wrinkled bark and branched crown. The leaves are 10–15 cm long and 1 cm broad, with the margins recurved. The tree grows in wet mountain forests in the northern part of New Caledonia, mostly on slate and gneiss rocks at an altitude of 400–1600 m.

Austrotaxus was first described in 1922, but its taxonomic position still remains uncertain: it has fairly sharp characters of both Podocarpaceae and Taxaceae. That is why it was even proposed to separate austrotaxus under a different family. Spicate austrotaxus is not found under cultivation. All its natural habitats should be placed under protection.

FAMILY WINTERACEAE

Taxonomically, the family Winteraceae is closest to the family Magnoliaceae, from which all the flowering plants are believed to have originated. This exclusively tropical family is represented by trees and shrubs, which maintain primitive characters and the floral structure. The family comprises 7 genera and 190 species. The monotypic genus *Degeneria* is separated by some authors under the independent family Degeneraceae, and they consider it to be endemic to the Fiji floristic regions.

Fijian degeneria (*D. vitiensis* J.W. Bailey et A.C. Smith) is a 'living fossil', discovered in 1934 in Fiji. It was found by the American botanist A. Smith in deep forests of Vanua Levu Island. This plant did not grow beyond 15 m in height, had pinnately veined leathery leaves and strong, large fruits resembling cucumbers; each fruit had longitudinal furrows (Plate 60). The discoverer of the species could not determine its affinity with the existing species because, on the one hand, it did not resemble any of the known plants in the world, and on the other, the tree did not have flowers, without which it is impossible to determine the genus or family. In 1941, another botanist, O. Degner, found a similar tree on the neighboring Viti Levu Island. The branches of this tree were covered with flowers which were structurally very similar to the flowers of Magnoliaceae. The plant received the generic name *Degeneria* in honor of O. Degner. The species of degeneria are characterized by the presence of flat and broad stamens and carpels, which are only slightly fused on the lower side. Conservation of the plant of this genus has great scientific importance for studying the phylogeny of flowering plants.

The present status of the Fijian degeneria is not known. Possibly, it is being conserved on one of the reserve territories.

The species of the genus *Bubbia*—**long-leaved bubbia** (*B. longifolia* A.C. Smith)—has similar floral structure to that of degeneria, but its wood does not resemble that of conifers, which is a primitive character. In spite of its ancient origin, bubbia has maintained a fairly wide area of distribution, covering New Guinea, eastern Australia, New Caledonia, and Lord Howe Island, where it grows in mountain forests. The reduction in the area of these forests because of human activity also affects the status of their constituent species, including bubbia. The survival of this species depends on the conservation of the environment.

8

Islands of the Atlantic and Indian Oceans

314 The flora of most islands differs by its unique composition and a large percentage of the flora, as a rule, consists of endemics.

CANARY ISLANDS (ATLANTIC OCEAN)

Macronesia or the Lucky Islands, which received the first name because of their mildly warm climate, are located off the southwestern coasts of Europe and northwestern coasts of Africa. These islands form a separate floristic region; their flora is associated with the Mediterranean flora but each archipelago of this region has its own unique and specific property. The flora of the Canary Islands is distinguished by a high percentage of endemics, but many species have already become extinct under the influence of human activities (development of tourism, introduction of plants, forest fellings, and reclamation of land for agriculture). Out of the 95 flowering plant species endemic to Tenerife Island, 9 are endangered, 18 have been included under category of vulnerable species, and 35 have become rare. On Fuerteventura Island, 5 out of 17 endemics are endangered, 3 are vulnerable, and 6 are rare. Out of the 56 endemic species of the Grand Canary Island, 2 have already disappeared, 8 are endangered, 9 are vulnerable, and 21 are rare. Out of the 33 endemic species of Palm Island, 8 are endangered and 10 have become rare. Eleven plant species of the Canary Islands have been included in the Red Data Book of IUCN.

FAMILY ASTERACEAE

Juno's centaury (*Centaurea junoniana* Svent.) is an arboreal perennial plant, 30–100 cm tall with numerous shoots. The flowers are pale, pink-lilac. The only habitat of centaury is the extreme south of Palm Island, near the Tepegia vulcano, where a few hundred plants has survived in the crevices of isolated rocks surrounded by dry lava. The vulcano is 1 km from the plant population and, during its last eruption in 1971, spewed lava on the rocks. The species may completely become extinct if the rocks are flooded with lava during regular eruptions.

FAMILY BORAGINACEAE

The **pine-shaped viper's bugloss** (*Echium pininana* Webb. et Berthel.), a giant grass (monocarpic plant) living for a few years or more (Plate 61), grows in the laurel forests on basalt rocks. The plant is known from 3 habitats northeast of the city of Las Palmas (Grand Canary Island). In 1972, only 2 adult plants were found in one habitat and 35 plants and seedlings at a distance of 1 km from this place. Only a few plants are known from the second habitat, and the condition of the third population is unknown. The plant is highly ornamental and is being destroyed because of collection and transfer of many plants to gardens.

FAMILY CRASSULACEAE

Bolle's aichryson (*Aichryson bollei* Webb. ex C. Bolle), a 20–50 cm tall annual or biennial herb densely covered with glandular hairs, is a rare species in laurel forests at an altitude of 1600 m. The yellow flowers are arranged in corymbs. This species is endemic to the Canary Islands, where it grows in the center and northwest of Palm Island, Azores, and Madeira. It is found in dark, humid places, along the gorges, on volcanic soil. The **noble aeonium** [*Aeonium nobile* (Praeger) Praeger] is a herbaceous succulent (monocarpic) with strong, simple stems arising from a rosette of yellowish fleshy leaves with reddish tinge. The flowers are numerous, copper to pale red. The plant grows along the northern coast of Palm Island and on other Canary Islands, on steep desert cliffs facing south and on rocks at an altitude of 120–800 m. It was earlier found in abundance but has now become rare as a result of land clearing for agriculture, cattle grazing, and excessive collection for gardens.

FAMILY ERICACEAE

The endemic **Canary strawberry tree** (*Arbutus canariensis* Veillard ex Duhamel) is found in evergreen laurel forests. This is a tree up to 15 m tall with brown bark, pealing off in the form of flakes; leaves are oblong-lanceolate (Plate 61). On the Canary Islands, it is found very rarely. Only 5 habitats are known; it has almost disappeared, and possibly has become completely extinct, from Palm Island (one habitat), from Gomera (2 habitats), and from Grand Canary Island (one habitat with 3 plants). It is still found quite commonly on Hperfo Island. The species is cultivated in the botanical garden on Grand Canary Island.

FAMILY EUPHORBIACEAE

Handia spurge (*Euphorbia handiensis* Burchard) is a cactus-like succulent, 50–100 m tall, often highly branched. It grows at a small distance from the coast, in the lower part of the valley south of the Gandia peninsula (southwest of Fuerteventura), on sand or stone at an altitude of 150 m, forming unique colonies. At present, it is found in 3 valleys where dense thickets of spurge existed earlier, but now it is already rare and endangered.

Other endangered species, also grow here: *Pulicaria burchardii* Hutch. (only 4 plants known), *Argyranthemum winteri* (Svent.) Humphries, and *Echium handiense* Svent.

FAMILY FABACEAE

The almost extinct Berthelot's **bird's-foot trefoil** (*Lotus berthelotii* Masferrer) is a small, highly branched shrub on Tenerife Island on a cliff under the canopy of pine forest at an altitude of 700–1200 m. Its flowers are bright scarlet to dark red (Plate 62). Even in 1884, it was recorded as a very rare plant. Its collection for transfer to gardens was mainly responsible for almost complete depletion of its populations. At present, this plant is known only from 2 habitats in Tenerife: one near Orotawa and another on the south of the island.

FAMILY GLOBULARIACEAE

The endemic species **Askan's globe daisy** (*Globularia ascanii* D. Bramwell et Kunkel), a small procumbent shrub with woody stem and slender branches, is endangered. The inflorescence is dense, terminal, consisting of 4–6 almost globose heads; flowers are numerous, milky white with bluish-purple tips. Two small populations of this plant are known in the northwest of Grand Canary Island, in the Camadiba forests, on a steep basalt cliff at an altitude of 950–1200 m, where it grows in a Canary pine forest.

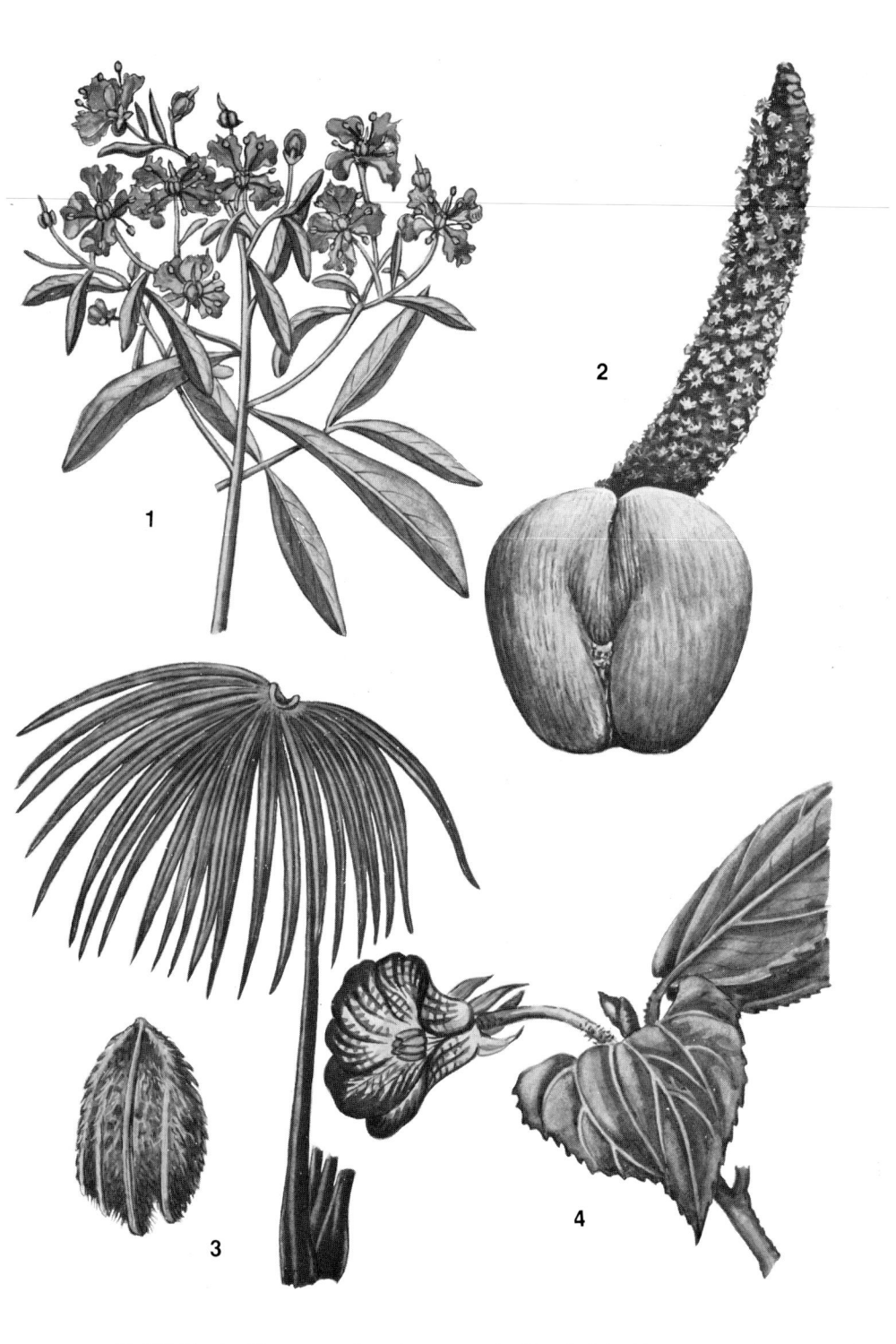

PLATE 63. 1—Pinnate rue; 2—twin coconut; 3—Verschaffelt's Bourbon palm; 4—St. Helena trochetia.

PLATE 64. 1—Lemon aerangis; 2—elephant faham; 3—Verschaffelt's palm; 4—sesquipedal faham.

FAMILY LILIACEAE

The **dragon tree** (*Dracaena draco* (L.) L.) is endangered. It grows to a height of 5–10 m forming umbellate crown and a strong silvery gray trunk which repeatedly branches after flowering and carries spreading clusters of branches. The leaves are hard, sword-like, glaucous, forming dense rosettes at the ends of branches. The inflorescence is in the form of dense terminal spikes (Plate 62). A dark red sticky sap, called dragon's blood, is discharged when the tree is cut. It is widely used for medicinal purposes and in magical rituals. The berries are edible and resemble sweet cherry in taste. The dragon tree grows on stony volcanic rocks at an altitude of 500 m. Its distribution is restricted to the Canary Islands—Gomera, Grand Canary, Palm, Tenerife—and the islands of Cape Green—Brava, Fogo, Santo Antano, Sao Nicolau, Sao Tiago (possibly, it has disappeared here), Sao Vicente (found in 1935, now on the verge of extinction), Madeira (a single population of a few plant)—and the Porto Santo Islands (the plant has already become extinct here).

At present, small scattered populations of the dragon tree are known, mainly on inaccessible precipices. The tree has poor regeneration ability, although it produces fruits abundantly. The seedlings appear in very small numbers and are either dug out by the collectors or eaten by domestic animals.

FAMILY PLUMBAGINACEAE

The only habitat of the **tree sea-lavender** (*Limonium arborescens* (Brouss.) Kuntze) is along the northern coast of Tenerife Island. This is a shrub up to 180 cm tall with poorly branched trunk in the upper part. Inflorescence is terminal and large. This giant coastal sea-lavender is exceptionally rare due to cattle grazing and its collection for ornamental gardening. For many years, it was considered to be extinct, but a small group of plants was subsequently found on Tenerife.

The flora of the coastal cliffs on the Canary Islands also includes many endemic species. Another sea-lavender species, *L. fruticans* (Webb) Kuntze, grows on a cliff in the Tepo region; species from other families are also endangered: *Centaurea canariensis* Willd., *Tolpis crassiuscula* Svent., *Vieraea laevigata* Webb et Berthel., and *Argyranthemum coronopifolium* (Willd.) Webb ex Schultz Bip.

FAMILY RUTACEAE

317 The endemic species **pinnate rue** (*Ruta pinnata* L. f.) is found at an altitude of 150–600 m on dry hills and volcanic discharges on Tenerife Island. This is a shrub 1.5–2 m tall with light green pinnate leaves. Yellow flowers are borne in groups of 4–5 in inflorescences at the ends of branches (Plate 63). The plant is known from only 6 habitats and each consists of a small population with about 100–150 plants. The plant is fast disappearing as a result of reclamation of lands for agriculture and cattle grazing; now it is surviving only on steep slopes.

ST. HELENA ISLAND (ATLANTIC OCEAN)

About 30 endemic species of angiosperms are known on the island, of which 10 have already disappeared and 15 are endangered. When the island was discovered, about 400 years ago, it was covered entirely with dense forests. Some scientists believe that at that time its flora comprised not less than 100 endemic species, but many plants recorded during the botanical survey of 1805–1810 are already unknown. Now its flora consists mainly of introduced plants, which have been responsible for the disappearance of local plants. The last shelter for a few endemic species is the Central Range of the island with an altitude of about 600 m. The cabbage trees (or woody daisy) are the most wonderful plants in the flora of St. Helena.

FAMILY ASTERACEAE

The **woody daisy** (*Psiadia rotundifolia* Hook. f.) was discovered in 1888. This is a fairly tall tree with spreading branches, at the ends of which small, petiolate, serrate leaves are borne that leave distinct scars on the branches after falling. The flowers are clustered in dense inflorescences, resembling flowers, on the basis of which the name has been derived. Now this species has completely become extinct, as has the **she cabbage tree** (*Senecio prenanthiflorus* Benth. et Hook. f.), a slender, erect tree with bright lilac young stems and leaves and inflorescences of white flowers. It grew at an altitude of 600–780 m. Possibly, a few trees might have still survived in thickets on the slope near the Diana peak.

Bastard cabbage tree (*Melanodendron integrifolium* DC.) is a beautiful spreading tree. Large, glossy dark green leaves are borne at the ends of its black and almost smooth branches. The inflorescences are greenish-white, resembling the cabbage tree. It is still found quite frequently in humid localities on the islands.

White cabbage tree (*Petrobium arboreum* R. Br.) is a slender, up to 6 m tall tree with thinner branches than the bastard cabbage tree and smaller, roundish leaves on lilac-colored petioles. The flowers are greenish-white. Earlier, it was the most widespread species on the island, but now it has become more rare than the bastard cabbage tree, although it grows together with it on the Central Range. It flowers and produces fruits, but young specimens have not been found on the island during recent years. The **he cabbage tree** (*Senecio leucadendron* Benth. et Hook. f.) reaches a height of 1.5–4.5 m, with smooth (glabrous) branches; leaves are pale green at the ends of branches. The tree flowers profusely but possibly produces sterile seeds. The white inflorescences of small flowers, resembling cabbage tree, are arranged so compactly that they resemble the inflorescence of cauliflower. Young trees are found very rarely.

The trees of genus *Commidendrum*—**toddy** (*C. robustum* DC.) and **scrub wood** (*C. rugosum* DC.)—were widely distributed on the island earlier, but the reserves of these species have sharply declined because of the goats damaging the undergrowth. About 200 plant specimens of toddy have survived near Longwood remnants of the Great Forest, covering a large part of the territory till 1720), and a few groups of trees in the mountains. About 200 trees of different ages are estimated to exist in the largest group near the Fairley peak.

Three plant species from St. Helena Island are included in the Red Data Book of IUCN, 2 of which are endangered.

FAMILY CAMPANULACEAE

Flax-leaved bluebell (*Wahlenbergia linifolia* A. DC.) has, possibly, already disappeared from nature. This is a shrub up to 1 m tall. A few drooping white flowers are borne at the tips of its branches. In 1875, the bluebell grew in abundance on the Central Range, at an altitude of 610–760 m. Its habitats were subsequently destroyed by goats or occupied by introduced plants. The survey of 1970 at the former habitats showed that only 5 plants have survived. These plants grow on rocky cliffs of the Central Range, sometimes as epiphytes on tree ferns. Other endemics of the island, including those now endangered, for example, the white cabbage tree, also grow along with it.

FAMILY GERANIACEAE

Cotyledonous stork's bill (*Pelargonium cotyledonis* (L.) L'Hérit.) is a perennial herb. For a greater part of the year, it had only a thick, crooked, leafless and woody chocolate-brown trunk, 3–5 cm thick and up to 30 cm tall, but a rosette of crimped, roundish leaves appears in May or June, and a slender shoot up to 20 cm tall bearing compound umbels of white flowers arises from the center of the rosette (Plate 62). The plant grows on stony cliffs at an altitude of 150–300 m in the southwestern part of the island. The reserves of the species are shrinking because the plants are being damaged by goats (only a few plants were found in 1970). The goat population has reduced on some parts of the island after 1950. As a result, the vegetation has been partly restored, and the population strength of *Pelargonium cotyledonis, Frankenia portulacaefolia*, and *Plantago robusta* has increased.

FAMILY STERCULIACEAE

St. Helena red trochetia [*Trochetia erythroxylon* (G. Forster) Benth.] is a very beautiful plant. It produces pale green ovate leaves, and the flowers are usually borne in pairs: they are initially pure white, then pink, then red (Plate 63). The reduction in its population strength is related to goat grazing and commercial exploitation of its wood in the past. In 1956, there was a single tree in nature and a few trees in gardens. The wild tree was found at an altitude of 300–450 m at the place of already receding forest. It grew with the **St. Helena ebony tree** (*T. melanoxylon* (Ayton f.) Benth.), which was earlier a common tree of the island and is now already extinct due to commercial exploitation of its bark for extracting tannins and to reduction in the forest area.

FALKLAND (MALVINAS) ISLANDS (ATLANTIC OCEAN)

The Falkland Islands are completely devoid of vegetation, which explains the persistent strong winds blowing here.

FAMILY PORTULACACEAE

Felton's rock purslane (*Calandrinia feltonii* Skottsb.) is an annual with numerous weak shoots. Its leaves are alternate, linear-scapulate. The clusters of red flowers are borne in the axils of upper leaves. Earlier, this plant was common in the northwestern part of West Falkland, where it grew on hills, covered with the short shrubby species *Empertum rubrum* on dry, warm, porous soils. Now it has disappeared from all wild plant populations. Felton's rock purslane has survived in the gardens of the island, but attempts to reintroduce it into nature have not yet been successful. The species is included in the Red Data Book of IUCN.

SOCOTRA ISLAND (INDIAN OCEAN)

About 3 percent of the genera and about 30 percent of the plant species are endemic to Socotra although some of them, may be found during surveys in Somalia and south of the Arabian peninsula. The flora of the island is in a critical condition. Out of 217 flowering plants, endemic to Socotra and Abd-el-Kuri Island, 132 are rare or endangered and 85 are already extinct.

FAMILY APOCYNACEAE

Socotra adenium [*Adenium obesum* (E. Forsk.) Roem. et Schult. ssp. *socotranum* Vierh.] has usually 2–3 swollen, extremely thick trunks arising from its base, whose tips are covered with a few hard leaves, usually falling before the appearance of red flowers. A poison is prepared from its milky sap. From a distance, the plant resembles monstrous growths on rocky slopes.

FAMILY BEGONIACEAE

The population of **Socotra begonia** (*Begonia socotrana* Hook. f.) is at a critically low level. This is a perennial herb with tuberous rhizome, rosettes of roundish leaves, and pink flowers. Now it is found only in the higher places that are inaccessible to goats. It is included in the Red Data Book of IUCN.

FAMILY CUCURBITACEAE

Socotra cucumber tree (*Dendrosicyos socotranus* Balf. f.) is of great botanical interest, as it is the only tree-like plant of the Cucurbitaceae. It is a knotty succulent, up to 7 m tall, with glabrous, swollen chalky white trunks up to 1 m thick. The trunks are swollen because of milky sap and are similar to the legs of an elephant. The tree bears a small spreading crown of a few slightly drooping branches at the apex, each of which terminates with a cluster of roundish or cordate, hard and curled leaves. The trunk tissue, whitish cellulose, is easily cut by a knife. Earlier this tree was very widely distributed on the island; it was found on limestone deposits. Now the reserves of this species have declined because of its use as camel fodder. It is included in the Red Data Book of IUCN.

FAMILY DIRACHMACEAE

Socotra dirachma (*Dirachma socotrana* Schweinf.) is a sweet-smelling shrub or small tree with white flowers. Only 30 adult plants of this species were found in 1967. Although they do produce flowers, practically no viable seeds are produced; the small undergrowth is destroyed by cattle. It is included in the Red Data Book of IUCN.

FAMILY EUPHORBIACEAE

The **Abd-el-Kuri spurge** (*Euphorbia abdelkuri* Balf. f.) growing on Abd-el-Kuri Island (in the Socotra Archipelago) is a leafless nonprickly succulent, usually growing up to a height of 2 m, and forming a dense pad of numerous erect unbranched stems. This is the only spurge species with yellow sap. It grows at an altitude of 230 m on granite gravel-covered slopes, almost without any vegetation. Its only habitat is in the main range of the island. In the past, it grew all along the slope on one side of this range at a height of 150–450 m, now 4 beds have survived at one location (data of 1967). The cause of decline in its population is not clear, as the domestic cattle do not touch this very poisonous plant. It is included in the Red Data Book of IUCN as an endangered species.

FAMILY FABACEAE

Most probably, the **silver-hair moneywort** (*Taverniera seracophylla* Balf. f.), an endemic plant of Socotra, has already become extinct. It is a small shrub covered with white-silky hairs. The species was found for the first time in 1880 at the northern coast of the island, at 2 places, but a thorough search for this plant in 1967 did not yield any results. This moneywort grew on the sands of coastal plains, but these places have suffered heavily from cattle grazing. The plant is included in the Red Data Book of IUCN.

FAMILY LILIACEAE

The **squarrose aloe** (*Aloe squarrosa* Baker), a basally branching perennial with grassy red flowers and spotted leaves, is endangered. It is interesting for ornamental use. Only one population of this species is known in the northwest of the island, on the limestone slope facing north, covered with lichens, at an altitude of 300 m. The population at present is at a critically low level. Sixteen endemic plants also grew here together with the squarrose aloe; 10 of them were not found during the survey of the place in 1967, among them, *Babiana socotrana* Hook. f., *Euphorbia obcordata* Balf. f., and *Helichrysum arachnoides* Balf. f. Possibly, these plants have already disappeared. Only a single plant each of *Teucrium balfourii* Vierh. and *Xylocalyx aculeolatus* S. Carter was found among the 5 endangered species.

Cinnabar dragon tree (*Dracaena cinnabari* Balf. f.), a relict tree, is one of the most unusual trees in the world. Its crown resembles an umbrella turned upside down. Like the famous canary dragon tree, it produces blood-red sap or resin (the dragon's blood), which is an item of trade. The tree is very beautiful, growing on suspended cliffs.

FAMILY MORACEAE

Giant dorstenia (*Dorstenia gigas* Schweinf.) is a knotty succulent, sometimes a tree (Plate 60). It is fairly widely distributed on the island, but found in small groups. It is included in the Red Data Book of IUCN.

FAMILY PUNICACEAE

Like aloe, the **wild pomegranate** (*Punica protopunica* Balf. f.) is included in the Red Data Book of IUCN. This is a tree up to 5 m tall, with elliptical or roundish leaves. As a close relative of the cultivated pomegranate, it is of great value as a genetic resource. In 1880, wild pomegranate was found in the upper belt of the granite mountains, where it grew in abundance on a limestone plateau. In 1953, it was recorded as the main component on the limestone slopes in the northeastern part of the island, but only 4 old trees were found in 1967 at a long distance from each other. Regeneration of these trees is not taking place; when they die, the species may become extinct.

SEYCHELLES (INDIAN OCEAN)

The Seychelles have an endemic family (Medusagynaceae), 13 endemic monotypic genera, and 72 endemic plant species.

FAMILY ASCLEPIADACEAE

Schimper's toxocarpus (*Toxocarpus schimperianus* Hemsley) is a slender creeper with leathery leaves and white or pale pink flowers. It is found only on the Seychelles in a small protected area near the coast.

FAMILY CUCURBITACEAE

Sublittoral peponium (*Peponium sublitorale* C. Jeffrey et J.S. Page) is a rare plant of Aldabra Island (not more than 1000 scattered plants are known). It is a twining or creeping 2 m long plant with broadly ovate leaves and yellow flowers. It is included in the Red Data Book of IUCN.

FAMILY DIPTEROCARPACEAE

The **Seychelles copal tree** (*Vateria seychellarum* Dyer) reaches 25–30 m in height and up to 2 m in diameter. Its leaves are leathery and elliptical; flowers are white or yellow borne in axillary racemes. Five habitats of the species are known in rainforests of Mahe Island, but the population with about 50 trees is at a critically low level. The population in the territory has declined because of tree felling (it yields valuable wood).

FAMILY MEDUSAGYNACEAE

Opposite-leaved medusa tree (*Medusagyne oppositifolia* Baker) is a shrub or small tree with dense umbellate crown. The leaves are leathery, elliptical; flowers are white or pink (Plate 62). It grows in crevices between granite rocks at an altitude of 220 m, where about 10 plants are known. It is endemic to the island.

FAMILY PALMAE

The pride of the Seychelles is the world famous **Seychelles lodoicea, sea coconut** or **twin coconut** [*Lodoicea maldivica* (J.F. Gmelin) Pers.]. This is a tree with slender columnar trunks, up to 30 m tall, and a crown of 12–20 pale yellowish-green, fan-shaped leaves. The plant is dioecious. The fruits, 40–50 cm, long, resemble 2 fused fruits of coconut; the weight of a single fruit is 10–20 kg, and a single tree bears up to 70 fruits (Plate 63), which keep hanging for 5–8 years. This palm is found rarely and in fewer numbers on the Praslin and Curieuse islands; it has already disappeared from Round Island. Now the species is protected by law.

MADAGASCAR (INDIAN OCEAN)

About 70–80 percent of the plants in the Madagascar flora are endemic. The impoverishment of the flora became clear even at the beginning of the 20th century, and this process is continuing. Many wonderful species are close to extinction as a result of destruction of their habitats due to cattle grazing or fire.

FAMILY APOCYNACEAE

The **leathery catharanthus** (*Catharanthus coriaceus* Markgraf), a small, erect shrub with hard opposite leaves and solitary reddish-violet flowers, is endangered. It is known from a few locations in the center of the island, west and southwest of Antananarivo, in a low deciduous sclerophyllous forest. The population is very small. The species is included in the Red Data Book of the IUCN.

FAMILY CAESALPINIACEAE

The **royal poinciana** or **flamboyant tree** (*Delonix regia* Rafin), a tree with broad umbellate crown and leaves up to 50 cm long, is a rare species. During flowering, the entire plant is covered with large, flame-red flowers (Plate 61).

FAMILY DIDIEREACEAE

The family Didiereaceae is included in the Appendix of CITES. It is endemic to Madagascar, its members being found only in its southern and southeastern parts. These are trees or shrubs with their trunks and branches covered with spines, externally resembling cacti. The unisexual flowers are borne in cymose inflorescence or groups of inflorescences. The family comprises 4 genera and 11 species; the genus *Alluaudia* has 6, *Alluaudiopsis* 3, *Didierea* 2, and *Decaryia* 1 species. The Didiereaceae plants grow in xerophytic forests or thickets of prickly shrubs, often in communities with leafless spurges on limestone or stony or sandy soils, up to an altitude of 900–1200 m.

FAMILY EUPHORBIACEAE

All the succulent spurges of Madagascar are protected, and among them are several rare species, for example, **Cape Ste. Marie spurge** (*Euphorbia capsaintemariensis* Raunt.), one of the most interesting succulents. It is found only in southern Madagascar, on Cape Ste. Marie. It forms a small crown of prickly branches bearing minute, glossy green leaves with wavy margins. The plant is covered with small orange-yellow flowers round the year.

The **bright spurge** (*E. splendens* Boj. et Hook.)—a widely distributed succulent xerophytic shrub—is found in the southwestern and western parts of the island.

FAMILY LILIACEAE

The **bright aloe** (*Aloe laeta* Berger), a small plant with bluish-gray leaves and crimson-red flowers, is a rare species. In the extreme south, south of Ambowmbe, a few plants of **Suzanna's aloe** (*A. suzannae* Dlary) have survived.

FAMILY ORCHIDACEAE

Among the orchids, the faham tea plant genus (*Angraecum*), comprising about 125 species, is represented particularly widely on Madagascar. These are mostly small plants (2–3 cm tall) and a few large plants. The **elephant faham** (*A. eburneum* Bory) is a large plant with well-developed aerial roots arising from erect stems. The leaves are arranged in 2 rows, leathery, and lorate. The inflorescence is a multiflowered cluster of sessile greenish-white flowers; the labellum is broadly ovate, open, pure white, with a light green spur at the base (Plate 64). The **sesquipedal faham** (*A. sesquipedale* Thou.) is also a large orchid. This plant has densely leafy erect stem and greenish chocolate-brown swollen roots (Plate 64).

The genus *Aerangis* includes more than 35 epiphytic species. The **jointed aerangis** (*A. articulata* Schltr.) and **modest aerangis** (*A. modesta* Schltr.) with snow-white flowers and **lemon aerangis** (*A. citrina* Schltr.) with yellowish-white flowers are particularly beautiful (Plate 64).

FAMILY PALMAE

Palm trees from 18 genera are found on Madagascar, 12 of which are endemic to the island. Many species are endangered, particularly on the eastern coast, where the forests are greatly destroyed. For example, the population of *Ravenea robustor* Jumella et Perr. is reduced to a few plants; *R. latisecta* Jumella and *Beccariophoenix madagascariensis* Jumella et Perr. are already extinct or are close to extinction. *Masoala madagascariensis* Jumella, *Sindroa longisquama* Jumella, and *Marojejya insignis* Humbert, members of monotypic genera, are known from 1–3 habitats, and their population strength is extremely low. The only habitat of the endemic palm *Neodypsis decaryi* Jum. is located in a reserve 40 km northwest of Fort Dauphin.

THE MASCARENE ISLANDS (INDIAN OCEAN)

All the 34 endemic plants of Rodrigues Island are either almost extinct or endangered. The earlier descriptions of the island indicate that it was covered with forests or savannas, but now the natural vegetation has been conserved on small areas as a result of land clearing for agriculture, felling of trees, and goat and cattle grazing. About 119 weeds have appeared on the island since 1874, which are replacing the endemic flora. A similar situation exists on Mauritius. On Réunion Island, where high mountains exist (up to 3069 m), the flora has degraded to a lesser extent, although even here 24 endemic species have become rare or are now endangered (the reasons are the same as on the other islands, besides the harmful effect of introduced plants). Six plant species of the Mascarene Islands are included in the Red Data Book of IUCN.

FAMILY AMARYLLIDACEAE

Mauritian lily (*Crinum mauritianum* Lodd.) is a bulbous perennial with pseudostems formed by the leaf bases. Its peduncles are up to 1 m long bearing 4–12 large white flowers with purple tinge (Plate 61). At present, 100 plants of this species are known in warm standing water and along the Mauritian coast.

FAMILY APOCYNACEAE

Earlier, the **Madagascar bitter wood** (*Carissa xylopicron*), a prickly shrub with opposite leathery leaves and funnel-shaped flowers in few-flowered inflorescenses, was found on Réunion Island. Its fruits are berries. Now this bitter wood has possibly disappeared.

FAMILY EBENACEAE

Persimmon (*Diospyros* sp. nov. Richardson) is a tree with a height of 8 m and thickness 30 cm. The only tree (about 100 years old) is found in southwest Mauritius. It is a female tree, which will die without reproducing if a male tree is not found. Out of the 17 persimmon species of the Mascarene Islands, about 10 species either have become extinct or are now endangered.

FAMILY EUPHORBIACEAE

Caustic wild orange (*Drypetes caustica* (Frapp. ex Cordem.) Airy Shaw.) is a tree up to 20 m tall with porous light gray bark and dense dull foliage. The tree is dioecious. In the past, it served as a source of timber. Now 2 trees are growing in Mauritius and 12 on Réunion, but they are scattered in different parts of the island, in low and medium altitude forests up to a height of 800 m. This species is endangered.

FAMILY LYTHRACEAE

Tetrataxis salicifolia Thouars ex Tul. is a tree up to 12 m tall with dense foliage and gray squamate bark. The leaves are ovate and elliptical-ovate. Seven trees are known in southern Mauritius, where they grow on a steep slope in a protected virgin forest.

FAMILY MYRSINACEAE

Badula crassa A. DC. is a sparsely branched shrub or a small tree with thick juicy and brittle branches, bearing glossy elliptical-ovate leaves at their tips. It has survived only on Réunion Island (3 plants are known on the fringe of a high mountain forest); earlier it was also found in Mauritius, but now it does not exist there any more.

FAMILY PALMAE

Five palm genera are found on Mauritius. Only a few plants of Loddiges' Bourbon palm (*Latania loddigesii* Mart.), which was once abundant in the coastal savannas of Mauritius, has survived till now. Only 4 or 5 plants of *Hyophorbe verschaffeltii* H. Wendl. (Plate 64) and a few plants of *Latania verschaffeltii* Lem. (Plate 63) are known on Rodrigues Island. Endangered species on Round Island are *Dictyosperma album* H. Wendl. et Drude (a few plants), *Hyophorbe amaricaulis* Mart. (one plant), *H. lagenicaulis* (L.H. Bailey) H.E. Moore, and *H. vaughanii* L.H. Bailey (4 plants). It must be noted, however, that all these palms are widely cultivated on these islands.

FAMILY RUTACEAE

Paniculate toothache tree (*Zanthoxylon paniculatum* Balf. f.) is a deciduous, aromatic tree with dark gray bark, thick branches, and pinnate leaves (flowers not found). Two trees have been conserved in the southwestern part of Rodrigues Island; they produce seeds but seedlings have not been found.

JUAN FERNANDEZ ISLANDS (PACIFIC OCEAN)

The Juan Fernandez Islands are an archipelago of volcanic islands in the Pacific Ocean, located 600 km west of Chile. They belong to the Juan Fernandez region of the Holarctic kingdom. The island flora is of great interest because of its unusual evolution and geographic associations. Besides, it includes many giant plants, especially among Asteraceae and Umbelliferae. At present, most of the endemic species are endangered due to destruction of the forests as a result of introduction of sheep, goats, cattle, rabbits, and rats on the islands. The local plants are being replaced by introduced species. Six plant species of these islands are included in the Red Data Book of IUCN.

FAMILY FABACEAE

Two species of sophora must be protected on Robinson Crusoe Island. The **masafueran sophora** [*Sophora masafuerana* (Phil.) Sottsb.] is known from several locations on the island and is endangered due to grazing by goats. This is a tree, one or a few meters tall, with spreading and erect trunks and thin hairy branches bearing pinnate leaves. Forests with this species were cleared before the 20th century for sandalwood plantations.

Juan Fernandez sophora [*S. fernandeziana* (Phil.) Scottsb.] is a vulnerable species. It is a tree up to 10 m tall with hard, dark-colored wood and pinnate leaves. In the beginning of the 20th century, this plant was found in the lower belt of evergreen mountain forests, at an altitude of 200–450 m, but always very locally. The populations of this species have significantly declined as a result of forest clearing as well as grazing by goats.

FAMILY LACTORIDACEAE

The family Lactoridaceae consists of only one endemic monotypic genus *Lactoris*, with the species **Fernandez lactoris** (*L. fernandeziana* Phil.), whose trees were rarely found in the high altitudes of wet forests of Robinson Crusoe (Mas a Tierra) Island, at an altitude higher than 500 m. In 1954–1955, only 10 plants from as many habitats were found, and in 1965 only 3 flowering plants remained. Possibly, this lactoris was earlier distributed more widely. Reduction in the population strength of the species is related to the introduction of goats, sheep, and cattle, as well as plants of the blackberry genus.

Fernandez lactoris is a highly branched shrub up to 1 m tall, but with one more developed stem. The small flowers are borne in groups of 2–4 on reduced axillary inflorescences.

FAMILY PALMAE

The **southern juania** (*Juania australis* (Mart.) Drude), an erect palm up to 15 m tall, with green and smooth glossy trunk covered with whitish or brownish leaf scars and a crown of 18 pinnate leaves, is among the rare species of the island. The leaves reach a length of more than 1.3 m. The flowers are minute, white, unisexual (the plant is dioecious), and borne in dense branched panicles up to 1 m long; fruits are spherical, 15–18 mm in diameter, orange-red. It is found only on Robinson Crusoe Island, where it forms small colonies from Puerto Francois to Puerto Ingles, on steep slopes and along the ridges in the lower and upper forests. In 1965, its only population consisted of 500–1000 plants. At present, about 2000 young palms and about 4000 plants of undergrowth are known. The reserves of the species are shrinking as a result of felling (the tips of trunks with spreading leaves are edible). The palm flora also suffered in the 19th century from cattle grazing. Now this *Juania* species is protected by law. All its habitats are on the territory of the national park.

All the species of the Juan Fernandez Islands described are included in the Red Data Book of IUCN.

FAMILY RANUNCULACEAE

Possibly, the **goat crowsfoot** (*Ranunculus caprarum* Scottsb.), an erect, coarse, hairy herb with the stem up to 75 cm tall and short rhizome, and a rosette of hairy, roundish-cordate leaves, has already become extinct or is endangered. It was collected for the first time in 1917; later the plant was not found. This crowsfoot grew in rocky Kasas ravines on Alexander Selkirk Island at an altitude of 1200–1300 m. Possibly, it was destroyed by goats.

FAMILY SANTALACEAE

Most probably, the **Juan Fernandez sandalwood** (*Santalum fernandezianum* Phil.), a semiparasite, has become extinct. It is usually less than 10 m tall with opposite pairs of glossy, dark green, slightly succulent, oblong leaves and terminal pyramidal panicles of dense, whitish, fleshy flowers. The distribution of this species was confined to the forests on Robinson Crusoe Island. The last tree (an old 9 m tall plant) was found in 1908, but it died in 1916. The species is known since 1624. At that time, its trees grew in abundance and were exploited intensively for aromatic and valuable wood and were transported to Peru. In 1740, the species became very rare. Its extinction was also accelerated by the introduction of goats on the island, which grazed the young seedlings and prevented regeneration.

The extinction of any species, subspecies, variety, or even individual plant population in any country leads to a general impoverishment of flora of the earth and its germplasm. Therefore, the attempts of any country to conserve the rare plants in natural or artificial conditions have not only a national but also international importance. Saving the plants from extinction would help in conserving the beauty of the earth for our future generations.

Index of Latin Names of Plants*

Abies gracilis 111
— guatemalensis 224
— kawakamii 111
— nebrodensis 59
— pinsapo 59
Acacia albida 154
— anomala 248
— aphylla 248
— galpinii 155
— giraffae 155
— menzelii 249
— peuce 249
— pubescens 249
— raddeana 15
Acanthaceae 129, 209
Aceras anthropophorum 53
Achillea lanulosa ssp. megacephala 198, 200, 202
Achyranthes arborescens 281
— mangarevica 280
— marchionica 281
Acidanthera divina 148
Acmopyle pancheri 307
— sahniana 307
Aconitum noveboracense 200
Acradenia franklinii 273
Acrocomia chunta 224
Adansonia digitata 134
Adenia pechuelii 161
Adenium obesum var. multiflorum 130
— ssp. socotranum 319
Adenochilus nortonii 256
Adiantaceae 209
Adiantum lorentzii 209
Adonis distorta 68
Aechmea dichlamydea 212
Aeonium nobile 315
Aerangis articulata 323

— citrina 323
— modesta 323
Aërides crispum 101
— maculosus 106
— multiflorum 101
Agalinis stenophylla 206
Agathis robusta 228
Agavaceae 169
Agave arizonica 170
Agropyrum kingianum 291
Aichryson bollei 314
Ailanthus fordii 121
Akebia chingshuiensis 94
Alberta magna 166
Albizia glabrior 87
Alliaceae 95
Allium caespitosum 95
— pumilum 95
Aloe allooides 150
— angelica 150
— arborescens 150
— bainesii 150
— castanea 150
— dolomitica 151
— fibrosa 151
— laeta 323
— littoralis 151
— marlothii 151
— parviflora 152
— pillansii 152
— polyphylla 151
— sessiliflora 152
— squarrosa 320
— striatula 152
— variegata 152
Alstroemeria pelegrina 210
Alstroemeriaceae 209
Alyssum fastigiatum 25

---

*Reproduced from the Russian original. Original Russian page numbers have been given in the left-hand margin of the English translation—General Editor.

Amaranthaceae 210, 281
Amaranthus haughtii 210
Amaryllis aviflora 210
— santacatarina 210
— scopulorum 210
Amaryllidaceae 6, 210, 324
Amborella trichopoda 281
Amborellaceae 281
Ammopiptanthus mongolicus 87
— nanus 87
Amorphophallus titanus 120
Anacardiaceae 129, 170
Anchusa crispa 22
Ancistrocactus tobuschii 177
Andersonia grandiflora 239
Androsace brevis 64
— ciliata 65
— helvetica 65
Andryala levitomentosa 12
Angelica heterocarpa 8
Angraecum distichum 157
— eburneum 323
— eichlerianum 157
— sesquipedale 323
Anguloa clowesii 218
— ruckeri 218
Anigozanthos gabrielae 245
— humilis 245
— kalbariensis 245
— manglesii 245
— preissii 245
— pulcherrimus 245
Annonaceae 171
Anoectochilus brevilabris 101
— sikkimensis 101
Anselia confusa 157
Antennaria nordhageniana 13
Anthemis gerardiana 13
Antrophyum annetii 168
Aotus carinata 240
Apiaceae 8
Apocynaceae 76, 130, 281, 319, 322, 324
Aquilegia thalictrifolia 68
Araceae 77
Araliaceae 77, 172, 282
Araucaria araucana 210
— bidwillii 228
Araucariaceae 210, 228
Arbutus canariensis 315
Arctostaphylos densiflora 187
Areca concinna 107
Argania sideroxylon 167

Argemona arizonica 196
— pleiacantha var. pinnatisecta 196
Argyranthemum coronopifolium 316
— winteri 315
Ariocarpus agavoides 177
— scapharostrus 178
Arisaema ternatipartium 77
Aristolochiaceae 172
Armeria arcuata 60
— maritima ssp. interior 198
— pseudarmeria 60
— rouyana 60
— soleirolii 60
Artemisia granatensis 13
— insipida 14
— molinieri 14
Asclepiadaceae 79, 131, 323
Asimina rugelii 171
— tetramera 172
Aspleniaceae 11
Asplenium jahandiezii 11
Aster pyrenaeus 14
Asteraceae 12, 79, 133, 173, 229, 282, 314, 317
Astragalus arnacantha 38
— beatleyae 188
— peterfii 39
— phoenix 188
— physocalyx 39
— pseudopurpureus 39
— roemeri 39
Astrophytum capricorne 177
Asystasia africana 129
Athrotaxis laxifolia 276
Atraphaxis muschketowii 177
Atropa komarovii 121
Austrobaileya maculata 230
Austrobaileyaceae 230
Austrocedrus chilensis 213
Austrotaxus spicata 312
Aztekium ritterii 178

Babiana socotrana 320
Badula crassa 324
Baikiaea plurijuga 147
Balanophora involucrata 80
— taylori 285
Balanophoraceae 80, 285
Balmea stormae 201
Banksia coccinea 264
— goodii 264
— menziesii 264

— tricuspis 264
Beccariophoenix madagascariensis 323
Begonia socotrana 319
Begoniaceae 319
Beilschmiedia berteroana 216
— miersii 216
— tarairi 301
Bellis bernardii 14
Berardia subacaulis 14
Berberidaceae 80, 173, 211
Berberis littoralis 211
— sonnei 173
Beta nana 35
Betula alleghaniensis 174
— lenta 174
— pumila var. glandulifera 174
— uber 174
Betulaceae 174
Bidaria cuspidata 79
Bidens hendersonensis 284
Bignoniaceae 80, 133, 285
Biscutella neustriaca 25
— sclerocarpa 25
Blaeria mannii 169
Boerlagiodendron sp. 290
Bombacaceae 134, 211
Boraginaceae 22, 286, 314
Boronia affinis 272
— edwardsii 273
— keysii 272
— megastigma 273
— rivularis 272
Boscia 134
Boswellia carteri 134
Bowenia serrulata 278
— spectabilis 278
Brachycome mulleri 229
Brassavola perrinii 218
— subulifolia 218
Brassia verrucosa 218
Brassica balearica 25
— macrocarpa 25
Brassicaceae 24, 81, 175
Bromeliaceae 211
Bromus interruptus 62
Bubbia longifolia 313
Buddleia incana 217
— longifolia 217
Buglossoides calabra 22
— glandulosa 22
Bulbophyllum barbigerum 157
— carreyanum 157

— cocoinum 157
— globuriforme 256
— minutissimum 256
— lageriforme 257
Bupleurum kakiskale 8
Burkea africana 146
Burmanniaceae 176
Burnettia cuneata 257
Burretiokentia hapala 302
Burseraceae 134
Byblidaceae 231
Byblis gigantea 231
— liniflora 231

Cactaceae 176, 212
Caesalpinia echinata 213
— spinosa 213
Caesalpiniaceae 135, 213, 322
Caladenia bicalliata 259
— gladiolata 259
— multiclavia 258
Calandrinia alba 225
— crenata 225
— feltonii 319
— paniculata 225
— ruizii 225
Caleana minor 302
Calendula suffruticosa ssp. maritima 14
Callitris oblonga 237
Calocedrus formosana 82
Calochilus richae 257
Calochone acuminata 166
Calycanthaceae 231
Calyptronoma rivalis 223
Calytrix laricina 249
Camellia crapnelliana 125
— granthamiana 125
Campanula isophylla 30
— thyrsoides 31
— transsilvanica 31
Campanulaceae 30, 232, 318
Campynema lineare 257
Capparaceae 134
Caralluma distincta 132
— tubiformis 132
Carica chilensis 213
Caricaceae 213
Carissa xylopicron 324
Carlina cirsioides 15
— diae 15
— onopordifolia 15
Carmichaelia exsul 287

— prona 287
Caryophyllaceae 31, 82
Caryopteris mongolica 128
Caryota no 108
Cassytha nodiflora 248
— racemosa 248
— tepperana 248
Castanea ozarkensis 189
Castilleja ludoviciana 206
Casuarina bicuspidata 233
— chamaecyparis 233
— fibrosa 232
— ramosissima 233
Casuarinaceae 232
Catharanthus coriaceus 322
Cathaya argyrophylla 113
— nanchuanensis 113
Cattleya aclandiae 219
— bowringiana 219
— persivaliana 219
— skinneri 219
— trianae 219
Cedrus brevifolia 114
— libani 113
Celmisia morganii 283
— philocremna 283
Centaurea canariensis 316
— corymbosa 16
— junoniana 314
— micracantha 16
— ragusina 16
Centaurium enclusensis 42
— namophilum 188, 189
— rigualii 42
— triphyllum 41
Cephaëlis ipecacuanha 225
Cephalanthera 53
Cephalaria radiata 37
Cephalotaceae 233
Cephalotus follicularis 233
Ceratolobus glaucescens 108
Ceratonia oreothauma 88
Ceratozamia miqueliana 209
Cereus robinii 212
Ceropegia elegans var. gardneri 79
Ceroxylon andicola 224
— quinquiense 224
Chamaecyparis formosensis 83
— obtusa var. formosana 83
Chamaedorea tuerckheimii 223
Championia reticulata 90
Cheirodendron trigynum 282

— — var. rochii 282
— — var. subcordatum 282
Chenopodiaceae 34
Chiloglottis formicifera 302
Chloanthaceae 234
Chloanthes coccinea 235
Chlorophora excelsa 167
Chlorophytum brasiliense 216
Chondropetalum acockii 165
Chordospartium stevensonii 287
Chrysosallidosperma smithii 223
Chunia bucklandioides 91
Cinchona succirubra 225
Cirsium bourgaeanum 16
— valentinum 16
Cissus uter 167
Cistaceae 35, 136, 184
Cistus albanicus 35
— palhinae 35
Claytonia bellidifolia 199
— lanceolata var. chrysantha 199
Clianthus puniceus 288
Cliffortia acockii 164
Cochlearia aragonensis 26
— polonica 26
Colchicum arenarium 49
Colpothrinax cookii 224
Combretaceae 136
Combretum imberbe 136
Commidendrum robustum 318
— rugosum 318
Comperia comperana 53
Conospermum debile 265
— incurvum 265
— scaposum 265
Conradina verticillata 191
Coprosma acutifolia 310
Cordeauxia edulis 145
Cordyline kaspar 294
Corynanthe dolychocarpa 166
Correa decumbens 273
Corybas aconitoflorus 258
— carinatus 102
— despectans 258
— fimbriatus 258
— fornicatus 102
— unguiculatus 258
— vinosus 102
Crassulaceae 35, 184, 314
Crepis suffreniana 17
Crinum mauritianum 324
Crocus albiflorus 46

— heuffelianus 46
— maesiacus 46
— olivieri 46
— versicolor 46
Crowea angustifolia 273
— exalata 274
Crowea saligna 274
Cruciferae 24
Cryptandra alpina 271
— nudiflora 271
Cryptanthemis slateri 255
Cryptostylos hunteriana 258
Cucurbitaceae 319, 321
Culcita villosa 237
Cunninghamia konishii 124
— lanceolata 123
Cupressaceae 82, 137, 185, 213, 237, 286
Cupressus arizonica 185
— — var. nevadensis 185
— — var. stephensonii 185
— dupreziana 137
— goveniana var. abramsiana 185
— macrocarpa 185
Cyathea camerooniana 139
— capensis 139
— celebica 237
— cunninghamii 237
— dregei 138
— felina 237
— flaccida 214
— manniana 139
Cyatheaceae 138, 214, 237
Cycadaceae 83, 139
Cycas revoluta 83
Cyclamen coum 66
— hederifolium 65
— persicum 66
— purpurascens 65
Cyclobalanopsis hypophaea 89
Cymbidium aloifolium 103
— giganteum 102
— grandiflorum 102
Cynara algarbiensis 17
Cypella osteniana 215
— pusilla 215
Cyperaceae 141
Cyperus papirus ssp. hadidii 141
Cyphophcenix elegans 303
— nucele 302
Cypripedium calceolus 54
— guttatum 54
— macranthon 54

Cytisus aeolicus 40
— emeriflorus 39

Dacrydium fonkii 224
Dactylostalix ringens 103
Dactylorhiza 55
Dampiera diversifolia 244
Daphne arbuscula 73
— baksanica 126
— petraea 74
— rodriguezii 73
Darlingtonia californica 203
Darwinia macrostegia 250
— thomasii 250
Davalliaceae 142
Degeneraceae 313
Degeneria vitiensis 313
Degenia velebitica 26
Delonix regia 322
Dendrobium biggibum 259
— chrysanthum 104
— densiflorum 104
— mabelae 104
— pauciflorum 103
Dendrosicyos socotranus 319
Deschampsia mackenziana 198, 200, 202
Dianthus callizonus 32
— glacialis 32
— nitidus 32
— urumoffii 31
Diastella buekii 163
Dichapetalaceae 142
Dicksonia berteriana 214
— esellowiana 214
— externa 214
— karsteniana 214
— youngiae 237
Dicksoniaceae 237
Dicliptera dodsonii 209
Dicrastylidaceae 234
Dicrastylis beveridgei 235
— costelloi 235
— doranii 235
— verticillata 235
Dictyosperma album 325
Didiciea cunninghamii 104
— japonica 104
Didiereaceae 322
Didymocarpus floccocus 89
— humboldtianus 90
— zeylanicus 90
Dilleniaceae 238

Dionaea muscipula 186
Dionysia aretiodes 118
— hissarica 118
— involucrata 118
— kossinskyi 118
— mira 119
Dioon edule 208
— mejaei 227
— spinulosum 208
Dioscorea caucasica 84
— deltoides 84
— elephantipes 84
— hemicrypta 84
— montana 84
— sylvatica 84
Dioscoreaceae 83
Diospyros oppositifolia 85
Dipcadi turkestanicum 95
Diplocaulobium masonii 259
Diplomeris hirsuta 104
— pulchella 104
Dipsacaceae 37
Dipterocarpaceae 85, 321
Dirachma socotrana 320
Dirachmaceae 320
Disa uniflora 158
Discocactus horstii 212
Diuris fastidiosa 259
— palustris 260
— pedunculata 260
Dodonea ericoides 275
— eriocarpa 311
— microzyga 275
Doronicum cataractarum 17
Dorstenia gigas 321
Draba haynaldii 26
— ladina 26
— loiseleurii 27
Dracaena cinnabari 321
— draco 316
— ombet 152
— usambarensis 153
Drosera regia 142
Droseraceae 142, 186
Dryandra formosa 265
— polycephala 265
— pulchella 266
Dryobalanops aromatica 85
Drypetes caustica 324
Ducampopinus krempfii 114
Dudleya traskiae 184

Ebenaceae 85, 142, 324
Echinocactus grusonii 178
Echinocereus engelmannii 179
— — var. howei 179
— — var. munzii 179
— — var. purpureus 179
— lindsayi 179
— reichenbachii 179
— — var. albertii 179
Echium handiense 315
— pininana 314
Eleorchis japonica 105
Elingamita johnsonii 285, 299
Elliotia racemosa 187
Elytranthe adamsii 294
Encephalartos altensteinii 140
— caffer 140
— hildebrandtii 140
— laevifolius 141
— natalensis 140
— paucidentatus 141
— transvenosus 140
— umbeluziensis 140
— villosus 141
Engelghardia pterocarpa 216
Entandophragma spicatum 154
Epacridaceae 238
Epacris barbata 239
Ephippianthus sawadanus 105
— schmidtii 105
Epiblema grandiflorum 260
Epidendrum mutelianum 219
Epigaea asiatica 86
— gaultherioides 85
Epipactis dunensis 54
Equisetaceae 214
Equisetum giganteum 214
Eremurus lachnostegius 96
— roseolus 96
Erica chrysocodon 143
— jasminiflora 143
Ericaceae 85, 143, 187, 214, 315
Eriostemon obovalis 274
Erodium astragaloides 42
Erucastrum palustre 27
Eryngium alpinum 9
— maritimum 9
— spinalba 9
Erysimum capitatum var. angustifolium 194
— pieninicum 27
Erythroxylaceae 214

Erythroxylon coca 214
Eucalyptus amygdalina 258
— argophloia 251
— carnabyi 251
— cladocalyx 253
— cneorifolia 253
— cordata 253
— crenulata 251
— curtisii 251
— froggattii 252
— obliqua 258
— rhodantha 252
— risdonii 252
— steedmanii 252
Euchlaena perennis 196
Euclea pseudebenis 142
Eumorphia swaziensis 133
Euphorbia abdelkuri 320
— angularis 145
— cameronii 143
— capsaintemariensis 323
— curvirama 144
— grandicornis 145
— grandidens 145
— handiensis 315
— horrhida 145
— mammillaris 145
— meloformis 145
— obcordata 320
— obesa 145
— proballyana 144
— tirucallii 145
— triangularis 145
— virosa 145
— wakefieldii 144
— waterbergensis 144
Euphorbiaceae 143, 286, 315, 320, 324
Euphrasia calida 71
— dunensis 71
Eupomatia bennettii 240
Eupomatiaceae 240
Eurypetalum unijugum 135
Euterpe edulis 224
Exocarpos luteolus 310

Fabaceae 38, 87, 145, 188, 214, 240, 287, 315, 320, 325
Fagaceae 89, 189, 214, 243
Falcatifolium angustum 117
— taxoides 308
Ferocactus acanthodes var. eastwoodiae 179

— glaucescens 180
Ferulago capillaris 9
Filitia africana 129
Fitzroya cupressoides 213
Forsythia europaea 52
Fothergilla gardeni 190
Frankenia bracteata 243
— parvula 243
— portulacaefolia 318
Frankeniaceae 243
Franklandia fucifolia 266
— triaristata 266
Franklinia alatamaha 207
Frerea indica 79
Freziera forerorum 226
Fritillaria arabica 96
— involucrata 50
— liliacea 192
— meleagris 50
Fuchsia procumbens 301
Fumaria jankae 41
— occidentalis 41
Fumariaceae 41

Galeola cassythoides 260
Gastrolobium hamulosum 241
— pyramidale 241
Gaultheria sphagnicola 214
Gaussia attenuata 223
— prinseps 223
Geleznovia verrucosa 274
Gentiana ligustica 42
— purpurea 42
Gentianaceae 41, 189
Geraniaceae 42, 292, 318
Geranium cuneatum 292
Gerardia stenophylla 206
Gesneriaceae 43, 89
Gigasiphon macrosiphon 135
Gillediodendron glandulosum 135
Gilliesia graminea 217
Ginkgo biloba 90
Ginkgoaceae 90
Gladiolus aureus 148
Gleditsia caspia 88
Globularia ascanii 315
— cambessedesii 44
— incanescens 44
Globulariaceae 44, 315
Glomeropitcairnia erectiflora 212
— penduliflora 212
Glossocalyx brevipes 155

Gomortega keule 215
Gomortegaceae 215
Goodenia chambersii 244
— quadrilocularis 244
Goodeniaceae 243, 292
Gossypium tomentosum 295
Grevillea dryandroides 267
— flexuosa 266
— rogersii 267
— scabra 267
— thyrsoides 267
— treueriana 267
Greyia sutherlandii 154
Griffinia concinna 210
— hyacinthina 210
— liboniana 210
Grindelia fraxino-pratensis 188
Grossulariaceae 44
Guaduella bedermannii 161
— macrostachys 161
Guibourtia coleosperma 146
Gunnera hamiltonii 292
Gunneraceae 292
Gymnocarpos przewalskii 82
Gypsophila papillosa 32

Haberlea rhodopensis 43
Haemodoraceae 245
Hakea cristata 268
— cycloptera 268
— obtusa 268
Hamamelidaceae 91, 190
Haworthia batteniae 153
Hebe breviracemosa 311
— dieffenbachii 312
— cupressoides 312
— speciosa 311
Hedycarya rivularis 299
Heimerliodendron brunonianum 300
Helianthemum sphaerocalyx 136
Helichrysum arachnoides 321
— dimorphum 283
— ericetum 229
— milliganii 229
Hemiandra gardneri 247
Hemigenia glabrescens 247
Heracleum minimum 9
Heritiera longipetiolata 290, 312
Hernandia bivalis 246
— labyrinthica 290
Hernandiaceae 246
Hesperis oblongifolia 27

Hexastylis naniflora 173
— speciosa 173
Hibbertia lasiopus 238
— leptopus 238
Hibiscadelphus distans 296
Hibiscadelphus giffardianus 295
— hualalaiensis 296
— wilderianus 296
Hibiscus insularis 291, 296
Hippomane mancinella 223
Holarrhena mitis 76
Hubbardia heptaneuron 116
Hudsonia montana 184
Hutera leptocarpa 28
— rupestris 27
Hymenostemma pseudanthemis 17
Hyophorbe amaricaulis 325
— lagenicaulis 325
— vaughanii 325
— verschaffeltii 325
Hypericaceae 45
Hypericum caprifolium 45
— haplophylloides 45
— nummularium 45
Hyphaene thebaica 158, 159
— ventricosa 159

Iberis fontoqueri 28
— sampaiana 28
Icacinaceae 293
Idiospermum australiense 231
Incarvillea semiretschenskia 81
Ionopsidium acaule 28
— savianum 29
Iriartea ventricosa 223
Iriartella ferreyrae 223
Iridaceae 45, 91, 148, 190, 215, 246
Iridodictyum winogradowii 93
Iris atrofusca 92
— camillae 93
— elegantissima 92
— grant-duffi 92
— iberica 93
— lortetii 91
— nazarena 92
— serotina 46
— winogradowii 93
Isatis platyloba 28
Isoëtaceae 46, 190
Isoëtes boryana 47
— delilei 47
— lithophylla 191

— louisianensis 190
— malinverniana 47
— tenuissima 47
Isophysis tasmanica 246
Isopogon adenanthoides 268
— latifolius 268
— tridens 268
Isotria medeoloides 194
Itaya amicorum 223
Ivesia eremica 188

Jacksonia foliosa 241
Jerdonia indica 90
Johannesteijsmannia altifrons 109
— lanceolata 109
Johannesteijsmannia magnifica 109
— perakensis 109
Juania australis 326
Jubaea spectabilis 223
Jubaeopsis caffra 160
Juglandaceae 216
Juglans neotropica 216
Juniperus bermudica 213
— procera 137
Juno magnifica 94
Jurinea fontqueri 18
— tzar-ferdinandii 18

Kennedia glabrata 242
— macrophylla 242
Keteleeria davidiana 112
— fortunei 112
Knautia basaltica 37
— foreziensis 37
— godetii 37
— nevadensis 37
— rupicola 38
Kniphofia splendida 154
— umbrina 153
Kokia cookei 296
— kauaiensis 297

Labiatae 47
Lachnostachys verbascifolia 236
Lactoridaceae 325
Lactoris fernandeziana 325
Lactuca livida 18
— longidentata 18
Laelia dayana 220
— jongheana 220
— praestans 220

— purpurata 219
— sincorana 220
Laevenworthia torulosa 175
Lafuentea rotundifolia 71
Lambertia echinata 269
Lamiaceae 47, 191, 247, 293
Lamyropsis microcephala 18
Lannea spinosa 129
Lardizabalaceae 94
Laserpitium longiradiatum 9
Lasiopetalum bracteatum 276
Latania loddigesii 325
— verschaffeltii 325
Launaea cervicornis 19
Lauraceae 149, 191, 247
Lebronnecia kokioides 297
Leontopodium alpinum 19
Lepidozamia hopei 279
— peroffskiana 279
Leptospermum ericoides 286
— floridum 254
Lereschia thomasii 10
Leschenaultia chlorantha 244
— superba
Lesquerella densipila 175
Leucadendron argenteum 164
— verticillatum 164
Leucanthemum corsicum 19
Leucojum longifolium 6
— nicaeënse 6
— valentinum 6
Leucopogon interruptus 239
— obtectus 240
Leuzea longifolia 20
— rhaponticoides 20
Lewisia cantelowii 199
— cotyledon 199
— oppositifolia 199
— tweedyi 199
Libertia tricocca 216
Ligusticum albanicum 10
— corsicum 10
— lucidum ssp. huteri 10
Liliaceae 49, 95, 149, 192, 216, 294, 316, 320, 323
Lilium albanicum 51
— bulbiferum 51
— carniolicum 51
— croceum 50
— jankae 51
— philippinense 50
— rhodopaeum 50

— speciosum 97
Limonium arborescens 316
— asterotrichum 61
— fruticans 316
— paradoxum 61
— recurvum 61
— transwallianum 61
Linaria algarviana 72
— hellenica 72
Liparis swenssonii 261
Lithodora nitida 23
— oleifolia 23
Livistona mariae 262
Lodoicea maldivica 322
Loganiaceae 217
Lonchocarpus capassa 147
Loranthaceae 294
Lotus berthelotii 315
Lycaste suaveolens 220
Lychnis nivalis 32
Lyperanthus forrestii 261
— nigricans 260
Lysimachia minoricensis 66
Lythraceae 324

Macropidia fuliginosa 246
Macrozamia communis 279
— macdonnellii 279
— moorei 279
— riedlei 279
Magnolia campbellii 98
— kachirachirai 99
Magnoliaceae 98
Malacothamnus clementinus 193
Malvaceae 193, 295
Mandragora turcomanica 122
Marattia salicina 298
Marattiaceae 298
Marojejya insignis 323
Masdevallia bella 221
— chimaera 220
— coccinea 221
Masoala madagascariensis 323
Mastigostyla mirabilis 216
Mastixiodendron pilosum 310
— robustum 310
Maxburretia rupicola 110
Medemia argun 158
Medusagynaceae 322
Medusagyne oppositifolia 322
Medusandra richardsina 169
Megadenia bardunovii 81

Melaleuca elachophylla 254
— polycephala 254
Melanodendron integrifolium 317
Melastomataceae 217
Meliaceae 154, 217
Melianthaceae 154
Melocactus amethystinus 212
— bahiensis 212
— pruinosus 212
Mentzelia leucophylla 188
Metasequoia glyptostroboides 125
Metrosideros carminea 300
Mezonevron kavaiensis 288
Micranthocereus auri-azureus 213
Microberlinia bisulcata 135
— brazzavillensis 135
Microcycas calocoma 227
Micromeria pulegium 48
Microstrobos fitzgeraldii 263
— niphophyllus 263
Mimosa lanuginosa 217
Mimosaceae 154, 217, 248
Mimusops balata 226
Minuartia cataractarum 33
— olonensis 33
Mirabilis macfarlanei 193
— pudica 193
Moehringia grisebachii 33
Moltkia doerfleri 23
Monimiaceae 155, 299
Moraceae 321
Moraea loubseri 148
Myosotidium hortensia 286
Myosotis colensoi 309
— corsicana 24
— ruscinonensis 23
— soleirolii 24
— traversei var. cinerascens 309
Myristica fragrans 99
Myristicaceae 99
Myroxylon pereira 214
Myrsinaceae 299, 324
Myrtaceae 249, 300

Narcissus calceolus 8
— cuatraecasasii 7
— hedraeanthus 7
— longispathus 7
— scaberulus 7
— tazetta 7
Naufraga balearica 10
Nenga gajah 110

Neocallitropsis araucarioides 286
Neostapfia 196
Neoveitchia storckii 303
Neowawraea phyllanthoides 286
Nepenthaceae 99
Nepenthes khasiana 101
— rafflesiana 100
— rajah 100
Nepeta sphaciotica 48
Newcastelia dixonii 236
— chrysophylla 236
— chrysotricha 236
— cladotricha 236
Nigritella nigra 55
Nolina interrata 192
Nothofagus alessandrii 215
— glauca 215
— moorei 243
Notospartium carmichaeliae 289
— torulosum 289
Notylia bicolor 221
Nyctaginaceae 193, 300

Obregonia denegrii 180
Ochroma lagopus 211
Ocotea bulluta 149
— glaziovii 216
Octoknemaceae 156
Oenanthe conioides 10
Oenothera deltoides ssp. howellii 194
Okoubaka aubrevillei 156
Olea capensis 156
— laperrinai 156
Oleaceae 51, 156
Oleandra annetti 142
Olearia ericoides 230
Omphalodes littoralis 24
Onagraceae 193, 300
Oncidium krameranum 221
— henekenii 221
— ornithorhynchum 221
— papilio 221
— varicosum 221
Onosma bubanii 24
— tornensis 24
— vaudensis 24
Ophrys apifera 55
Orchidaceae 52, 101, 156, 194, 218, 256, 301, 323
Orchis 56
Orcuttia mucronata 196
Oreomunnea pterocarpa 216

Orothamnus zeyheri 162
Ourisia integrifolia 275
Oxytropis deflexa ssp. norvegica 40

Pachycormus discolor 170
Pachypodium bispinosum 131
— lealii 131
— namaquanum 130
— rutenbergianum 131
— saunderersii 131
Paeonia cambessedesii 56
Paeoniacae 56
Palaeocyanus crassifolius 20
Palmae 57, 107, 158, 195, 222, 262, 302, 322, 323, 325, 326
Panax ginseng 78, 172
— quinquefolius 78, 172
Papaver bracteatum 111
— laestadianum 58
— walpolei 196
Papaveraceae 58, 110, 196
Paphiopedilum druryi 106
— fairieanum 106
— sukhakulii 107
— villosum 106
Paraberlinia bifoliata 136
Paradisea liliastrum 51
— stenantha 216
Parasitaxus ustus 308
Passiflora herbertiana ssp. insulaehowei 305
Passifloraceae 161, 305
Pedicularis furbishiae 205
Pediocactus bradyi 181
— knowltonii 180
— peeblesianus var. peeblesianus 181
— sileri 181
— winkleri 180
Pelagodoxa henryana 304
Pelargonium cotyledonis 318
Pelecyphora aselliformis 181
— pseudopectinatus 181
— strobiliformis 181
Peltophorum africanum 145
Pennantia baylisiana 285, 293
Peperomia atocongona 224
— breviramula 306
— crystallina 224
— kraemeri 306
— kusaiensis 306
— limaensis 224
— non-hispidula 224
— pseudo-galapagensis 224

— seleri 224
— umbelliformis 224
Peponium sublitorale 321
Persea borbonia 191
— — var. humilis 191
— linguae 216
— theobromifolia 216
Persoonia brachystylis 269
— marginata 269
Petagnia saniculifolia 10
Petchia ceylanica 76
Petrobium arboreum 317
Petrophile biloba 269
— multisecta 269
Petroselinum segetum 11
Phaius tancarvillae 261
Pheballum montanum 274
— ralstonii 275
Phoenix dactylifera 158
— theophrasti 57
Phyllostegia hirsuta 293
Picea koyamai 112
Pilgerodendron uviferum 213
Pimelea physodes 277
Pimpinella bicknellii 11
Pinaceae 59, 111, 224
Pinus aycahuite 224
— brutia 115
— — ssp. eldarica 115
— — ssp. pityusa 115
— bungeana 116
Piper guahamensis 306
— hosokawea 306
Piperaceae 224, 306
Pisonia brunoniana 300
Pitcairnia feliciana 169
Pittosporaceae 306
Pittosporum dallii 307
Plagianthus betulinus var. chatamica 298
Plantago robusta 318
Plumbaginaceae 59, 198, 316
Poaceae 61, 116, 161, 196
Podocarpaceae 116, 161, 224, 263, 307
Podocarpus andina 225
— falcatus 161
— milanjianus 161
— nubigenus 225
— parlatorei 224
— salignus 225
Podolepis monticola 230
Polygalla cowellii 225
Polygalaceae 225

Polygonaceae 63, 117
Pomaderis grandis 271
Portulaca pilosissima 225
Portulacaceae 199, 225, 319
Potentilla robbinsiana 201
Prasophyllum concinnum 261
— diversifolium 261
— validum 261
Primula allionii 67
— hirsuta 67
— palinuri 66
— scotica 67
Primulaceae 64, 118
Prionotes cerinthoides 239
Pritchardia macrocarpa 304
— munroi 305
Prosopis atacamensis 218
Prosopis chilensis 218
— tamarugo 218
Prostanthera cryptandroides 247
Protea angolensis 162
Proteaceae 162, 264
Protolirion sakuraii 97
Prunus maritima 200
— gravesii 200
— ramburii
Pseudophoenix ekmanii 222
— vinifera 223
Pseudolarix kaempferi 113
Pseudotsuga wilsoniana 116
Psiadia rotundifolia 317
Psilotaceae 165
Psilotum nudum 165
Pteralyxia caumiana 281
Pterocarpus angolensis 148
Ptilotrichum pyrenaicum 29
Ptychosema pusillum 242
Puelia acuminata 161
Pulicaria burchardii 315
Pultenea calycina 242
— trifida 242
— viscidula 242
Punica protopunica 321
Punicaceae 321

Quercus georgiana 189
— graciliformis 189
— hinckleyi 189
— oglethorpensis 189
— parvula 189
— shumardii 189
— — var. acerifolia 489

— tardifolia 189
— tomentella 189
Quillaja saponaria 225

Rafflesia arnoldii 119
— hasseltii 120
Rafflesiaceae 119
Ramonda myconi 43
— nathaliae 43
— serbica 43
Ranunculaceae 67, 200, 308, 326
Ranunculus carparum 326
— pauciflorus 308
— godleyanus 309
— hyperboreus 200
Ranzania japonica 80
Raphia australis 160
Rauwolfia caffra 131
— serpentina 77
Ravenea latisecta 323
— robustor 323
Renanthera imschootiana 107
Restio acockii 166
— duthieae 164
Restionaceae 165, 309
Rhamnaceae 270
Rhapidophyllum hystrix 195
Rheum rhaponticum 63
Rhizanthella gardneri 255
Rhizanthes lowii 120
Rhododendron dalnousiae 86
— edgeworthii 87
— stenophyllum 86
— taxoides 87
— thomsonii 86
— vanderbiltianum 87
Rhus kearneyi 171
— lancea 130
Rhynchosinapis wrightii 29
Ribes sardoum 44
Richea scoparia 239
Rosaceae 120, 200, 225
Roystonea elata 195
Rubiaceae 166, 201, 225, 310
Rumex rothschildianus 117
Ruta corsica 68
— pinnata 317
Rutaceae 68, 166, 271, 316, 325

Salicaceae 202
Salix brachycarpa var. psammophila 198, 200, 202
— silicola 202
— turnorii 198, 200, 202
— tyrrellii 198, 200, 202
Salvia candelabrum 49
Sanguisorba magnifica 120
Santalaceae 310, 326
Santalum fernandezianum 326
Santolina elegans 21
— oblongifolia 20
— viscosa 21
Sapindanaceae 275, 311
Sapotaceae 167, 226
Sarcochilus fitzgeraldii 261
— hartmannii 261
Sarracenia alabamensis ssp. alabamensis 204
— flava 205
— jonesii 204
— oreophila 205
Sarraceniaceae 202
Saussurea lappa 80
Saxegothaea conspicua 225
Saxifraga biternata 69
— cochlearis 70
— facchinii 69
— florulenta 70
— mutata ssp. demissa 70
— vendellii 69
Saxifragaceae 69
Scaevola kilaueae 292
— parvifolia 245
Schizaea pennula 226
Schizaeaceae 226
Scholtzia uberiflora 254
Sciadopitys verticillata 124
Sclerocactus contortus 182
— glaucus 183
— mesae-verdae 183
— polyancistrus 182
— terrae-canyonae 182
Sclerocarya caffra 130
Scrophulariaceae 70, 205, 275, 311
Sedum kostovii 36
— pruinatum 36
— stefco 36
— wilkommianum 36
— zollikoferi 36
Sempervivum pittonii 36
— wulfenii 36

Senecio cambrensis 21
— leucadendron 317
— lopezii 21
— multicorymbosus 133
— newcombei 173
— prenanthiflorus 317
Serianthes nelsonii 289
Serratula bulgarica 21
Serruria ciliata 163
— florida 163
— roxburghii 163
Seseli degenii 11
Seymeria havardii 206
Silene diclinis 33
— holzmannii
— velutina 34
— viscariopsis
Simaroubia glauca 226
Simaroubaceae 121, 226
Sindroa longisquama 323
Sinofranchetia chinensis 94
Sisymbrium matritense 29
Sobralia xantholeuca 222
Socratea sabazarii 223
Solanaceae 121
Sophora chrysophylla 290
— fernandeziana 325
— masafuerana 325
— toromiro 290
Sorbaria olgae 121
Speea humilis 217
Spiranthes aestivalis 56
— romanzoffiana 56
Sporodanthus traversii 309
Stackhousia annua 276
Stackhousiaceae 275
Stangeria eriopus 139
Stanhopea oculata 222
Stellaria arenicola 198, 200, 202
Stelleropsis caucasica 126
Stenanthemum pimeloides 271
Stenogyne crenata 293
Sterculiaceae 276, 312, 318
Stereospermum kunthianum 133
Stilbocarpa robusta 282
Stipa bavarica 63
— danubialis 63
Streblorrhiza speciosa 288, 291
Swainsona viridis 243
Swallenia alexandre 197
Swietenia macrophylla 217
— mahagoni 217

Taiwania cryptomerioides 124
Tanacetum huronense var. floccosum 198, 200, 202
Tapura africana 142
Tasmannia insipida 278
— purpurascens 278
— stipitata 278
Taverniera seracophylla 320
Taxaceae 206, 312
Taxodiaceae 123, 276
Taxus floridana 207
Tecomanthe speciosa 285
Telekia speciosissima 22
Telopea mongaensis 270
Tetraplasandra lydgatei 282
Tetrataxis salicifolia 324
Tetratheca deltoidea 277
— glandulosa 277
— gunnii 277
— halmanturina 277
— remota 277
Teucrium balfourii 321
Theaceae 125, 207, 226
Thelocactus bicolor var. bolansis 183
— — var. flavidispinus 183
— heterochromus 184
— rinconensis 183
— tulensis 184
Thelymitra epipactoides 262
Thismia americana 176
— rodwayi 176
Thomasia montana 276
Thymelaeaceae 72, 126, 277
Tibouchina chamaecistus 217
Tigridia chiapensis 190
— hintonii 190
— purpusii 190
Tolpis crassiuscula 316
Torreya taxifolia 207
Toxocarpus schimperianus 321
Tremandraceae 277
Trichipteris tryonorum 214
Trigonobalanus doichangensis 89
— verticillata 89
Trilepidia adamsii 294
Trimezia caerulea 215
— candida 215
— glauca 215
— humulis 215
— lutea 215
— minuta 215
— northiana 215

— sabinii 215
— silvestris 215
— steyermarkii 215
— variegata 215
Trochetia erythroxylon 318
— melanoxylon 319
Trochodendraceae 127
Trochodendron aralioides 127
Tsuga blaringhemii 112
— calcarea 112
— forrestii 112
Tulipa albertii 98
— celsiana 51
— greigii 98
— kaufmanniana 98
— ostrovskiana 98
— rhodopea 51
— sylvestris 51
— thracica 51
— urumoffii 51

Ulex densus 40
Ulmaceae 127
Ulmus wallichiana 127
Umbelliferae 8
Umtisa listerana 146
Urceolina 210
Urticaceae 167

Vancouveria chrysantha 174
— hexandra 174
— planipetala 174
Vanda coerulea 107
— tricolor 262
— whitaena 262
Vanilla fragrans 222
— odorata 222
— pompona 222
Vateria seychellarum 321
Vella pseudocytisus 30
Vepris glandulosa 167
Verbascum anisophyllum 72
Verbenaceae 128
Verticordia chrysantha 255
— grandis 254
— nitens 255
Vieraea laevigata 316
Vigna owahuensis 291
— sandwicensis 292

Viola cryana 75
— hispida 74
Viola jaubertiana 74
Violaceae 74
Vitaceae 167
Vittaria schaeferi 168
Vittariaceae 168
Vriesia broadwayi 212
— johnstonii 212

Wahlenbergia brockiei 309
— linifolia 318
— saxicola 232
Watsonia strictiflora 164
Welwitschia bainesii 168
— mirabilis 168
Welwitschiaceae 168
Whitesloanea crassa 132
Widdringtonia cedarbergennis 138
Wikstroemia capitellata 127
Willughbeia cirrhifera 76
Winteraceae 278, 313
Wissmannia carinensis 158

Xanthocercis zambesiaca 147
Xeronema callistemon 294
— moorei 294
Xylocalyx aculeolatus
Xylomelum angustifolium 270
— occidentale 270
— pyriforme 270

Yoania australis 301

Zamia angustifolia 227
— floridana 208
— integrifolia 227
— kickxii 227
— latifolialata 227
— media 227
— multifoliata 227
— ottonis 227
— poeppigiana 227
— pseudo-parasitica 227
— pumila 227
— silicea 226
Zamiaceae 208, 226, 278
Zanthoxylon paniculatum 325
Zenkerella citrina 136
Zizania texana 197
Zombia antillarum 223

Printed in India